Undergraduate Lecture Notes in Physics

Undergraduate Lecture Notes in Physics (ULNP) publishes authoritative texts covering topics throughout pure and applied physics. Each title in the series is suitable as a basis for undergraduate instruction, typically containing practice problems, worked examples, chapter summaries, and suggestions for further reading.

ULNP titles must provide at least one of the following:

- An exceptionally clear and concise treatment of a standard undergraduate subject.
- A solid undergraduate-level introduction to a graduate, advanced, or non-standard subject.
- A novel perspective or an unusual approach to teaching a subject.

ULNP especially encourages new, original, and idiosyncratic approaches to physics teaching at the undergraduate level.

The purpose of ULNP is to provide intriguing, absorbing books that will continue to be the reader's preferred reference throughout their academic career.

Series editors

More information about this series at http://www.springer.com/series/8917

Giovanni Landi · Alessandro Zampini

Linear Algebra and Analytic Geometry for Physical Sciences

Giovanni Landi
University of Trieste
Trieste
Italy

Alessandro Zampini
INFN Sezione di Napoli
Napoli
Italy

ISSN 2192-4791 ISSN 2192-4805 (electronic)
Undergraduate Lecture Notes in Physics
ISBN 978-3-319-78360-4 ISBN 978-3-319-78361-1 (eBook)
https://doi.org/10.1007/978-3-319-78361-1

Library of Congress Control Number: 2018935878

Printed on acid-free paper

This Springer imprint is published by the registered company Springer International Publishing AG part of Springer Nature
The registered company address is: Gewerbestrasse 11, 6330 Cham, Switzerland

To our families

Contents

Introduction

This book originates from a collection of lecture notes that the first author prepared at the University of Trieste with Michela Brundu, over a span of fifteen years, together with the more recent one written by the second author. The notes were meant for undergraduate classes on linear algebra, geometry and more generally basic mathematical physics delivered to physics and engineering students, as well as mathematics students in Italy, Germany and Luxembourg.

The book is mainly intended to be a self-contained introduction to the theory of finite-dimensional vector spaces and linear transformations (matrices) with their spectral analysis both on Euclidean and Hermitian spaces, to affine Euclidean geometry as well as to quadratic forms and conic sections.

Many topics are introduced and motivated by examples, mostly from physics. They show how a definition is natural and how the main theorems and results are first of all plausible before a proof is given. Following this approach, the book presents a number of examples and exercises, which are meant as a central part in the development of the theory. They are all completely solved and intended both to guide the student to appreciate the relevant formal structures and to give in several cases a proof and a discussion, within a geometric formalism, of results from physics, notably from mechanics (including celestial) and electromagnetism.

Being the book intended mainly for students in physics and engineering, we tasked ourselves not to present the mathematical formalism *per se*. Although we decided, for clarity's sake of our readers, to organise the basics of the theory in the classical terms of *definitions* and the main results as *theorems* or *propositions*, we do often not follow the standard sequential form of *definition—theorem—corollary—example* and provided some two hundred and fifty solved problems given as exercises.

Chapter 1 of the book presents the Euclidean space used in physics in terms of applied vectors with respect to orthonormal coordinate system, together with the operation of scalar, vector and mixed product. They are used both to describe the motion of a point mass and to introduce the notion of vector field with the most relevant differential operators acting upon them.

Chapters 2 and 3 are devoted to a general formulation of the theory of finite-dimensional vector spaces equipped with a scalar product, while the Chaps. 4 –6 present, via a host of examples and exercises, the theory of finite rank matrices and their use to solve systems of linear equations.

These are followed by the theory of linear transformations in Chap. 7. Such a theory is described in Chap. 8 in terms of the Dirac's Bra-Ket formalism, providing a link to a geometric–algebraic language used in quantum mechanics.

The notion of the diagonal action of an endomorphism or a matrix (the problem of diagonalisation and of reduction to the Jordan form) is central in this book, and it is introduced in Chap. 9.

Again with many solved exercises and examples, Chap. 10 describes the spectral theory for operators (matrices) on Euclidean spaces, and (in Chap. 11) how it allows one to characterise the rotations in classical mechanics. This is done by introducing the Euler angles which parameterise rotations of the physical three-dimensional space, the notion of angular velocity and by studying the motion of a rigid body with its inertia matrix, and formulating the description of the motion with respect to different inertial observers, also giving a characterisation of polar and axial vectors.

Chapter 12 is devoted to the spectral theory for matrices acting on Hermitian spaces in order to present a geometric setting to study a finite level quantum mechanical system, where the time evolution is given in terms of the unitary group. All these notions are related with the notion of Lie algebra and to the exponential map on the space of finite rank matrices.

In Chap. 13, we present the theory of quadratic forms. Our focus is the description of their transformation properties, so to give the notion of signature, both in the real and in the complex cases. As the most interesting example of a non-Euclidean quadratic form, we present the Minkowski spacetime from special relativity and the Maxwell equations.

In Chaps. 14 and 15, we introduce through many examples the basics of the Euclidean affine linear geometry and develop them in the study of conic sections, in Chap. 16, which are related to the theory of Kepler motions for celestial body in classical mechanics. In particular, we show how to characterise a conic by means of its eccentricity.

A reader of this book is only supposed to know about number sets, more precisely the natural, integer, rational and real numbers and no additional prior knowledge is required. To try to be as much self-contained as possible, an appendix collects a few basic algebraic notions, like that of group, ring and field and maps between them that preserve the structures (homomorphisms), and polynomials in one variable. There are also a few basic properties of the field of complex numbers and of the field of (classes of) integers modulo a prime number.

Trieste, Italy Giovanni Landi
Napoli, Italy Alessandro Zampini
May 2018

Chapter 1
Vectors and Coordinate Systems

The notion of a *vector*, or more precisely of a *vector applied at a point*, originates in physics when dealing with an observable quantity. By this or simply by *observable*, one means anything that can be measured in the physical space—the space of physical events— via a suitable measuring process. Examples are the velocity of a point particle, or its acceleration, or a force acting on it. These are characterised at the point of application by a *direction*, an *orientation* and a *modulus* (or *magnitude*). In the following pages we describe the physical space in terms of points and applied vectors, and use these to describe the physical observables related to the motion of a point particle with respect to a coordinate system (a reference frame). The geometric structures introduced in this chapter will be more rigorously analysed in the next chapters.

1.1 Applied Vectors

We refer to the common intuition of a physical space made of points, where the notions of *straight* line between two points and of the length of a segment (or equivalently of distance of two points) are assumed to be given. Then, a vector v can be denoted as

$$v = B - A \qquad \text{or} \qquad v = AB,$$

where A, B are two points of the physical space. Then, A is the point of application of v, its direction is the straight line joining B to A, its orientation the one of the arrow pointing from A towards B, and its modulus the real number $\|B - A\| = \|A - B\|$, that is the length (with respect to a fixed unit) of the segment AB.

G. Landi and A. Zampini, *Linear Algebra and Analytic Geometry for Physical Sciences*, Undergraduate Lecture Notes in Physics,
https://doi.org/10.1007/978-3-319-78361-1_1

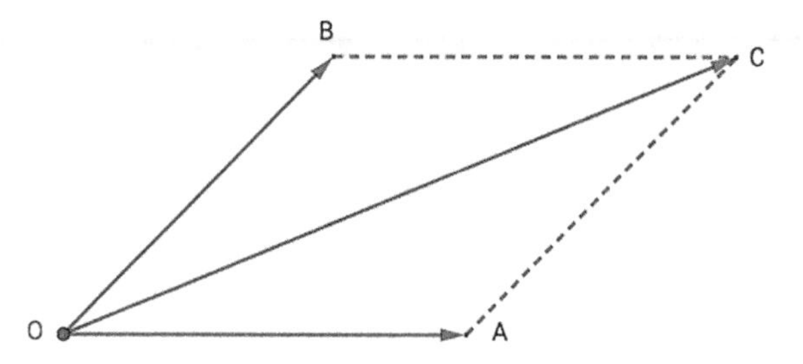

Fig. 1.1 The parallelogram rule

If $\mathcal{S}$ denotes the usual three dimensional physical space, we denote by

$$\mathcal{W}^3 = \{B - A \mid A, B \in \mathcal{S}\}$$

the collection of all applied vectors at any point of $\mathcal{S}$ and by

$$\mathcal{V}^3_A = \{B - A \mid B \in \mathcal{S}\}$$

the collection of all vectors applied at A in $\mathcal{S}$. Then

$$\mathcal{W}^3 = \bigcup_{A \in \mathcal{S}} \mathcal{V}^3_A.$$

Remark 1.1.1 Once fixed a point O in $\mathcal{S}$, one sees that there is a bijection between the set $\mathcal{V}^3_O = \{B - O \mid B \in \mathcal{S}\}$ and $\mathcal{S}$ itself. Indeed, each point B in $\mathcal{S}$ uniquely determines the element $B - O$ in $\mathcal{V}^3_O$, and each element $B - O$ in $\mathcal{V}^3_O$ uniquely determines the point B in $\mathcal{S}$.

It is well known that the so called *parallelogram rule* defines in $\mathcal{V}^3_O$ a sum of vectors, where

$$(A - O) + (B - O) = (C - O),$$

with C the fourth vertex of the parallelogram whose other three vertices are A, O, B, as shown in Fig. 1.1.

The vector $\mathbf{0} = O - O$ is called the *zero vector* (or *null* vector); notice that its modulus is zero, while its direction and orientation are undefined.

It is evident that $\mathcal{V}^3_O$ is closed with respect to the notion of sum defined above. That such a sum is associative and abelian is part of the content of the proposition that follows.

Proposition 1.1.2 *The datum* $(\mathcal{V}^3_O, +, \mathbf{0})$ *is an abelian group.*

Proof Clearly the zero vector $\mathbf{0}$ is the neutral (identity) element for the sum in $\mathcal{V}^3_O$, that added to any vector leave the latter unchanged. Any vector $A - O$ has an inverse

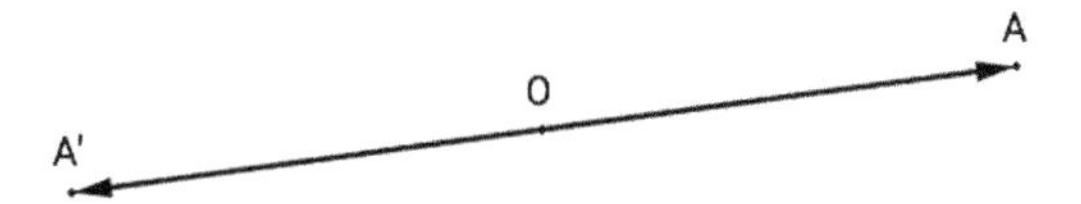

Fig. 1.2 The opposite of a vector: $A^{'} - O = -(A - O)$

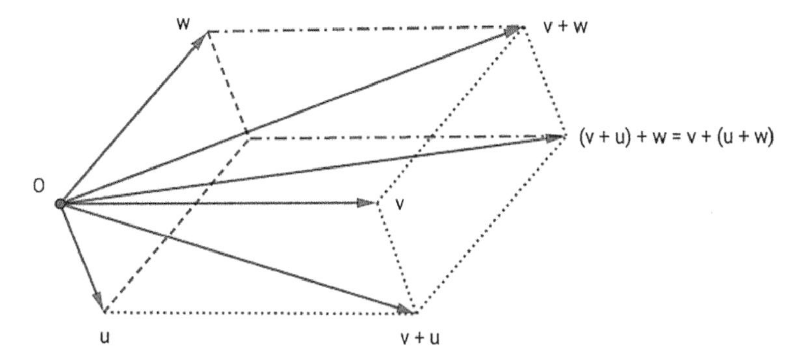

Fig. 1.3 The associativity of the vector sum

with respect to the sum (that is, any vector has an opposite vector) given by $A' - O$, where A' is the symmetric point to A with respect to O on the straight line joining A to O (see Fig. 1.2).

From its definition the sum of two vectors is a commutative operation. For the associativity we give a pictorial argument in Fig. 1.3. □

There is indeed more structure. The physical intuition allows one to consider multiples of an applied vector. Concerning the collection $\mathcal{V}_O^3$, this amounts to define an operation involving vectors applied in O and real numbers, which, in order not to create confusion with vectors, are called (real) *scalars*.

Definition 1.1.3 Given the scalar $\lambda \in \mathbb{R}$ and the vector $A - O \in \mathcal{V}_O^3$, the *product by a scalar*

$$B - O = \lambda(A - O)$$

is the vector such that:

(i) A, B, O are on the same (straight) line,
(ii) $B - O$ and $A - O$ have the same orientation if $\lambda > 0$, while $A - O$ and $B - O$ have opposite orientations if $\lambda < 0$,
(iii) $\|B - O\| = |\lambda| \, \|A - O\|$.

The main properties of the operation of product by a scalar are given in the following proposition.

Proposition 1.1.4 *For any pair of scalars* $\lambda, \mu \in \mathbb{R}$ *and any pair of vectors* $A - O,\, B - O \in \mathcal{V}_O^3$*, it holds that:*

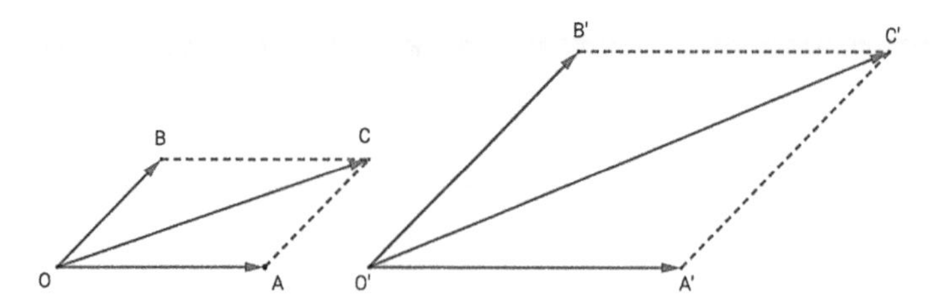

Fig. 1.4 The scaling $\lambda(C - O) = (C' - O)$ with $\lambda > 1$

1. $\lambda(\mu(A - O)) = (\lambda\mu)(A - O)$,
2. $1(A - O) = A - O$,
3. $\lambda((A - O) + (B - O)) = \lambda(A - O) + \lambda(B - O)$,
4. $(\lambda + \mu)(A - O) = \lambda(A - O) + \mu(A - O)$.

Proof 1. Set $C - O = \lambda(\mu(A - O))$ and $D - O = (\lambda\mu)(A - O)$. If one of the scalars λ, μ is zero, one trivially has $C - O = \mathbf{0}$ and $D - O = \mathbf{0}$, so Point 1. is satisfied. Assume now that $\lambda \neq 0$ and $\mu \neq 0$. Since, by definition, both C and D are points on the line determined by O and A, the vectors $C - O$ and $D - O$ have the same direction. It is easy to see that $C - O$ and $D - O$ have the same orientation: it will coincide with the orientation of $A - O$ or not, depending on the sign of the product $\lambda\mu \neq 0$. Since $|\lambda\mu| = |\lambda||\mu| \in R$, one has $\|C - O\| = \|D - O\|$.

2. It follows directly from the definition.
3. Set $C - O = (A - O) + (B - O)$ and $C' - O = (A' - O) + (B' - O)$, with $A' - O = \lambda(A - O)$ and $B' - O = \lambda(B - O)$.
   We verify that $\lambda(C - O) = C' - O$ (see Fig. 1.4).
   Since OA is parallel to OA' by definition, then BC is parallel to $B'C'$; OB is indeed parallel to OB', so that the planar angles $\widehat{OBC}$ and $\widehat{OB'C'}$ are equal. Also $\lambda(OB) = OB'$, $\lambda(OA) = OA'$, and $\lambda(BC) = B'C'$. It follows that the triangles OBC and $OB'C'$ are similar: the vector OC is then parallel OC' and they have the same orientation, with $\|OC'\| = \lambda \|OC\|$. From this we obtain $OC' = \lambda(OC)$.
4. The proof is analogue to the one in point 3. □

What we have described above shows that the operations of sum and product by a scalar give $\mathcal{V}_O^3$ an algebraic structure which is richer than that of abelian group. Such a structure, that we shall study in detail in Chap. 2, is called in a natural way *vector space*.

1.2 Coordinate Systems

The notion of coordinate system is well known. We rephrase its main aspects in terms of vector properties.

Definition 1.2.1 Given a line r, a *coordinate system* Λ on it is defined by a point $O \in r$ and a vector $\mathbf{i} = A - O$, where $A \in r$ and $A \neq O$.

The point O is called the *origin* of the coordinate system, the norm $\|A - O\|$ is the *unit of measure* (or *length*) of Λ, with $\mathbf{i}$ the basis *unit vector*. The orientation of $\mathbf{i}$ is the *orientation* of the coordinate system Λ.

A coordinate system Λ provides a bijection between the points on the line r and $\mathbb{R}$. Any point $P \in r$ singles out the real number x such that $P - O = x\mathbf{i}$; viceversa, for any $x \in \mathbb{R}$ one has the point $P \in r$ defined by $P - O = x\mathbf{i}$. One says that P has coordinate x, and we shall denote it by $P = (x)$, with respect to the coordinate system Λ that is also denoted as $(O; x)$ or $(O; \mathbf{i})$.

Definition 1.2.2 Given a plane α, a coordinate system Π on it is defined by a point $O \in \alpha$ and a pair of non zero distinct (and not having the same direction) vectors $\mathbf{i} = A - O$ and $\mathbf{j} = B - O$ with $A, B \in \alpha$, and $\|A - O\| = \|B - O\|$.

The point O is the origin of the coordinate system, the (common) norm of the vectors $\mathbf{i}, \mathbf{j}$ is the unit length of Π, with $\mathbf{i}, \mathbf{j}$ the basis *unit vectors*. The system is oriented in such a way that the vector $\mathbf{i}$ coincides with $\mathbf{j}$ after an anticlockwise rotation of angle ϕ with $0 < \phi < \pi$. The line defined by O and $\mathbf{i}$, with its given orientation, is usually referred to as a the *abscissa axis*, while the one defined by O and $\mathbf{j}$, again with its given orientation, is called *ordinate axis*.

As before, it is immediate to see that a coordinate system Π on α allows one to define a bijection between points on α and *ordered* pairs of real numbers. Any $P \in \alpha$ uniquely provides, via the parallelogram rule (see Fig. 1.5), the ordered pair $(x, y) \in \mathbb{R}^2$ with $P - O = x\mathbf{i} + y\mathbf{j}$; conversely, for any given ordered pair $(x, y) \in \mathbb{R}^2$, one defines $P \in \alpha$ as given by $P - O = x\mathbf{i} + y\mathbf{j}$.

With respect to Π, the elements $x \in \mathbb{R}$ and $y \in \mathbb{R}$ are the *coordinates* of P, and this will be denoted by $P = (x, y)$. The coordinate system Π will be denoted $(O; \mathbf{i}, \mathbf{j})$ or $(O; x, y)$.

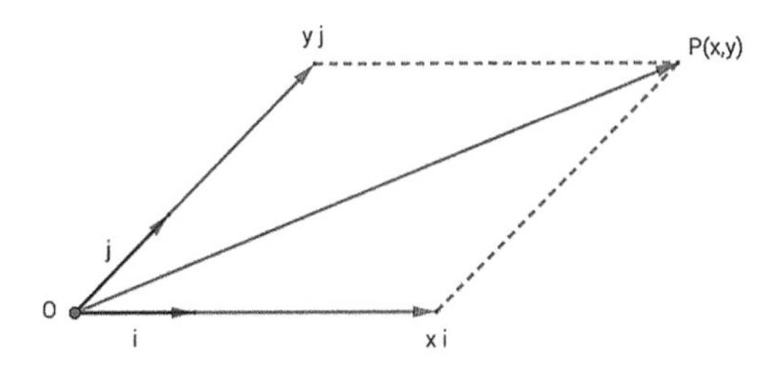

Fig. 1.5 The bijection $P(x, y) \leftrightarrow P - O = x\mathbf{i} + y\mathbf{j}$ in a plane

Definition 1.2.3 A coordinate system $\Pi = (O; \mathbf{i}, \mathbf{j})$ on a plane α is called an *orthogonal cartesian coordinate system* if $\phi = \pi/2$, where ϕ is as before the width of the anticlockwise rotation under which $\mathbf{i}$ coincides with $\mathbf{j}$.

In order to introduce a coordinate system for the physical three dimensional space, we start by considering three unit-length vectors in $\mathcal{V}^3_O$ given as $\mathbf{u} = U - O$, $\mathbf{v} = V - O$, $\mathbf{w} = W - O$, and we assume the points O, U, V, W not to be on the same plane. This means that any two vectors, $\mathbf{u}$ and $\mathbf{v}$ say, determine a plane which does not contain the third point, say W. Seen from W, the vector $\mathbf{u}$ will coincide with $\mathbf{v}$ under an anticlockwise rotation by an angle that we denote by $\widehat{\mathbf{uv}}$.

Definition 1.2.4 An ordered triple $(\mathbf{u}, \mathbf{v}, \mathbf{w})$ of unit vectors in $\mathcal{V}^3_O$ which do not lie on the same plane is called *right-handed* if the three angles $\widehat{\mathbf{uv}}$, $\widehat{\mathbf{vw}}$, $\widehat{\mathbf{wu}}$, defined by the prescription above are smaller than π. Notice that the order of the vectors matters.

Definition 1.2.5 A coordinate system Σ for the space $\mathcal{S}$ is given by a point $O \in \mathcal{S}$ and three non zero distinct (and not lying on the same plane) vectors $\mathbf{i} = A - O$, $\mathbf{j} = B - O$ and $\mathbf{k} = C - O$, with $A, B, C \in \mathcal{S}$, and $\|A - O\| = \|B - O\| = \|C - O\|$ and $(\mathbf{i}, \mathbf{j}, \mathbf{k})$ giving a right-handed triple.

The point O is the *origin* of the coordinate system, the common length of the vectors $\mathbf{i}, \mathbf{j}, \mathbf{k}$ is the unit measure in Σ, with $\mathbf{i}, \mathbf{j}, \mathbf{k}$ the basis *unit vectors*. The line defined by O and $\mathbf{i}$, with its orientation, is the abscissa axis, that defined by O and $\mathbf{j}$ is the ordinate axis, while the one defined by O and $\mathbf{k}$ is the quota axis.

With respect to the coordinate system Σ, one establishes, via $\mathcal{V}^3_O$, a bijection between ordered triples of real numbers and points in $\mathcal{S}$. One has

$$P \quad \leftrightarrow \quad P - O \quad \leftrightarrow \quad (x, y, z)$$

with $P - O = x\mathbf{i} + y\mathbf{j} + z\mathbf{k}$ as in Fig. 1.6. The real numbers x, y, z are the *components* (or *coordinates*) of the applied vector $P - O$, and this will be denoted by $P = (x, y, z)$. Accordingly, the coordinate system will be denoted by $\Sigma = (O; \mathbf{i}, \mathbf{j}, \mathbf{k}) = (O; x, y, z)$. The coordinate system Σ is called cartesian orthogonal if the vectors $\mathbf{i}, \mathbf{j}, \mathbf{k}$ are pairwise orthogonal.

By writing $\mathbf{v} = P - O$, it is convenient to denote by v_x, v_y, v_z the components of $\mathbf{v}$ with respect to a cartesian coordinate system Σ, so to have

$$\mathbf{v} = v_x\mathbf{i} + v_y\mathbf{j} + v_z\mathbf{k}.$$

In order to simplify the notations, we shall also write this as

$$\mathbf{v} = (v_x, v_y, v_z),$$

implicitly assuming that such components of $\mathbf{v}$ refer to the cartesian coordinate system $(O; \mathbf{i}, \mathbf{j}, \mathbf{k})$. Clearly the components of a given vector $\mathbf{v}$ depend on the particular coordinate system one is using.

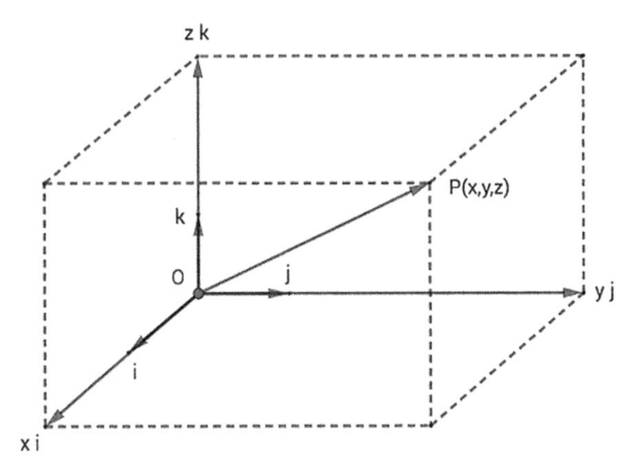

Fig. 1.6 The bijection $P(x, y, z) \leftrightarrow P - O = x\mathbf{i} + y\mathbf{j} + z\mathbf{k}$ in the space

Exercise 1.2.6 One has

1. The zero (null) vector $\mathbf{0} = O - O$ has components $(0, 0, 0)$ with respect to any coordinate system whose origin is O, and it is the only vector with this property.
2. Given a coordinate system $\Sigma = (O; \mathbf{i}, \mathbf{j}, \mathbf{k})$, the basis unit vectors have components
$$\mathbf{i} = (1, 0, 0)\ , \quad \mathbf{j} = (0, 1, 0)\ , \quad \mathbf{k} = (0, 0, 1).$$
3. Given a coordinate system $\Sigma = (O; \mathbf{i}, \mathbf{j}, \mathbf{k})$ for the space $\mathcal{S}$, we call *coordinate plane* each plane determined by a pair of axes of Σ. We have $\mathbf{v} = (a, b, 0)$, with $a, b \in \mathbb{R}$, if $\mathbf{v}$ is on the plane xy, $\mathbf{v}' = (0, b', c')$ if $\mathbf{v}'$ is on the plane yz, and $\mathbf{v}'' = (a'', 0, c'')$ if $\mathbf{v}''$ is on the plane xz.

Example 1.2.7 The motion of a point mass in three dimensional space is described by a map $t \in \mathbb{R} \mapsto \mathbf{x}(t) \in \mathcal{V}^3_O$ where t represents the *time* variable and $\mathbf{x}$(t) is the position of the point mass at time t. With respect to a coordinate system $\Sigma = (O; x, y, z)$ we then write

$$\mathbf{x}(t) = (x(t), y(t), z(t)) \qquad \text{or equivalently} \qquad \mathbf{x}(t) = x(t)\mathbf{i} + y(t)\mathbf{j} + z(t)\mathbf{k}.$$

The corresponding *velocity* is a vector applied in $\mathbf{x}(t)$, that is $\mathbf{v}(t) \in \mathcal{V}^3_{\mathbf{x}(t)}$, with components

$$\mathbf{v}(t) = (v_x(t), v_y(t), v_z(t)) = \frac{\mathrm{d}\mathbf{x}(t)}{\mathrm{d}t} = (\frac{\mathrm{d}x}{\mathrm{d}t}, \frac{\mathrm{d}y}{\mathrm{d}t}, \frac{\mathrm{d}z}{\mathrm{d}t}),$$

while the acceleration is the vector $\mathbf{a}(t) \in \mathcal{V}^3_{\mathbf{x}(t)}$ with components

$$\mathbf{a}(t) = \frac{\mathrm{d}\mathbf{v}(t)}{\mathrm{d}t} = (\frac{\mathrm{d}x^2}{\mathrm{d}t^2}, \frac{\mathrm{d}^2y}{\mathrm{d}t^2}, \frac{\mathrm{d}^2z}{\mathrm{d}t^2}).$$

One also uses the notations

$$\mathbf{v} = \frac{\mathrm{d}\mathbf{x}}{\mathrm{d}t} = \dot{\mathbf{x}} \qquad \text{and} \qquad \mathbf{a} = \frac{\mathrm{d}^2\mathbf{x}}{\mathrm{d}t^2} = \dot{\mathbf{v}} = \ddot{\mathbf{x}}.$$

In the newtonian formalism for the dynamics, a *force* acting on the given point mass is a vector applied in $\mathbf{x}(t)$, that is $\mathbf{F} \in \mathcal{V}^3_{\mathbf{x}(t)}$ with components $\mathbf{F} = (F_x, F_y, F_z)$, and the *second law of dynamics* is written as

$$m\,\mathbf{a} = \mathbf{F}$$

where $m > 0$ is the value of the *inertial mass* of the moving point mass. Such a relation can be written component-wise as

$$m\,\frac{\mathrm{d}^2x}{\mathrm{d}t^2} = F_x, \qquad m\,\frac{\mathrm{d}^2y}{\mathrm{d}t^2} = F_y, \qquad m\,\frac{\mathrm{d}^2z}{\mathrm{d}t^2} = F_z.$$

A coordinate system for $\mathcal{S}$ allows one to express the operations of sum and product by a scalar in $\mathcal{V}^3_O$ in terms of elementary algebraic expressions.

Proposition 1.2.8 *With respect to the coordinate system* $\Sigma = (O; \mathbf{i}, \mathbf{j}, \mathbf{k})$, *let us consider the vectors* $\mathbf{v} = v_x\mathbf{i} + v_y\mathbf{j} + v_z\mathbf{k}$ *and* $\mathbf{w} = w_x\mathbf{i} + w_y\mathbf{j} + w_z\mathbf{k}$, *and the scalar* $\lambda \in \mathbb{R}$. *One has:*

(1) $\mathbf{v} + \mathbf{w} = (v_x + w_x)\mathbf{i} + (v_y + w_y)\mathbf{j} + (v_z + w_z)\mathbf{k}$,
(2) $\lambda\mathbf{v} = \lambda v_x\mathbf{i} + \lambda v_y\mathbf{j} + \lambda v_z\mathbf{k}$.

Proof (1) Since $\mathbf{v} + \mathbf{w} = (v_x\mathbf{i} + v_y\mathbf{j} + v_z\mathbf{k}) + (w_x\mathbf{i} + w_y\mathbf{j} + w_z\mathbf{k})$, by using the commutativity and the associativity of the sum of vectors applied at a point, one has

$$\mathbf{v} + \mathbf{w} = (v_x\mathbf{i} + w_x\mathbf{i}) + (v_y\mathbf{j} + w_y\mathbf{j}) + (v_z\mathbf{k} + w_z\mathbf{k}).$$

Being the product distributive over the sum, this can be regrouped as in the claimed identity.

(2) Along the same lines as (1). □

Remark 1.2.9 By denoting $\mathbf{v} = (v_x, v_y, v_z)$ and $\mathbf{w} = (w_x, w_y, w_z)$, the identities proven in the proposition above are written as

$$(v_x, v_y, v_z) + (w_x, w_y, w_z) = (v_x + w_x, v_y + w_y, v_z + w_z),$$
$$\lambda(v_x, v_y, v_z) = (\lambda v_x, \lambda v_y, \lambda v_z).$$

This suggests a generalisation we shall study in detail in the next chapter. If we denote by $\mathbb{R}^3$ the set of ordered triples of real numbers, and we consider a pair of elements (x_1, x_2, x_3) and (y_1, y_2, y_3) in $\mathbb{R}^3$, with $\lambda \in \mathbb{R}$, one can introduce a sum of triples and a product by a scalar:

$$\begin{aligned}(x_1, x_2, x_3) + (y_1, y_2, y_3) &= (x_1 + y_1, x_2 + y_2, x_3 + y_3), \\ \lambda(x_1, x_2, x_3) &= (\lambda x_1, \lambda x_2, \lambda x_3).\end{aligned}$$

1.3 More Vector Operations

In this section we recall the notions—originating in physics—of scalar product, vector product and mixed products.

Before we do this, as an elementary consequence of the Pythagora's theorem, one has the following (see Fig. 1.6)

Proposition 1.3.1 *Let* $\mathbf{v} = (v_x, v_y, v_z)$ *be an arbitrary vector in* $\mathcal{V}_O^3$ *with respect to the cartesian orthogonal coordinate system* $(O; \mathbf{i}, \mathbf{j}, \mathbf{z})$. *One has*

$$\|\mathbf{v}\| = \sqrt{v_x^2 + v_y^2 + v_z^2}.$$

Definition 1.3.2 Let us consider a pair of vectors $\mathbf{v}, \mathbf{w} \in \mathcal{V}_O^3$. The *scalar product* of $\mathbf{v}$ and $\mathbf{w}$, denoted by $\mathbf{v} \cdot \mathbf{w}$, is the real number

$$\mathbf{v} \cdot \mathbf{w} = \|\mathbf{v}\| \ \|\mathbf{w}\| \cos\alpha$$

with $\alpha = \widehat{\mathbf{vw}}$ the plane angle defined by $\mathbf{v}$ and $\mathbf{w}$. Since $\cos\alpha = \cos(-\alpha)$, for this definition one has $\cos\widehat{\mathbf{vw}} = \cos\widehat{\mathbf{wv}}$.

The definition of a scalar product for vectors in $\mathcal{V}_O^2$ is completely analogue.

Remark 1.3.3 The following properties follow directly from the definition.

(1) If $\mathbf{v} = \mathbf{0}$, then $\mathbf{v} \cdot \mathbf{w} = 0$.
(2) If $\mathbf{v}$, $\mathbf{w}$ are both non zero vectors, then

$$\mathbf{v} \cdot \mathbf{w} = 0 \iff \cos\alpha = 0 \iff \mathbf{v} \perp \mathbf{w}.$$

(3) For any $\mathbf{v} \in \mathcal{V}_O^3$, it holds that:

$$\mathbf{v} \cdot \mathbf{v} = \|\mathbf{v}\|^2$$

and moreover

$$\mathbf{v} \cdot \mathbf{v} = 0 \iff \mathbf{v} = 0.$$

(4) From (2), (3), if $(O; \mathbf{i}, \mathbf{j}, \mathbf{k})$ is an orthogonal cartesian coordinate system, then

$$\mathbf{i} \cdot \mathbf{i} = \mathbf{j} \cdot \mathbf{j} = \mathbf{k} \cdot \mathbf{k} = 1, \qquad \mathbf{i} \cdot \mathbf{j} = \mathbf{j} \cdot \mathbf{k} = \mathbf{k} \cdot \mathbf{i} = 0.$$

Proposition 1.3.4 *For any choice of* $\mathbf{u}, \mathbf{v}, \mathbf{w} \in \mathcal{V}_O^3$ *and* $\lambda \in \mathbb{R}$, *the following identities hold.*

(i) $\mathbf{v} \cdot \mathbf{w} = \mathbf{w} \cdot \mathbf{v}$,
(ii) $(\lambda\mathbf{v}) \cdot \mathbf{w} = \mathbf{v} \cdot (\lambda\mathbf{w}) = \lambda(\mathbf{v} \cdot \mathbf{w})$,
(iii) $\mathbf{u} \cdot (\mathbf{v} + \mathbf{w}) = \mathbf{u} \cdot \mathbf{v} + \mathbf{u} \cdot \mathbf{w}$.

Proof (i) From the definition one has

$$\mathbf{v} \cdot \mathbf{w} = \|\mathbf{v}\| \ \|\mathbf{w}\| \cos \widehat{\mathbf{v}\mathbf{w}} = \|\mathbf{w}\| \ \|\mathbf{v}\| \cos \widehat{\mathbf{w}\mathbf{v}} = \mathbf{w} \cdot \mathbf{v}.$$

(ii) Setting $a = (\lambda\mathbf{v}) \cdot \mathbf{w}$, $b = \mathbf{v} \cdot (\lambda\mathbf{w})$ and $c = \lambda(\mathbf{v} \cdot \mathbf{w})$, from the Definition 1.3.2 and the properties of the norm of a vector, one has

$$\begin{aligned} a &= (\lambda\mathbf{v}) \cdot \mathbf{w} = \|\lambda\mathbf{v}\| \ \|\mathbf{w}\| \cos\alpha' = |\lambda| \|\mathbf{v}\| \ \|\mathbf{w}\| \cos\alpha' \\ b &= \mathbf{v} \cdot (\lambda\mathbf{w}) = \|\mathbf{v}\| \ \|\lambda\mathbf{w}\| \cos\alpha'' = \|\mathbf{v}\| \ |\lambda| \|\mathbf{w}\| \cos\alpha'' \\ c &= \lambda(\mathbf{v} \cdot \mathbf{w}) = \lambda(\|\mathbf{v}\| \ \|\mathbf{w}\| \cos\alpha) = \lambda\|\mathbf{v}\| \ \|\mathbf{w}\| \cos\alpha \end{aligned}$$

where $\alpha' = \widehat{(\lambda\mathbf{v})\mathbf{w}}$, $\alpha'' = \widehat{\mathbf{v}(\lambda\mathbf{w})}$ and $\alpha = \widehat{\mathbf{v}\mathbf{w}}$. If $\lambda = 0$, then $a = b = c = 0$. If $\lambda > 0$, then $|\lambda| = \lambda$ and $\alpha = \alpha' = \alpha''$; from the commutativity and the associativity of the product in $\mathbb{R}$, this gives that $a = b = c$. If $\lambda < 0$, then $|\lambda| = -\lambda$ and $\alpha' = \alpha'' = \pi - \alpha$, thus giving $\cos\alpha' = \cos\alpha'' = -\cos\alpha$. These read $a = b = c$.

(iii) We sketch the proof for parallel $\mathbf{u}, \mathbf{v}, \mathbf{w}$. Under this condition, the result depends on the relative orientations of the vectors. If $\mathbf{u}, \mathbf{v}, \mathbf{w}$ have the same orientation, one has

$$\begin{aligned} \mathbf{u} \cdot (\mathbf{v} + \mathbf{w}) &= \|\mathbf{u}\| \ \|\mathbf{v} + \mathbf{w}\| \\ &= \|\mathbf{u}\|(\|\mathbf{v}\| + \|\mathbf{w}\|) \\ &= \|\mathbf{u}\| \ \|\mathbf{v}\| + \|\mathbf{u}\| \ \|\mathbf{w}\| \\ &= \mathbf{u} \cdot \mathbf{v} + \mathbf{u} \cdot \mathbf{w}. \end{aligned}$$

If $\mathbf{v}$ and $\mathbf{w}$ have the same orientation, which is not the orientation of $\mathbf{u}$, one has

$$\begin{aligned} \mathbf{u} \cdot (\mathbf{v} + \mathbf{w}) &= -\|\mathbf{u}\| \ \|\mathbf{v} + \mathbf{w}\| \\ &= -\|\mathbf{u}\|(\|\mathbf{v}\| + \|\mathbf{w}\|) \\ &= -\|\mathbf{u}\| \ \|\mathbf{v}\| - \|\mathbf{u}\| \ \|\mathbf{w}\| \\ &= \mathbf{u} \cdot \mathbf{v} + \mathbf{u} \cdot \mathbf{w}. \end{aligned}$$

We leave the reader to explicitly prove the other cases. □

By expressing vectors in $\mathcal{V}_O^3$ in terms of an orthogonal cartesian coordinate system, the scalar product has an expression that will allow us to define the scalar product of vectors in the more general situation of euclidean spaces.

Proposition 1.3.5 *Given* $(O; \mathbf{i}, \mathbf{j}, \mathbf{k})$, *an orthogonal cartesian coordinate system for* $\mathcal{S}$; *with vectors* $\mathbf{v} = (v_x, v_y, v_z)$ *and* $\mathbf{w} = (w_x, w_y, w_z)$ *in* $\mathcal{V}_O^3$, *one has*

$$\mathbf{v} \cdot \mathbf{w} = v_x w_x + v_y w_y + v_z w_z.$$

Proof With $\mathbf{v} = v_x\mathbf{i} + v_y\mathbf{j} + v_z\mathbf{k}$ and $\mathbf{w} = w_x\mathbf{i} + w_y\mathbf{j} + w_z\mathbf{k}$, from Proposition 1.3.4, one has

$$\begin{aligned} \mathbf{v} \cdot \mathbf{w} = &\ (v_x\mathbf{i} + v_y\mathbf{j} + v_z\mathbf{k}) \cdot (w_x\mathbf{i} + w_y\mathbf{j} + w_z\mathbf{k}) \\ = &\ v_x w_x\mathbf{i} \cdot \mathbf{i} + v_y w_x\mathbf{j} \cdot \mathbf{i} + v_z w_x\mathbf{k} \cdot \mathbf{i} \\ + &\ v_x w_y\mathbf{i} \cdot \mathbf{j} + v_y w_y\mathbf{j} \cdot \mathbf{j} + v_z w_y\mathbf{k} \cdot \mathbf{j} + \ v_x w_z\mathbf{i} \cdot \mathbf{k} + v_y w_z\mathbf{j} \cdot \mathbf{k} + v_z w_z\mathbf{k} \cdot \mathbf{k}. \end{aligned}$$

The result follows directly from (4) in Remark 1.3.3, that is $\mathbf{i} \cdot \mathbf{j} = \mathbf{j} \cdot \mathbf{k} = \mathbf{k} \cdot \mathbf{i} = 0$ as well as $\mathbf{i} \cdot \mathbf{i} = \mathbf{j} \cdot \mathbf{j} = \mathbf{k} \cdot \mathbf{k} = 1$. □

Exercise 1.3.6 With respect to a given cartesian orthogonal coordinate system, consider the vectors $\mathbf{v} = (2, 3, 1)$ and $\mathbf{w} = (1, -1, 1)$. We verify they are orthogonal. From (2) in Remark 1.3.3 this is equivalent to show that $\mathbf{v} \cdot \mathbf{w} = 0$. From Proposition 1.3.5, one has $\mathbf{v} \cdot \mathbf{w} = 2 \cdot 1 + 3 \cdot (-1) + 1 \cdot 1 = 0$.

Example 1.3.7 If the map $\mathbf{x}(t) : \mathbb{R} \ni t \mapsto \mathbf{x}(t) \in \mathcal{V}_O^3$ describes the motion (notice that the range of the map gives the trajectory) of a point mass (with mass m), its *kinetic energy* is defined by

$$T = \frac{1}{2} m \, \|\mathbf{v}(t)\|^2.$$

With respect to an orthogonal coordinate system $\Sigma = (O; \mathbf{i}, \mathbf{j}, \mathbf{k})$, given $\mathbf{v}(t) = (v_x(t), v_y(t), v_z(t))$ as in the Example 1.2.7, we have from the Proposition 1.3.5 that

$$T = \frac{1}{2} m \, (v_x^2 + v_y^2 + v_z^2).$$

Also the following notion will be generalised in the context of euclidean spaces.

Definition 1.3.8 Given two non zero vectors $\mathbf{v}$ and $\mathbf{w}$ in $\mathcal{V}_O^3$, the *orthogonal projection* of $\mathbf{v}$ along $\mathbf{w}$ is defined as the vector $\mathbf{v}_\mathbf{w}$ in $\mathcal{V}_O^3$ given by

$$\mathbf{v}_\mathbf{w} = \frac{\mathbf{v} \cdot \mathbf{w}}{\|\mathbf{w}\|^2} \, \mathbf{w}.$$

As the first part of Fig. 1.7 displays, $\mathbf{v}_\mathbf{w}$ is parallel to $\mathbf{w}$.

From the identities proven in Proposition 1.3.4 one easily has

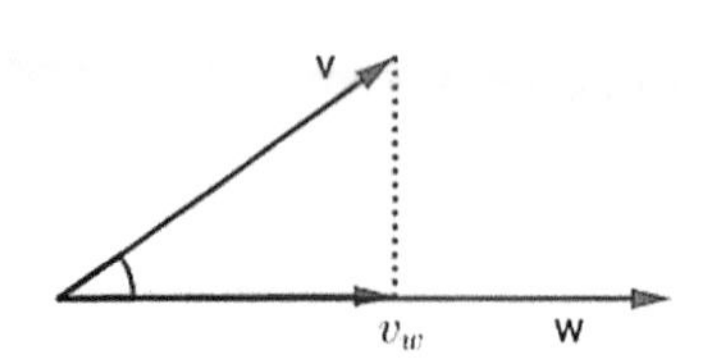

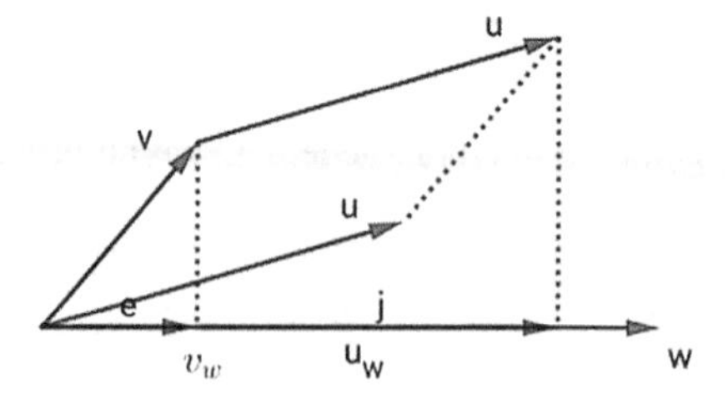

Fig. 1.7 Orthogonal projections

Proposition 1.3.9 *For any* $\mathbf{u}, \mathbf{v}, \mathbf{w} \in \mathcal{V}_O^3$, *the following identities hold:*

(a) $(\mathbf{u} + \mathbf{v})_\mathbf{w} = \mathbf{u}_\mathbf{w} + \mathbf{v}_\mathbf{w}$,
(b) $\mathbf{v} \cdot \mathbf{w} = \mathbf{v}_\mathbf{w} \cdot \mathbf{w} = \mathbf{w}_\mathbf{v} \cdot \mathbf{v}$.

The point (a) is illustrated by the second part of the Fig. 1.7.

Remark 1.3.10 The scalar product we have defined is a map

$$\sigma : \mathcal{V}_O^3 \times \mathcal{V}_O^3 \longrightarrow \mathbb{R}, \qquad \sigma(\mathbf{v}, \mathbf{w}) = \mathbf{v} \cdot \mathbf{w}.$$

Also, the scalar product of vectors on a plane is a map $\sigma : \mathcal{V}_O^2 \times \mathcal{V}_O^2 \longrightarrow \mathbb{R}$.

Definition 1.3.11 Let $\mathbf{v}, \mathbf{w} \in \mathcal{V}_O^3$. The *vector product* between $\mathbf{v}$ and $\mathbf{w}$, denoted by $\mathbf{v} \wedge \mathbf{w}$, is defined as the vector in $\mathcal{V}_O^3$ whose modulus is

$$\|\mathbf{v} \wedge \mathbf{w}\| = \|\mathbf{v}\| \; \|\mathbf{w}\| \sin \alpha,$$

where $\alpha = \widehat{\mathbf{v}\mathbf{w}}$, with $0 < \alpha < \pi$ is the angle defined by $\mathbf{v}$ e $\mathbf{w}$; the direction of $\mathbf{v} \wedge \mathbf{w}$ is orthogonal to both $\mathbf{v}$ and $\mathbf{w}$; and its orientation is such that $(\mathbf{v}, \mathbf{w}, \mathbf{v} \wedge \mathbf{w})$ is a right-handed triple as in Definition 1.2.4.

Remark 1.3.12 The following properties follow directly from the definition.

(i) if $\mathbf{v} = \mathbf{0}$ then $\mathbf{v} \wedge \mathbf{w} = \mathbf{0}$,
(ii) if $\mathbf{v}$ and $\mathbf{w}$ are both non zero then

$$\mathbf{v} \wedge \mathbf{w} = 0 \quad \Longleftrightarrow \quad \sin \alpha = 0 \quad \Longleftrightarrow \quad \mathbf{v} \parallel \mathbf{w},$$

(one trivially has $\mathbf{v} \wedge \mathbf{v} = 0$),
(iii) if $(O; \mathbf{i}, \mathbf{j}, \mathbf{k})$ is an orthogonal cartesian coordinate system, then

$$\mathbf{i} \wedge \mathbf{j} = \mathbf{k} = -\mathbf{j} \wedge \mathbf{i}, \qquad \mathbf{j} \wedge \mathbf{k} = \mathbf{i} = -\mathbf{k} \wedge \mathbf{j}, \qquad \mathbf{k} \wedge \mathbf{i} = \mathbf{j} = -\mathbf{i} \wedge \mathbf{k}.$$

We omit to prove the following proposition.

Proposition 1.3.13 *For any* $\mathbf{u}, \mathbf{v}, \mathbf{w} \in \mathcal{V}_O^3$ *and* $\lambda \in \mathbb{R}$, *the following identities holds:*

(i) $\mathbf{v} \wedge \mathbf{w} = -\mathbf{w} \wedge \mathbf{v}$,
(ii) $(\lambda \mathbf{v}) \wedge \mathbf{w} = \mathbf{v} \wedge (\lambda \mathbf{w}) = \lambda(\mathbf{v} \wedge \mathbf{w})$
(iii) $\mathbf{u} \wedge (\mathbf{v} + \mathbf{w}) = \mathbf{u} \wedge \mathbf{v} + \mathbf{u} \wedge \mathbf{w}$,

Exercise 1.3.14 With respect to a given cartesian orthogonal coordinate system, consider in $\mathcal{V}_O^3$ the vectors $\mathbf{v} = (1, 0, -1)$ e $\mathbf{w} = (-2, 0, 2)$. To verify that they are parallel, we recall the abov e result (ii) in the Remark 1.3.12 and compute, using the Proposition 1.3.15, that $\mathbf{v} \wedge \mathbf{w} = 0$.

Proposition 1.3.15 *Let* $\mathbf{v} = (v_x, v_y, v_z)$ *and* $\mathbf{w} = (w_x, w_y, w_z)$ *be elements in* $\mathcal{V}_O^3$ *with respect to a given cartesian orthogonal coordinate system. It is*

$$\mathbf{v} \wedge \mathbf{w} = (v_y w_z - v_z w_y, v_z w_x - v_x w_z, v_x w_y - v_y w_x).$$

Proof Given the Remark 1.3.12 and the Proposition 1.3.13, this comes as an easy computation. □

Remark 1.3.16 The vector product defines a map

$$\tau : \mathcal{V}_O^3 \times \mathcal{V}_O^3 \longrightarrow \mathcal{V}_O^3, \qquad \tau(\mathbf{v}, \mathbf{w}) = \mathbf{v} \wedge \mathbf{w}.$$

Clearly, such a map has no meaning on a plane.

Example 1.3.17 By slightly extending the Definition 1.3.11, one can use the vector product for additional notions coming from physics. Following Sect. 1.1, we consider vectors $\mathbf{u}, \mathbf{w}$ as elements in $\mathcal{W}^3$, that is vectors applied at arbitrary points in the physical three dimensional space $\mathcal{S}$, with components $\mathbf{u} = (u_x, u_y, u_z)$ and $\mathbf{w} = (w_x, w_y, w_z)$ with respect to a cartesian orthogonal coordinate system $\Sigma = (O; \mathbf{i}, \mathbf{j}, \mathbf{k})$. In parallel with Proposition 1.3.15, we define $\tau : \mathcal{W}^3 \times \mathcal{W}^3 \to \mathcal{W}^3$ as

$$\mathbf{u} \wedge \mathbf{w} = (u_y w_z - u_z w_y, \; u_z w_x - u_x w_z, \; u_x w_y - u_y w_x).$$

If $\mathbf{u} \in \mathcal{V}_{\mathbf{x}}^3$ is a vector applied at $\mathbf{x}$, its *momentum* with respect to a point $\mathbf{x}' \in \mathcal{S}$ is the vector in $\mathcal{W}^3$ defined by

$$\mathbf{M} = (\mathbf{x} - \mathbf{x}') \wedge \mathbf{u}.$$

In particular, if $\mathbf{u} = \mathbf{F}$ is a force acting on a point mass in $\mathbf{x}$, its momentum is $\mathbf{M} = (\mathbf{x} - \mathbf{x}') \wedge \mathbf{F}$.

If $\mathbf{x}(t) \in \mathcal{V}_O^3$ describes the motion of a point mass (with mass $m > 0$), whose velocity is $\mathbf{v}(t)$, then its corresponding *angular momentum* with respect to a point $\mathbf{x}'$ is defined by

$$\mathbf{L}_{\mathbf{x}'}(t) = (\mathbf{x}(t) - \mathbf{x}') \wedge m\mathbf{v}(t).$$

Exercise 1.3.18 The angular momentum is usually defined with respect to the origin of the coordinate system Σ, giving $\mathbf{L}_O(t) = \mathbf{x}(t) \wedge m\mathbf{v}(t)$. If we consider a circular uniform motion

$$\mathbf{x}(t) = \Big(x(t) = r\cos(\omega t),\ \ y(t) = r\sin(\omega t),\ \ z(t) = 0\Big),$$

with $r > 0$ the radius of the trajectory and $\omega \in \mathbb{R}$ the angular velocity, then

$$\mathbf{v}(t) = \Big(v_x(t) = -r\omega\sin(\omega t),\ \ y(t) = r\omega\cos(\omega t),\ \ v_z(t) = 0\Big)$$

so that

$$\mathbf{L}_O(t) = (0, 0, mr\omega).$$

Thus, a circular motion on the xy plane has angular momentum along the z axis.

Definition 1.3.19 Given an *ordered* triple $\mathbf{u}, \mathbf{v}, \mathbf{w} \in \mathcal{V}^3_O$, their *mixed product* is the real number

$$\mathbf{u} \cdot (\mathbf{v} \wedge \mathbf{w}).$$

Proposition 1.3.20 *Given a cartesian orthogonal coordinate system in $\mathcal{S}$ with $\mathbf{u} = (u_x, u_y, u_z)$, $\mathbf{v} = (v_x, v_y, v_z)$ and $\mathbf{w} = (w_x, w_y, w_z)$ in $\mathcal{V}^3_O$, one has*

$$\mathbf{u} \cdot (\mathbf{v} \wedge \mathbf{w}) = u_x(v_y w_z - v_z w_y) + u_y(v_z w_x - v_x w_z) + u_z(v_x w_y - v_y w_x).$$

Proof It follows immediately by Propositions 1.3.5 and 1.3.15. □

In the space $\mathcal{S}$, the vector product between $\mathbf{u} \wedge \mathbf{w}$ is the area of the parallelogram defined by $\mathbf{u}$ and $\mathbf{v}$, while the mixed product $\mathbf{u} \cdot (\mathbf{v} \wedge \mathbf{w})$ give the volume of the parallelepiped defined by $\mathbf{u}, \mathbf{v}, \mathbf{w}$.

Proposition 1.3.21 *Given $\mathbf{u}, \mathbf{v}, \mathbf{w} \in \mathcal{V}^3_O$.*

1. *Denote $\alpha = \widehat{\mathbf{v}\mathbf{w}}$ the angle defined by $\mathbf{v}$ and $\mathbf{w}$. Then, the area A of the parallelogram whose edges are $\mathbf{u}$ and $\mathbf{v}$, is given by*

$$A = \|\mathbf{v}\|\ \|\mathbf{w}\| \sin\alpha = \|\mathbf{v} \wedge \mathbf{w}\|.$$

2. *Denote $\theta = \widehat{\mathbf{u}(\mathbf{v} \wedge \mathbf{w})}$ the angle defined by $\mathbf{u}$ and $\mathbf{v} \wedge \mathbf{w}$. Then the volume V of the parallelepiped whose edges are $\mathbf{u}, \mathbf{v}, \mathbf{w}$, is given by*

$$V = A\|\mathbf{u}\| \cos\theta = \|\mathbf{u} \cdot \mathbf{v} \wedge \mathbf{w}\|.$$

Proof The claim is evident, as shown in the Figs. 1.8 and 1.9. □

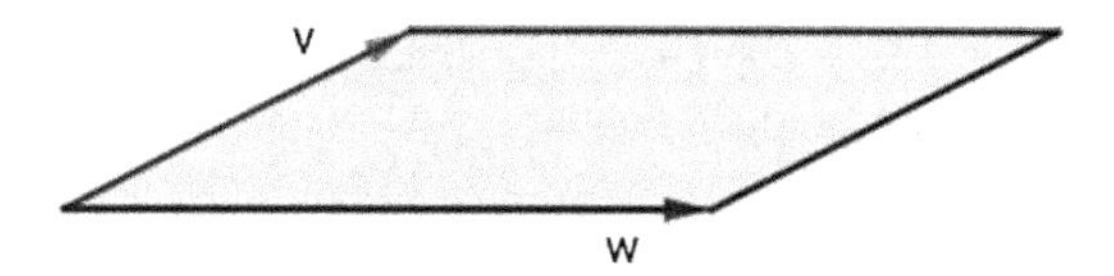

Fig. 1.8 The area of the parallelogramm with edges **v** and **w**

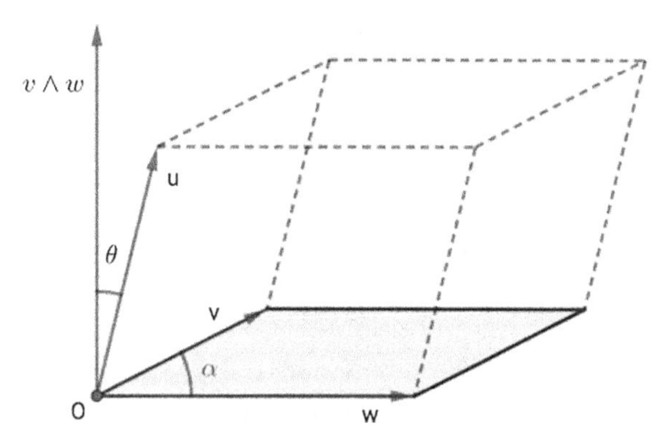

Fig. 1.9 The volume of the parallelogramm with edges **v**, **w**, **u**

1.4 Divergence, Rotor, Gradient and Laplacian

We close this chapter by describing how the notion of vector applied at a point also allows one to introduce a definition of a *vector field.*

The intuition coming from physics requires to consider, for each point $\mathbf{x}$ in the physical space $\mathcal{S}$, a vector applied at $\mathbf{x}$. We describe it as a map

$$\mathcal{S} \ni \mathbf{x} \quad \mapsto \quad \mathbf{A}(\mathbf{x}) \in \mathcal{V}^3_{\mathbf{x}}.$$

With respect to a given cartesian orthogonal reference system for $\mathcal{S}$ we can write this in components as $\mathbf{x} = (x_1, x_2, x_3)$ and $\mathbf{A}(\mathbf{x}) = (A_1(\mathbf{x}), A_2(\mathbf{x}), A_3(\mathbf{x}))$ and one can act on a vector field with partial derivatives (first order differential operators), $\partial_a = (\partial/\partial x_a)$ with $a = 1, 2, 3$, defined as usual by

$$\partial_a(x_b) = (\delta_{ab}), \qquad \text{with} \ \ \delta_{ab} = \begin{cases} 1 & \text{if } a = b \\ 0 & \text{if } a \neq b \end{cases}.$$

Then, (omitting the explicit dependence of $\mathbf{A}$ on $\mathbf{x}$) one defines

$$\begin{aligned} \operatorname{div} \mathbf{A} \;&=\; \sum_{k=1}^{3}(\partial_k A_k) \in \mathbb{R} \\ \operatorname{rot} \mathbf{A} \;&=\; (\partial_2 A_3 - \partial_3 A_2)\mathbf{i} \,+\, (\partial_3 A_1 - \partial_1 A_3)\mathbf{j} \,+\, (\partial_1 A_2 - \partial_2 A_1)\mathbf{k} \in \mathcal{V}^3_{\mathbf{x}}. \end{aligned}$$

By introducing the triple $\nabla = (\partial_1, \partial_2, \partial_3)$, such actions can be formally written as a scalar product and a vector product, that is

$$\begin{aligned}\operatorname{div}\mathbf{A} &= \nabla\cdot\mathbf{A}\\ \operatorname{rot}\mathbf{A} &= \nabla\wedge\mathbf{A}\,.\end{aligned}$$

Furthermore, if $f : \mathcal{S} \to \mathbb{R}$ is a real valued function defined on $\mathcal{S}$, that is a (real) *scalar field* on $\mathcal{S}$, one has the grad operator

$$\operatorname{grad} f = \nabla f = (\partial_1 f,\, \partial_2 f,\, \partial_3 f)$$

as well as the Laplacian operator

$$\nabla^2 f = \operatorname{div}(\nabla f) = \Big(\sum_{k=1}^{3} \partial_k\partial_k\Big) f = \partial_1^2 f + \partial_2^2 f + \partial_3^2 f\,.$$

Exercise 1.4.1 The properties of the mixed products yields a straightforward proof of the identity

$$\operatorname{div}(\operatorname{rot}\mathbf{A}) = \nabla\cdot(\nabla\wedge\mathbf{A}) = 0\,,$$

for any vector field **A**. On the other hand, a direct computation shows also the identity

$$\operatorname{rot}(\operatorname{grad} f) = \nabla\wedge(\operatorname{grad} f) = 0\,,$$

for any scalar field f.

Chapter 2
Vector Spaces

The notion of vector space can be defined over any field $\mathbb{K}$. We shall mainly consider the case $\mathbb{K} = \mathbb{R}$ and briefly mention the case $\mathbb{K} = \mathbb{C}$. Starting from our exposition, it is straightforward to generalise to any field.

2.1 Definition and Basic Properties

The model of the construction is the collection of all vectors in the space applied at a point with the operations of sum and multiplication by a scalar, as described in the Chap. 1.

Definition 2.1.1 A non empty set V is called a *vector space over* $\mathbb{R}$ (or a *real vector space* or an $\mathbb{R}$-*vector space*) if there are defined two operations,

(a) an internal one: a sum of vectors $s : V \times V \to V$,

$$V \times V \ni (v, v') \mapsto s(v, v') = v + v',$$

(b) an exterior one: the product by a scalar $p : \mathbb{R} \times V \to V$

$$\mathbb{R} \times V \ni (k, v) \mapsto p(k, v) = kv,$$

and these operations are required to satisfy the following conditions:

(1) There exists an element $0_V \in V$, which is neutral for the sum, such that $(V, +, 0_V)$ is an abelian group.
For any $k, k' \in \mathbb{R}$ and $v, v' \in V$ one has
(2) $(k + k')v = kv + k'v$
(3) $k(v + v') = kv + kv'$

G. Landi and A. Zampini, *Linear Algebra and Analytic Geometry for Physical Sciences*, Undergraduate Lecture Notes in Physics,
https://doi.org/10.1007/978-3-319-78361-1_2

(4) $k(k'v) = (kk')v$
(5) $1v = v$, with $1 = 1_{\mathbb{R}}$.

The elements of a vector space are called *vectors*; the element 0_V is the *zero* or *null* vector. A vector space is also called a *linear space*.

Remark 2.1.2 Given the properties of a group (see A.2.9), the null vector 0_V and the opposite $-v$ to any vector v are (in any given vector space) unique. The sums can be indeed simplified, that is $v + w = v + u \Longrightarrow w = u$. Such a statement is easily proven by adding to both terms in $v + w = v + u$ the element $-v$ and using the associativity of the sum.

As already seen in Chap. 1, the collections $\mathcal{V}_O^2$ (vectors in a plane) and $\mathcal{V}_O^3$ (vectors in the space) applied at the point O are real vector spaces. The bijection $\mathcal{V}_O^3 \longleftrightarrow \mathbb{R}^3$ introduced in the Definition 1.2.5, together with the Remark 1.2.9, suggest the natural definitions of sum and product by a scalar for the set $\mathbb{R}^3$ of ordered triples of real numbers.

Proposition 2.1.3 *The collection* $\mathbb{R}^3$ *of triples of real numbers together with the operations defined by*

I. $(x_1, x_2, x_3) + (y_1, y_2, y_3) = (x_1 + y_1, x_2 + y_2, x_3 + y_3)$, *for any* (x_1, x_2, x_3), $(y_1, y_2, y_3) \in \mathbb{R}^3$,
II. $a(x_1, x_2, x_3) = (ax_1, ax_2, ax_3)$, *for any* $a \in \mathbb{R}$, $(x_1, x_2, x_3) \in \mathbb{R}^3$,

is a real vector space.

Proof We verify that the conditions given in the Definition 2.1.1 are satisfied. We first notice that (a) and (b) are fullfilled, since $\mathbb{R}^3$ is closed with respect to the operations in I. and II. of sum and product by a scalar. The neutral element for the sum is $0_{\mathbb{R}^3} = (0, 0, 0)$, since one clearly has

$$(x_1, x_2, x_3) + (0, 0, 0) = (x_1, x_2, x_3).$$

The datum $(\mathbb{R}^3, +, 0_{\mathbb{R}^3})$ is an abelian group, since one has

- The sum $(\mathbb{R}^3, +)$ is associative, from the associativity of the sum in $\mathbb{R}$:

$$\begin{aligned}(x_1, x_2, x_3) &+ ((y_1, y_2, y_3) + (z_1, z_2, z_3)) \\ &= (x_1, x_2, x_3) + (y_1 + z_1, y_2 + z_2, y_3 + z_3) \\ &= (x_1 + (y_1 + z_1), x_2 + (y_2 + z_2), x_3 + (y_3 + z_3)) \\ &= ((x_1 + y_1) + z_1, (x_2 + y_2) + z_2, (x_3 + y_3) + z_3) \\ &= (x_1 + y_1, x_2 + y_2, x_3 + y_3) + (z_1, z_2, z_3) \\ &= ((x_1, x_2, x_3) + (y_1, y_2, y_3)) + (z_1, z_2, z_3).\end{aligned}$$

- From the identity

$$(x_1, x_2, x_3) + (-x_1, -x_2, -x_3) = (x_1 - x_1, x_2 - x_2, x_3 - x_3) = (0, 0, 0)$$

one has $(-x_1, -x_2, -x_3)$ as the opposite in $\mathbb{R}^3$ of the element (x_1, x_2, x_3).
- The group $(\mathbb{R}^3, +)$ is commutative, since the sum in $\mathbb{R}$ is commutative:

$$\begin{aligned}(x_1, x_2, x_3) + (y_1, y_2, y_3) &= (x_1 + y_1, x_2 + y_2, x_3 + y_3) \\ &= (y_1 + x_1, y_2 + x_2, y_3 + x_3) \\ &= (y_1, y_2, y_3) + (x_1, x_2, x_3).\end{aligned}$$

We leave to the reader the task to show that the conditions (1), (2), (3), (4) in Definition 2.1.1 are satisfied: for any $\lambda, \lambda' \in \mathbb{R}$ and any $(x_1, x_2, x_3), (y_1, y_2, y_3) \in \mathbb{R}^3$ it holds that

1. $(\lambda + \lambda')(x_1, x_2, x_3) = \lambda(x_1, x_2, x_3) + \lambda'(x_1, x_2, x_3)$
2. $\lambda((x_1, x_2, x_3) + (y_1, y_2, y_3)) = \lambda(x_1, x_2, x_3) + \lambda(y_1, y_2, y_3)$
3. $\lambda(\lambda'(x_1, x_2, x_3)) = (\lambda\lambda')(x_1, x_2, x_3)$
4. $1(x_1, x_2, x_3) = (x_1, x_2, x_3)$. □

The previous proposition can be generalised in a natural way. If $n \in \mathbb{N}$ is a positive natural number, one defines the n-th cartesian product of $\mathbb{R}$, that is the collection of ordered n-tuples of real numbers

$$\mathbb{R}^n = \{X = (x_1, \dots, x_n) \ : \ x_k \in \mathbb{R}\},$$

and the following operations, with $a \in \mathbb{R}$, $(x_1, \dots, x_n), (y_1, \dots, y_n) \in \mathbb{R}^n$:

In. $(x_1, \dots, x_n) + (y_1, \dots, y_n) = (x_1 + y_1, \dots, x_n + y_n)$
IIn. $a(x_1, \dots, x_n) = (ax_1, \dots, ax_n)$.

The previous proposition can be directly generalised to the following.

Proposition 2.1.4 *With respect to the above operations, the set $\mathbb{R}^n$ is a vector space over $\mathbb{R}$.*

The elements in $\mathbb{R}^n$ are called *n-tuples* of real numbers. With the notation $X = (x_1, \dots, x_n) \in \mathbb{R}^n$, the scalar x_k, with $k = 1, 2, \dots, n$, is the *k-th component* of the vector X.

Example 2.1.5 As in the Definition A.3.3, consider the collection of all polynomials in the indeterminate x and coefficients in $\mathbb{R}$, that is

$$\mathbb{R}[x] = \{f(x) = a_0 + a_1 x + a_2 x^2 + \cdots + a_n x^n \ : \ a_k \in \mathbb{R},\ n \geq 0\},$$

with the operations of sum and product by a scalar $\lambda \in \mathbb{R}$ defined, for any pair of elements in $\mathbb{R}[x]$, $f(x) = a_0 + a_1 x + a_2 x^2 + \cdots + a_n x^n$ and $g(x) = b_0 + b_1 x + b_2 x^2 + \cdots + b_m x^m$, component-wise by

Ip. $f(x) + g(x) = a_0 + b_0 + (a_1 + b_1)x + (a_2 + b_2)x^2 + \cdots$
IIp. $\lambda f(x) = \lambda a_0 + \lambda a_1 x + \lambda a_2 x^2 + \cdots + \lambda a_n x^n$.

Endowed with the previous operations, the set $\mathbb{R}[x]$ is a real vector space; $\mathbb{R}[x]$ is indeed closed with respect to the operations above. The null polynomial, denoted by $0_{\mathbb{R}[x]}$ (that is the polynomial with all coefficients equal zero), is the neutral element for the sum. The opposite to the polynomial $f(x) = a_0 + a_1 x + a_2 x^2 + \cdots + a_n x^n$ is the polynomial $(-a_0 - a_1 x - a_2 x^2 - \cdots - a_n x^n) \in \mathbb{R}[x]$ that one denotes by $-f(x)$. We leave to the reader to prove that $(\mathbb{R}[x], +, 0_{\mathbb{R}[x]})$ is an abelian group and that all the additional conditions in Definition 2.1.1 are fulfilled.

Exercise 2.1.6 We know from the Proposition A.3.5 that $\mathbb{R}[x]_r$, the subset in $\mathbb{R}[x]$ of polynomials with degree not larger than a fixed $r \in \mathbb{N}$, is closed under addition of polynomials. Since the degree of the polynomial $\lambda f(x)$ coincides with the degree of $f(x)$ for any $\lambda \neq 0$, we see that also the product by a scalar, as defined in IIp. above, is defined consistently on $\mathbb{R}[x]_r$. It is easy to verify that also $\mathbb{R}[x]_r$ is a real vector space.

Remark 2.1.7 The proof that $\mathbb{R}^n$, $\mathbb{R}[x]$ and $\mathbb{R}[x]_r$ are vector space over $\mathbb{R}$ relies on the properties of $\mathbb{R}$ as a field (in fact a ring, since the multiplicative inverse in $\mathbb{R}$ does not play any role).

Exercise 2.1.8 The set $\mathbb{C}^n$, that is the collection of ordered n-tuples of complex numbers, can be given the structure of a vector space over $\mathbb{C}$. Indeed, both the operations In. and IIn. considered in the Proposition 2.1.3 when intended for complex numbers make perfectly sense:

Ic. $(z_1, \ldots, z_n) + (w_1, \ldots, w_n) = (z_1 + w_1, \ldots, z_n + w_n)$
IIc. $c(z_1, \ldots, z_n) = (cz_1, \ldots, cz_n)$

with $c \in \mathbb{C}$, and $(z_1, \ldots, z_n), (w_1, \ldots, w_n) \in \mathbb{C}^n$.

The reader is left to show that $\mathbb{C}^n$ is a vector space over $\mathbb{C}$.

The space $\mathbb{C}^n$ can also be given a structure of vector space over $\mathbb{R}$, by noticing that the product of a complex number by a real number is a complex number. This means that $\mathbb{C}^n$ is closed with respect to the operations of (component-wise) product by a real scalar. The condition IIc. above makes sense when $c \in \mathbb{R}$.

We next analyse some elementary properties of general vector spaces.

Proposition 2.1.9 *Let V be a vector space over $\mathbb{R}$. For any $k \in \mathbb{R}$ and any $v \in V$ it holds that:*

(i) $0_{\mathbb{R}} v = 0_V$,
(ii) $k 0_V = 0_V$,
(iii) *if* $kv = 0_V$ *then it is either* $k = 0_{\mathbb{R}}$ *or* $v = 0_V$,
(iv) $(-k)v = -(kv) = k(-v)$.

Proof (i) From $0_{\mathbb{R}} v = (0_{\mathbb{R}} + 0_{\mathbb{R}})v = 0_{\mathbb{R}} v + 0_{\mathbb{R}} v$, since the sums can be simplified, one has that $0_{\mathbb{R}} v = 0_V$.
(ii) Analogously: $k0_V = k(0_V + 0_V) = k0_V + k0_V$ which yields $k0_V = 0_V$.

(iii) Let $k \neq 0$, so $k^{-1} \in \mathbb{R}$ exists. Then, $v = 1v = k^{-1}kv = k^{-1}0_V = 0_V$, with the last equality coming from (ii).

(iv) Since the product is distributive over the sum, from (i) it follows that $kv + (-k)v = (k + (-k))v = 0_{\mathbb{R}}v = 0_V$ that is the first equality. For the second, one writes analogously $kv + k(-v) = k(v - v) = k0_V = 0_V$ □

Relations (i), (ii), (iii) above are more succinctly expressed by the equivalence:

$$kv = 0_V \iff k = 0_{\mathbb{R}} \quad \text{or} \quad v = 0_V.$$

2.2 Vector Subspaces

Among the subsets of a real vector space, of particular relevance are those which inherit from V a vector space structure.

Definition 2.2.1 Let V be a vector space over $\mathbb{R}$ with respect to the sum s and the product p as given in the Definition 2.1.1. Let $W \subseteq V$ be a subset of V. One says that W is a *vector subspace* of V if the restrictions of s and p to W equip W with the structure of a vector space over $\mathbb{R}$.

In order to establish whether a subset $W \subseteq V$ of a vector space is a vector subspace, the following can be seen as *criteria*.

Proposition 2.2.2 *Let W be a non empty subset of the real vector space V. The following conditions are equivalent.*

(i) W is a vector subspace of V,

(ii) W is closed with respect to the sum and the product by a scalar, that is

(a) $w + w' \in W$, for any $w, w' \in W$,

(b) $kw \in W$, for any $k \in \mathbb{R}$ and $w \in W$,

(iii) $kw + k'w' \in W$, for any $k, k' \in \mathbb{R}$ and any $w, w' \in W$.

Proof The implications (i) $\Longrightarrow$ ii) and (ii) $\Longrightarrow$ (iii) are obvious from the definition.

(iii) $\Longrightarrow$ (ii): By taking $k = k' = 1$ one obtains (a), while to show point (b) one takes $k' = 0_{\mathbb{R}}$.

(ii) $\Longrightarrow$ (i): Notice that, by hypothesis, W is closed with respect to the sum and product by a scalar. Associativity and commutativity hold in W since they hold in V.

One only needs to prove that W has a neutral element 0_W and that, for such a neutral element, any vector in W has an opposite in W. If $0_V \in W$, then 0_V is the zero element in W: for any $w \in W$ one has $0_V + w = w + 0_V = w$ since $w \in V$; from ii, (b) one has $0_{\mathbb{R}}w \in W$ for any $w \in W$; from the Proposition 2.1.9 one has $0_{\mathbb{R}}w = 0_V$; collecting these relations, one concludes that $0_V \in W$. If $w \in W$, again from the Proposition 2.1.9 one gets that $-w = (-1)w \in W$. □

Exercise 2.2.3 Both $W = \{0_V\} \subset V$ and $W = V \subseteq V$ are *trivial* vector subspaces of V.

Exercise 2.2.4 We have already seen that $\mathbb{R}[x]_r \subseteq \mathbb{R}[x]$ are vector spaces with respect to the same operations, so we may conclude that $\mathbb{R}[x]_r$ is a vector subspace of $\mathbb{R}[x]$.

Exercise 2.2.5 Let $v \in V$ a non zero vector in a vector space, and let

$$\mathcal{L}(v) = \{av \ : \ a \in \mathbb{R}\} \subset V$$

be the collection of all multiples of v by a real scalar. Given the elements $w = av$ and $w' = a'v$ in $\mathcal{L}(v)$, from the equality

$$\alpha w + \alpha' w' = (\alpha a + \alpha' a')v \in \mathcal{L}(v)$$

for any $\alpha, \alpha' \in \mathbb{R}$, we see that, from the Proposition 2.2.2, $\mathcal{L}(v)$ is a vector subspace of V, and we call it the *(vector) line generated by* v.

Exercise 2.2.6 Consider the following subsets $W \subset \mathbb{R}^2$:

1. $W_1 = \{(x, y) \in \mathbb{R}^2 \ : \ x - 3y = 0\}$,
2. $W_2 = \{(x, y) \in \mathbb{R}^2 \ : \ x + y = 1\}$,
3. $W_3 = \{(x, y) \in \mathbb{R}^2 \ : \ x \in \mathbb{N}\}$,
4. $W_4 = \{(x, y) \in \mathbb{R}^2 \ : \ x^2 - y = 0\}$.

From the previous exercise, one sees that W_1 is a vector subspace since $W_1 = \mathcal{L}((3, 1))$. On the other hand, W_2, W_3, W_4 are not vector subspaces of $\mathbb{R}^2$. The zero vector $(0, 0) \notin W_2$; while W_3 and W_4 are not closed with respect to the product by a scalar, since, for example, $(1, 0) \in W_3$ but $\frac{1}{2}(1, 0) = (\frac{1}{2}, 0) \notin W_3$. Analogously, $(1, 1) \in W_4$ but $2(1, 1) = (2, 2) \notin W_4$.

The next step consists in showing how, given two or more vector subspaces of a real vector space V, one can define new vector subspaces of V via suitable operations.

Proposition 2.2.7 *The intersection $W_1 \cap W_2$ of any two vector subspaces W_1 and W_2 of a real vector space V is a vector subspace of V.*

Proof Consider $a, b \in \mathbb{R}$ and $v, w \in W_1 \cap W_2$. From the Propostion 2.2.2 it follows that $av + bw \in W_1$ since W_1 is a vector subspace, and also that $av + bw \in W_2$ for the same reason. As a consequence, one has $av + bw \in W_1 \cap W_2$. □

Remark 2.2.8 In general, the union of two vector subspaces of V is *not* a vector subspace of V. As an example, the Fig. 2.1 shows that, if $\mathcal{L}(v)$ and $\mathcal{L}(w)$ are generated by different $v, w \in \mathbb{R}^2$, then $\mathcal{L}(v) \cup \mathcal{L}(w)$ is not closed under the sum, since it does not contain the sum $v + w$, for instance.

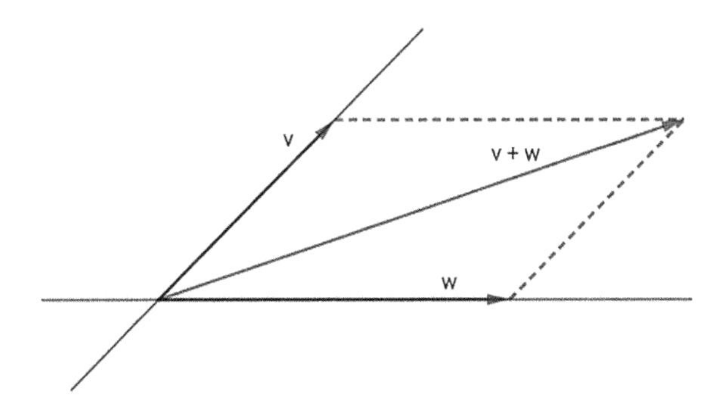

Fig. 2.1 The vector line $\mathcal{L}(v+w)$ with respect to the vector lines $\mathcal{L}(v)$ and $\mathcal{L}(w)$

Proposition 2.2.9 *Let W_1 and W_2 be vector subspaces of the real vector space V and let $W_1 + W_2$ denote*

$$W_1 + W_2 = \{v \in V \mid v = w_1 + w_2;\ w_1 \in W_1,\ w_2 \in W_2\} \subset V.$$

Then $W_1 + W_2$ is the smallest vector subspace of V which contains the union $W_1 \cup W_2$.

Proof Let $a, a' \in \mathbb{R}$ and $v, v' \in W_1 + W_2$; this means that there exist $w_1, w_1' \in W_1$ and $w_2, w_2' \in W_2$, so that $v = w_1 + w_2$ and $v' = w_1' + w_2'$. Since both W_1 and W_2 are vector subspaces of V, from the identity

$$av + a'v' = aw_1 + aw_2 + a'w_1' + a'w_2' = (aw_1 + a'w_1') + (aw_2 + a'w_2'),$$

one has $aw_1 + a'w_1' \in W_1$ and $aw_2 + a'w_2' \in W_2$. It follows that $W_1 + W_2$ is a vector subspace of V.

It holds that $W_1 + W_2 \supseteq W_1 \cup W_2$: if $w_1 \in W_1$, it is indeed $w_1 = w_1 + 0_V$ in $W_1 + W_2$; one similarly shows that $W_2 \subset W_1 + W_2$.

Finally, let Z be a vector subspace of V containing $W_1 \cup W_2$; then for any $w_1 \in W_1$ and $w_2 \in W_2$ it must be $w_1 + w_2 \in Z$. This implies $Z \supseteq W_1 + W_2$, and then $W_1 + W_2$ is the smallest of such vector subspaces Z. □

Definition 2.2.10 If W_1 and W_2 are vector subspaces of the real vector space V the vector subspace $W_1 + W_2$ of V is called the *sum* of W_1 e W_2.

The previous proposition and definition are easily generalised, in particular:

Definition 2.2.11 If $W_1, \dots, W_n$ are vector subspaces of the real subspace V, the vector subspace

$$W_1 + \cdots + W_n = \{v \in V \mid v = w_1 + \cdots + w_n;\ w_i \in W_i,\ i = 1, \dots, n\}$$

of V is the *sum* of $W_1, \dots, W_n$.

Definition 2.2.12 Let W_1 and W_2 be vector subspaces of the real vector space V. The sum $W_1 + W_2$ is called *direct* if $W_1 \cap W_2 = \{0_V\}$. A direct sum is denoted $W_1 \oplus W_2$.

Proposition 2.2.13 *Let W_1, W_2 be vector subspaces of the real vector space V. Their sum $W = W_1 + W_2$ is direct if and only if any element $v \in W_1 + W_2$ has a* unique *decomposition as $v = w_1 + w_2$ with $w_i \in W_i$, $i = 1, 2$.*

Proof We first suppose that the sum $W_1 + W_2$ is direct, that is $W_1 \cap W_2 = \{0_V\}$. If there exists an element $v \in W_1 + W_2$ with $v = w_1 + w_2 = w'_1 + w'_2$, and $w_i, w'_i \in W_i$, then $w_1 - w'_1 = w'_2 - w_2$ and such an element would belong to both W_1 and W_2. This would then be zero, since $W_1 \cap W_2 = \{0_V\}$, and then $w_1 = w'_1$ and $w_2 = w'_2$.

Suppose now that any element $v \in W_1 + W_2$ has a unique decomposition $v = w_1 + w_2$ with $w_i \in W_i$, $i = 1, 2$. Let $v \in W_1 \cap W_2$; then $v \in W_1$ and $v \in W_2$ which gives $0_V = v - v \in W_1 + W_2$, so the zero vector has a unique decomposition. But clearly also $0_V = 0_V + 0_V$ and being the decomposition for 0_V unique, this gives $v = 0_V$. □

These proposition gives a natural way to generalise the notion of direct sum to an arbitrary number of vector subspaces of a given vector space.

Definition 2.2.14 Let $W_1, \dots, W_n$ be vector subspaces of the real vector space V. The sum $W_1 + \cdots + W_n$ is called *direct* if any of its element has a unique decomposition as $v = w_1 + \cdots + w_n$ with $w_i \in W_i$, $i = 1, \dots, n$. The direct sum vector subspace is denoted $W_1 \oplus \cdots \oplus W_n$.

2.3 Linear Combinations

We have seen in Chap. 1 that, given a cartesian coordinate system $\Sigma = (O; \mathbf{i}, \mathbf{j}, \mathbf{k})$ for the space $\mathcal{S}$, any vector $\mathbf{v} \in \mathcal{V}^3_O$ can be written as $\mathbf{v} = a\mathbf{i} + b\mathbf{j} + c\mathbf{k}$. One says that $\mathbf{v}$ is a *linear combination* of $\mathbf{i}, \mathbf{j}, \mathbf{k}$. From the Definition 1.2.5 we also know that, given Σ, the components (a, b, c) are uniquely determined by $\mathbf{v}$. For this one says that $\mathbf{i}, \mathbf{j}, \mathbf{k}$ are *linearly independent*. In this section we introduce these notions for an arbitrary vector space.

Definition 2.3.1 Let $v_1, \dots, v_n$ be arbitrary elements of a real vector space V. A vector $v \in V$ is a *linear combination* of $v_1, \dots, v_n$ if there exist n scalars $\lambda_1, \dots, \lambda_n \in \mathbb{R}$, such that

$$v = \lambda_1 v_1 + \cdots + \lambda_n v_n.$$

The collection of all linear combinations of the vectors $v_1, \dots, v_n$ is denoted by $\mathcal{L}(v_1, \dots, v_n)$. If $I \subseteq V$ is an arbitrary subset of V, by $\mathcal{L}(I)$ one denotes the collection of all possible linear combinations of vectors in I, that is

$$\mathcal{L}(I) = \{\lambda_1 v_1 + \cdots + \lambda_n v_n \mid \lambda_i \in \mathbb{R},\ v_i \in I,\ n \geq 0\}.$$

The set $\mathcal{L}(I)$ is also called the *linear span* of I.

Proposition 2.3.2 *The space $\mathcal{L}(v_1, \ldots, v_n)$ is a vector subspace of V, called the space* generated *by $v_1, \ldots, v_n$ or the* linear span *of the vectors $v_1, \ldots, v_n$.*

Proof After Proposition 2.2.2, it is enough to show that $\mathcal{L}(v_1, \ldots, v_n)$ is closed for the sum and the product by a scalar. Let $v, w \in \mathcal{L}(v_1, \ldots, v_n)$; it is then $v = \lambda_1 v_1 + \cdots + \lambda_n v_n$ and $w = \mu_1 v_1 + \cdots + \mu_n v_n$, for scalars $\lambda_1, \ldots, \lambda_n$ and $\mu_1, \ldots, \mu_n$. Recalling point (2) in the Definition 2.1.1, one has

$$v + w = (\lambda_1 + \mu_1)v_1 + \cdots + (\lambda_n + \mu_n)v_n \in \mathcal{L}(v_1, \ldots, v_n).$$

Next, let $\alpha \in \mathbb{R}$. Again from the Definition 2.1.1 (point 4)), one has $\alpha v = (\alpha\lambda_1)v_1 + \cdots + (\alpha\lambda_n)v_n$, which gives $\alpha v \in \mathcal{L}(v_1, \ldots, v_n)$. □

Exercise 2.3.3 The following are two examples for the notion just introduced.

(1) Clearly one has $\mathcal{V}^2_O = \mathcal{L}(\mathbf{i}, \mathbf{j})$ and $\mathcal{V}^3_O = \mathcal{L}(\mathbf{i}, \mathbf{j}, \mathbf{k})$.
(2) Let $v = (1, 0, -1)$ and $w = (2, 0, 0)$ be two vectors in $\mathbb{R}^3$; it is easy to see that $\mathcal{L}(v, w)$ is a proper subset of $\mathbb{R}^3$. For example, the vector $\mathbf{u} = (0, 1, 0) \notin \mathcal{L}(v, w)$. If $\mathbf{u}$ were in $\mathcal{L}(v, w)$, there should be $\alpha, \beta \in \mathbb{R}$ such that

$$(0, 1, 0) = \alpha(1, 0, -1) + \beta(2, 0, 0) = (\alpha + 2\beta, 0, -\alpha).$$

No choice of $\alpha, \beta \in \mathbb{R}$ can satisfy this vector identity, since the second component equality would give $1 = 0$, independently of α, β.

It is interesting to explore which subsets $I \subseteq V$ yield $\mathcal{L}(I) = V$. Clearly, one has $V = \mathcal{L}(V)$. The example (1) above shows that there are proper subsets $I \subset V$ whose linear span coincides with V itself. We already know that $\mathcal{V}^2_O = \mathcal{L}(\mathbf{i}, \mathbf{j})$ and that $\mathcal{V}^3_O = \mathcal{L}(\mathbf{i}, \mathbf{j}, \mathbf{k})$: both $\mathcal{V}^3_O$ and $\mathcal{V}^2_O$ are generated by a *finite* number of (their) vectors. This is not always the case, as the following exercise shows.

Exercise 2.3.4 The real vector space $\mathbb{R}[x]$ is *not* generated by a finite number of vectors. Indeed, let $f_1(x), \ldots, f_n(x) \in \mathbb{R}[x]$ be arbitrary polynomials. Any $p(x) \in \mathcal{L}(f_1, \ldots, f_n)$ is written as

$$p(x) = \lambda_1 f_1(x) + \cdots + \lambda_n f_n(x)$$

with suitable $\lambda_1, \ldots, \lambda_n \in \mathbb{R}$. If one writes $d_i = \deg(f_i)$ and $d = \max\{d_1, \ldots, d_n\}$, from Remark A.3.5 one has that

$$\deg(p(x)) = \deg(\lambda_1 f_1(x) + \cdots + \lambda_n f_n(x)) \leq \max\{d_1, \ldots, d_n\} = d.$$

This means that any polynomial of degree $d + 1$ or higher is *not* contained in $\mathcal{L}(f_1, \ldots, f_n)$. This is the case for any *finite* n, giving a *finite* d; we conclude that, if n is finite, any $\mathcal{L}(I)$ with $I = (f_1(x), \ldots, f_n(x))$ is a proper subset of $\mathbb{R}[x]$ which can then *not* be generated by a *finite* number of polynomials.

On the other hand, $\mathbb{R}[x]$ is indeed the linear span of the infinite set

$$\{1, x, x^2, \dots, x^i, \dots\}.$$

Definition 2.3.5 A vector space V over $\mathbb{R}$ is said to be *finitely generated* if there exists a finite number of elements $v_1, \dots, v_n$ in V which are such that $V = \mathcal{L}(v_1, \dots, v_n)$. In such a case, the set $\{v_1, \dots, v_n\}$ is called a *system of generators* for V.

Proposition 2.3.6 *Let $I \subseteq V$ and $v \in V$. It holds that*

$$\mathcal{L}(\{v\} \cup I) = \mathcal{L}(I) \quad \iff \quad v \in \mathcal{L}(I).$$

Proof "$\Rightarrow$" Let us assume that $\mathcal{L}(\{v\} \cup I) = \mathcal{L}(I)$. Since $v \in \mathcal{L}(\{v\} \cup I)$, then $v \in \mathcal{L}(I)$.

"$\Leftarrow$" We shall prove the claim under the hypothesis that we have a finite system $\{v_1, \dots, v_n\}$. The inclusion $\mathcal{L}(I) \subseteq \mathcal{L}(\{v\} \cup I)$ is obvious. To prove the inclusion $\mathcal{L}(\{v\} \cup I) \subseteq \mathcal{L}(I)$, consider an arbitrary element $w \in \mathcal{L}(\{v\} \cup I)$, so that $w = \alpha v + \mu_1 v_1 + \cdots + \mu_n v_n$. By the hypothesis, $v \in \mathcal{L}(I)$ so it is $v = \lambda_1 v_1 + \cdots + \lambda_n v_n$. We can then write

$$w = \alpha(\lambda_1 v_1 + \cdots + \lambda_n v_n) + \mu_1 v_1 + \cdots + \mu_n v_n.$$

From the properties of the sum of vectors in V, one concludes that $w \in \mathcal{L}(v_1, \dots, v_n) = \mathcal{L}(I)$. □

Remark 2.3.7 From the previous proposition one has also the identity

$$\mathcal{L}(v_1, \dots, v_n, 0_V) = \mathcal{L}(v_1, \dots, v_n)$$

for any $v_1, \dots, v_n \in V$.

If I is a system of generators for V, the next question to address is whether I contains a *minimal* set of generators for V, that is whether there exists a set $J \subset I$ (with $J \neq I$) such that $\mathcal{L}(J) = \mathcal{L}(I) = V$. The answer to this question leads to the notion of *linear independence* for a set of vectors.

Definition 2.3.8 Given a collection $I = \{v_1, \dots, v_n\}$ of vectors in a real vector space V, the elements of I are called *linearly independent* on $\mathbb{R}$, and the system I is said to be *free*, if the following implication holds,

$$\lambda_1 v_1 + \cdots + \lambda_n v_n = 0_V \quad \Longrightarrow \quad \lambda_1 = \cdots = \lambda_n = 0_{\mathbb{R}}.$$

That is, if the *only* linear combination of elements of I giving the zero vector is the one whose coefficients are all zero.

Analogously, an infinite system $I \subseteq V$ is said to be *free* if any of its finite subsets is free.

The vectors $v_1, \dots, v_n \in V$ are said to be *linearly dependent* if they are not linearly independent, that is if there are scalars $(\lambda_1, \dots, \lambda_n) \neq (0, \dots, 0)$ such that $\lambda_1 v_1 + \cdots + \lambda_n v_n = 0_V$.

Exercise 2.3.9 It is clear that $\mathbf{i}, \mathbf{j}, \mathbf{k}$ are linearly independent in $\mathcal{V}_O^3$, while the vectors $v_1 = \mathbf{i} + \mathbf{j}$, $v_2 = \mathbf{j} - \mathbf{k}$ and $v_3 = 2\mathbf{i} - \mathbf{j} + 3\mathbf{k}$ are linearly dependent, since one computes that $2v_1 - 3v_2 - v_3 = 0$.

Proposition 2.3.10 *Let V be a real vector space and $I = \{v_1, \dots, v_n\}$ be a collection of vectors in V. The following properties hold true:*

(i) if $0_V \in I$, then I is not free,
(ii) I is not free if and only if one of the elements v_i is a linear combination of the other elements $v_1, \dots, v_{i-1}, v_{i+1}, \dots, v_n$,
(iii) if I is not free, then any $J \supseteq I$ is not free,
(iv) if I is free, then any J such that $J \subseteq I$ is free; that is any subsystem of a free system is free.

Proof i) Without loss of generality we suppose that $v_1 = 0_V$. Then, one has

$$1_{\mathbb{R}} v_1 + 0_{\mathbb{R}} v_2 + \cdots + 0_{\mathbb{R}} v_n = 0_V,$$

which amounts to say that the zero vector can be written as a linear combination of elements in I with a non zero coefficients.

(ii) Suppose I is not free. Then, there exists scalars $(\lambda_1, \dots, \lambda_n) \neq (0, \dots, 0)$ giving the combination $\lambda_1 v_1 + \cdots + \lambda_n v_n = 0_V$. Without loss of generality take $\lambda_1 \neq 0$; so λ_1 is invertible and we can write

$$v_1 = \lambda_1^{-1}(-\lambda_2 v_2 - \cdots - \lambda_n v_n) \in \mathcal{L}(v_2, \dots, v_n).$$

In order to prove the converse, we start by assuming that a vector v_i is a linear combination

$$v_i = \lambda_1 v_1 + \cdots + \lambda_{i-1} v_{i-1} + \lambda_{i+1} v_{i+1} + \cdots + \lambda_n v_n.$$

This identity can be written in the form

$$\lambda_1 v_1 + \cdots + \lambda_{i-1} v_{i-1} - v_i + \lambda_{i+1} v_{i+1} + \cdots + \lambda_n v_n = 0_V.$$

The zero vector is then written as a linear combination with coefficients not all identically zero, since the coefficient of v_i is -1. This amounts to say that the system I is not free.

We leave the reader to show the obvious points (iii) and (iv). □

2.4 Bases of a Vector Space

Given a real vector space V, in this section we determine its smallest possible systems of generators, together with their cardinalities.

Proposition 2.4.1 *Let V be a real vector space, with $v_1, \ldots, v_n \in V$. The following facts are equivalent:*

(i) the elements $v_1, \ldots, v_n$ are linearly independent,
(ii) $v_1 \neq 0_V$ and, for any $i \geq 2$, the vector v_i is not a linear combination of $v_1, \ldots, v_{i-1}$.

Proof The implication (i) $\Longrightarrow$ (ii) directly follows from the Proposition 2.3.10.

To show the implication (ii) $\Longrightarrow$ (i) we start by considering a combination $\lambda_1 v_1 + \cdots + \lambda_n v_n = 0_V$. Under the hypothesis, v_n is not a linear combination of $v_1, \ldots, v_{n-1}$, so it must be $\lambda_n = 0$: were it not, one could write $v_n = \lambda_n^{-1}(-\lambda_1 v_1 - \cdots - \lambda_{n-1} v_{n-1})$. We are then left with $\lambda_1 v_1 + \cdots + \lambda_{n-1} v_{n-1} = 0_V$, and an analogous reasoning leads to $\lambda_{n-1} = 0$. After $n - 1$ similar steps, one has $\lambda_1 v_1 = 0$; since $v_1 \neq 0$ by hypothesis, it must be (see 2.1.5) that $\lambda_1 = 0$. □

Theorem 2.4.2 *Any finite system of generators for a vector space V contains a free system of generators for V.*

Proof Let $I = \{v_1, \ldots, v_s\}$ be a system of generators for a real vector space V. Recalling the Remark 2.3.7, we can take $v_i \neq 0$ for any $i = 1, \ldots, s$. We define iteratively a system of subsets of I, as follows:

- take $I_1 = I = \{v_1, \ldots, v_s\}$,
- if $v_2 \in \mathcal{L}(v_1)$, take $I_2 = I_1 \setminus \{v_2\}$; if $v_2 \neq \mathcal{L}(v_1)$, take $I_2 = I_1$,
- if $v_3 \in \mathcal{L}(v_1, v_2)$, take $I_3 = I_2 \setminus \{v_3\}$; if $v_3 \neq \mathcal{L}(v_1, v_2)$, take $I_3 = I_2$,
- Iterate the steps above.

The whole procedure consists in examining any element in the starting $I_1 = I$, and deleting it if it is a linear combination of the previous ones. After s steps, one ends up with a chain $I_1 \supseteq \cdots \supseteq I_s \supseteq I$.

Notice that, for any $j = 2, \ldots, s$, it is $\mathcal{L}(I_j) = \mathcal{L}(I_{j-1})$. It is indeed either $I_j = I_{j-1}$ (which makes the claim obvious) or $I_{j-1} = I_j \cup \{v_j\}$, with $v_j \in \mathcal{L}(v_1, \ldots, v_{j-1}) \subseteq \mathcal{L}(I_{j-1})$; from Proposition 2.3.6, it follows that $\mathcal{L}(I_j) = \mathcal{L}(I_{j-1})$.

One has then $\mathcal{L}(I) = \mathcal{L}(I_1) = \cdots = \mathcal{L}(I_s)$, and I_s is a system of generators of V. Since no element in I_s is a linear combination of the previous ones, the Proposition 2.4.1 shows that I_s is free. □

Definition 2.4.3 Let V be a real vector space. An ordered system of vectors $I = (v_1, \ldots, v_n)$ in V is called a *basis* of V if I is a free system of generators for V, that is $V = \mathcal{L}(v_1, \ldots, v_n)$ and $v_1, \ldots, v_n$ are linearly independent.

Corollary 2.4.4 *Any finite system of generators for a vector space contains (at least) a basis. This means also that any finitely generated vector space has a basis.*

Proof It follows directly from the Theorem 2.4.2. □

Exercise 2.4.5 Consider the vector space $\mathbb{R}^3$ and the system of vectors $I = \{v_1, \ldots, v_5\}$ with

$$v_1 = (1, 1, -1), \quad v_2 = (-2, -2, 2), \quad v_3 = (2, 0, 1), \quad v_4 = (1, -1, 2), \quad v_5 = (0, 1, 1).$$

Following Theorem 2.4.2, we determine a basis for $\mathcal{L}(v_1, v_2, v_3, v_4, v_5)$.

- At the first step $I_1 = I$.
- Since $v_2 = -2v_1$, so that $v_2 \in \mathcal{L}(v_1)$, delete v_2 and take $I_2 = I_1 \setminus \{v_2\}$.
- One has $v_3 \notin \mathcal{L}(v_1)$, so keep v_3 and take $I_3 = I_2$.
- One has $v_4 \in \mathcal{L}(v_1, v_3)$ if and only if there exist $\alpha, \beta \in \mathbb{R}$ such that $v_4 = \alpha v_1 + \beta v_3$, that is $(1, -1, 2) = (\alpha + 2\beta, \alpha, -\alpha + \beta)$. By equating components, one has $\alpha = -1$, $\beta = 1$. This shows that $v_4 = -v_1 + v_3 \in \mathcal{L}(v_1, v_3)$; therefore delete v_4 and take $I_4 = I_3 \setminus \{v_4\}$.
- Similarly one shows that $v_5 \notin \mathcal{L}(v_1, v_3)$. A basis for $\mathcal{L}(I)$ is then $I_5 = I_4 = (v_1, v_3, v_5)$.

The next theorem characterises free systems.

Theorem 2.4.6 *A system $I = \{v_1, \ldots, v_n\}$ of vectors in V is free if and only if any element in $\mathcal{L}(v_1, \ldots, v_n)$ can be written in a* unique *way as a linear combination of the elements $v_1, \ldots, v_n$.*

Proof We assume that I is free and that $\mathcal{L}(v_1, \ldots, v_n)$ contains a vector, say v, which has two linear decompositions with respect to the vectors v_i:

$$v = \lambda_1 v_1 + \cdots + \lambda_n v_n = \mu_1 v_1 + \cdots + \mu_n v_n.$$

This identity would give $(\lambda_1 - \mu_1)v_1 + \cdots + (\lambda_n - \mu_n)v_n = 0_V$; since the elements v_i are linearly independent it would read

$$\lambda_1 - \mu_1 = 0, \quad \cdots \quad , \lambda_n - \mu_n = 0,$$

that is $\lambda_i = \mu_i$ for any $i = 1, \ldots, n$. This says that the two linear expressions above coincide and v is written in a unique way.

We assume next that any element in $\mathcal{L}(v_1, \ldots, v_n)$ as a unique linear decomposition with respect to the vectors v_i. This means that the zero vector $0_V \in \mathcal{L}(v_1, \ldots, v_n)$ has the unique decomposition $0_V = 0_{\mathbb{R}} v_1 + \cdots + 0_{\mathbb{R}} v_n$. Let us consider the expression $\lambda_1 v_1 + \cdots + \lambda_n v_n = 0_V$; since the linear decomposition of 0_V is unique, it is $\lambda_i = 0$ for any $i = 1, \ldots, n$. This says that the vectors $v_1, \ldots, v_n$ are linearly independent. □

Corollary 2.4.7 *Let $v_1, \ldots, v_n$ be elements of a real vector space V. The system $I = (v_1, \ldots, v_n)$ is a basis for V if and only if any element $v \in V$ can be written in a unique way as $v = \lambda_1 v_1 + \cdots + \lambda_n v_n$.*

Definition 2.4.8 Let $I = (v_1, \ldots, v_n)$ be a basis for the real vector space V. Any $v \in V$ is then written as a linear combination $v = \lambda_1 v_1 + \cdots + \lambda_n v_n$ is a unique way. The scalars $\lambda_1, \ldots, \lambda_n$ (which are uniquely determined by Corollary 2.4.7) are called the *components* of v on the basis I. We denote this by

$$v = (\lambda_1, \ldots, \lambda_n)_I.$$

Remark 2.4.9 Notice that we have taken a free system in a vector space V and a system of generators for V not to be ordered sets while on the other hand, the Definition 2.4.3 refers to a basis as an ordered set. This choice is motivated by the fact that it is more useful to consider the components of a vector on a given basis as an ordered array of scalars. For example, if $I = (v_1, v_2)$ is a basis for V, so it is $J = (v_2, v_1)$. But one considers I equivalent to J as systems of generators for V, not as bases.

Exercise 2.4.10 With $\mathcal{E} = (\mathbf{i}, \mathbf{j})$ and $\mathcal{E}' = (\mathbf{j}, \mathbf{i})$ two bases for $\mathcal{V}_O^2$, the vector $\mathbf{v} = 2\mathbf{i} + 3\mathbf{j}$ has the following components

$$\mathbf{v} = (2, 3)_{\mathcal{E}} = (3, 2)_{\mathcal{E}'}$$

when expressed with respect to them.

Remark 2.4.11 Consider the real vector space $\mathbb{R}^n$ and the vectors

$$\begin{aligned} e_1 &= (1, 0, \ldots, 0), \\ e_2 &= (0, 1, \ldots, 0), \\ &\vdots \\ e_n &= (0, 0, \ldots, 1). \end{aligned}$$

Since any element $v = (x_1, \ldots, x_n)$ can be uniquely written as

$$(x_1, \ldots, x_n) = x_1 e_1 + \cdots + x_n e_n,$$

the system $\mathcal{E} = (e_1, \ldots, e_n)$ is a basis for $\mathbb{R}^n$.

Definition 2.4.12 The system $\mathcal{E} = (e_1, \ldots, e_n)$ above is called the *canonical* basis for $\mathbb{R}^n$.

The canonical basis for $\mathbb{R}^2$ is $\mathcal{E} = (e_1, e_2)$, with $e_1 = (1, 0)$ and $e_2 = (0, 1)$; the canonical basis for $\mathbb{R}^3$ is $\mathcal{E} = (e_1, e_2, e_3)$, with $e_1 = (1, 0, 0)$, $e_2 = (0, 1, 0)$ and $e_3 = (0, 0, 1)$.

Remark 2.4.13 We have meaningfully introduced the notion of a canonical basis for $\mathbb{R}^n$. Our analysis so far should nonetheless make it clear that for an arbitrary vector space V over $\mathbb{R}$ there is no canonical choice of a basis. The exercises that follow indeed show that some vector spaces have bases which appear more natural than others, in a sense.

Exercise 2.4.14 We refer to the Exercise 2.1.8 and consider $\mathbb{C}$ as a vector space over $\mathbb{R}$. As such it is generated by the two elements 1 and i since any complex number can be written as $z = a + ib$, with $a, b \in \mathbb{R}$. Since the elements $1, \mathrm{i}$ are linearly independent over $\mathbb{R}$ they are a basis over $\mathbb{R}$ for $\mathbb{C}$.

As already seen in the Exercise 2.1.8, $\mathbb{C}^n$ is a vector space *both* over $\mathbb{C}$ *and* over $\mathbb{R}$. As a $\mathbb{C}$-vector space, $\mathbb{C}^n$ has canonical basis $\mathcal{E} = (e_1, \dots, e_n)$, where the elements e_i are given as in the Remark 2.4.11. For example, the canonical basis for $\mathbb{C}^2$ is $\mathcal{E} = (e_1, e_2)$, with $e_1 = (1, 0)$, $e_2 = (0, 1)$.

As a real vector space, $\mathbb{C}^n$ has no canonical basis. It is natural to consider for it the following basis $\mathcal{B} = (b_1, c_1 \dots, b_n, c_n)$, made of the $2n$ following elements,

$$\begin{aligned} b_1 &= (1, 0, \dots, 0), \quad c_1 = (\mathrm{i}, 0, \dots, 0), \\ b_2 &= (0, 1, \dots, 0), \quad c_2 = (0, \mathrm{i}, \dots, 0), \\ &\vdots \\ b_n &= (0, 0, \dots, 1), \quad c_n = (0, 0, \dots, \mathrm{i}). \end{aligned}$$

For $\mathbb{C}^2$ such a basis is $\mathcal{B} = (b_1, c_1, b_2, c_2)$, with $b_1 = (1, 0)$, $c_1 = (\mathrm{i}, 0)$, and $b_2 = (0, 1)$, $c_2 = (0, \mathrm{i})$.

Exercise 2.4.15 The real vector space $\mathbb{R}[x]_r$ has a natural basis given by all the monomials $(1, x, x^2, \dots, x^r)$ with degree less than r, since any element $p(x) \in \mathbb{R}[x]_r$ can be written in a unique way as

$$p(x) = a_0 + a_1 x + a_2 x^2 + \cdots a_r x^r,$$

with $a_i \in \mathbb{R}$.

Remark 2.4.16 We have seen in Chap. 1 that, by introducing a cartesian coordinate system in $\mathcal{V}_O^3$ and with the notion of components for the vectors, the vector space operations in $\mathcal{V}_O^3$ can be written in terms of operations among components. This fact is generalised in the following way.

Let $I = (v_1, \dots, v_n)$ be a basis for V. Let $v, w \in V$, with $v = (\lambda_1, \dots, \lambda_n)_I$ and $w = (\mu_1, \dots, \mu_n)_I$ the corresponding components with respect to I. We compute the components, with respect to I, of the vectors $v + w$. We have

$$\begin{aligned} v + w &= (\lambda_1 v_1 + \cdots + \lambda_n v_n) + (\mu_1 v_1 + \cdots + \mu_n v_n) \\ &= (\lambda_1 + \mu_1) v_1 + \cdots + (\lambda_n + \mu_n) v_n, \end{aligned}$$

so we can write

$$v + w = (\lambda_1 + \mu_1, \dots, \lambda_n + \mu_n)_I.$$

Next, with $a \in \mathbb{R}$ we also have

$$av = a(\lambda_1 v_1 + \cdots + \lambda_n v_n) = (a\lambda_1)v_1 + \cdots + (a\lambda_n)v_n,$$

so we can write

$$av = (a\lambda_1, \ldots, a\lambda_n)_I.$$

If $z = av + bw$ with, $z = (\xi_1, \ldots, \xi_n)_I$, it is immediate to see that

$$(\xi_1, \ldots, \xi_n)_I = (a\lambda_1 + b\mu_1, \ldots, a\lambda_n + b\mu_n)_I$$

or equivalently

$$\xi_i = a\lambda_i + b\mu_i, \qquad \text{for any} \quad i = 1, \ldots, n.$$

Proposition 2.4.17 *Let V be a vector space over $\mathbb{R}$, and $I = (v_1, \ldots, v_n)$ a basis for V. Consider a system*

$$w_1 = (\lambda_{11}, \ldots, \lambda_{1n})_I, \quad w_2 = (\lambda_{21}, \ldots, \lambda_{2n})_I, \quad \ldots, \quad w_s = (\lambda_{s1}, \ldots, \lambda_{sn})_I$$

of vectors in V, and denote $z = (\xi_1, \ldots, \xi_n)_I$. One has that

$$z = a_1 w_1 + \cdots + a_s w_s \quad \Longleftrightarrow \quad \xi_i = a_1\lambda_{1i} + \cdots + a_s\lambda_{si} \;\; \textit{for any} \;\; i = 1, \ldots, n.$$

The i-th component of the linear combination z of the vectors w_k, is given by the same linear combination of the i-th components of the vectors w_k.

Proof It comes as a direct generalisation of the previous remark. □

Corollary 2.4.18 *With the same notations as before, one has that*

(a) *the vectors $w_1, \ldots, w_s$ are linearly independent in V if and only if the corresponding n-tuples of components $(\lambda_{11}, \ldots, \lambda_{1n}), \ldots, (\lambda_{s1}, \ldots, \lambda_{sn})$ are linearly independent in $\mathbb{R}^n$,*
(b) *the vectors $w_1, \ldots, w_s$ form a system of generators for V if and only if the corresponding n-tuples of components $(\lambda_{11}, \ldots, \lambda_{1n}), \ldots, (\lambda_{s1}, \ldots, \lambda_{sn})$ generate $\mathbb{R}^n$.*

A free system can be completed to a basis for a given vector space.

Theorem 2.4.19 *Let V be a finitely generated real vector space. Any free finite system is contained in a basis for V.*

Proof Let $I = \{v_1, \ldots, v_s\}$ be a free system for the real vector space V. By the Corollary 2.4.4, V has a basis, that we denote $\mathcal{B} = (e_1, \ldots, e_n)$. The set $I \cup \mathcal{B} = \{v_1, \ldots, v_s, e_1, \ldots, e_n\}$ obviously generates V. By applying the procedure given in the Theorem 2.4.2, the first s vectors are not deleted, since they are linearly independent by hypothesis; the subsystem $\mathcal{B}'$ one ends up with at the end of the procedure will then be a basis for V that contains I. □

2.5 The Dimension of a Vector Space

The following (somewhat intuitive) result is given without proof.

Theorem 2.5.1 *Let V be a vector space over $\mathbb{R}$ with a basis made of n elements. Then,*

(i) any free system I in V contains at most *n elements,*
(ii) any system of generators for V has at least *n elements,*
(iii) any basis for V has n elements.

This theorem makes sure that the following definition is consistent.

Definition 2.5.2 If there exists a positive integer $n > 0$, such that the real vector space V has a basis with n elements, we say that V has *dimension* n, and write $\dim V = n$. If V is not finitely generated we set $\dim V = \infty$. If $V = \{0_V\}$ we set $\dim V = 0$.

Exercise 2.5.3 Following what we have extensively described above, it is clear that $\dim \mathcal{V}_O^2 = 2$ and $\dim \mathcal{V}_O^3 = 3$. Also $\dim \mathbb{R}^n = n$, with $\dim \mathbb{R} = 1$, and we have that $\dim \mathbb{R}[x] = \infty$ while $\dim \mathbb{R}[x]_r = r + 1$. Referring to the Exercise 2.4.14, one has that $\dim_{\mathbb{C}} \mathbb{C}^n = n$ while $\dim_{\mathbb{R}} \mathbb{C}^n = 2n$.

We omit the proof of the following results.

Proposition 2.5.4 *Let V be a n-dimensional vector space, and W a vector subspace of V. Then,* $\dim(W) \leq n$*, while* $\dim(W) = n$ *if and only if $W = V$.*

Corollary 2.5.5 *Let V be a n-dimensional vector space, and $v_1, \ldots, v_n \in V$. The following facts are equivalent:*

(i) the system $(v_1, \ldots, v_n)$ is a basis for V,
(ii) the system $\{v_1, \ldots, v_n\}$ is free,
(iii) the system $\{v_1, \ldots, v_n\}$ generates V.

Theorem 2.5.6 (Grassmann) *Let V a finite dimensional vector space, with U and W two vector subspaces of V. It holds that*

$$\dim(U + W) = \dim(U) + \dim(W) - \dim(U \cap W).$$

Proof Denote $r = \dim(U)$, $s = \dim(W)$ and $p = \dim(U \cap W)$. We need to show that $U + W$ has a basis with $r + s - p$ elements.

Let $(v_1, \ldots, v_p)$ be a basis for $U \cap W$. By the Theorem 2.4.19 such a free system can be completed to a basis $(v_1, \ldots, v_p, u_1, \ldots, u_{r-p})$ for U and to a basis $(v_1, \ldots, v_p, w_1, \ldots, w_{s-p})$ for W.

We then show that $I = (v_1, \ldots, v_p, u_1, \ldots, u_{r-p}, w_1, \ldots, w_{s-p})$ is a basis for the vector space $U + W$. Since any vector in $U + W$ has the form $u + w$, with $u \in U$ and $w \in W$, and since u is a linear combination of $v_1, \ldots, v_p, u_1, \ldots, u_{r-p}$, while w

is a linear combination of $v_1, \ldots, v_p, w_1, \ldots, w_{s-p}$, the system I generates $U + W$. Next, consider the combination

$$\alpha_1 v_1 + \cdots + \alpha_p v_p + \beta_1 u_1 + \cdots + \beta_{r-p} u_{r-p} + \gamma_1 w_1 + \cdots + \gamma_{s-p} w_{s-p} = 0_V.$$

Denoting for brevity $v = \sum_{i=1}^{p} \alpha_i v_i$, $u = \sum_{j=1}^{r-p} \beta_j u_j$ and $w = \sum_{k=1}^{s-p} \gamma_k w_k$, we write this equality as

$$v + u + w = 0_V,$$

with $v \in U \cap W$, $u \in U$, $w \in W$. Since $v, u \in U$, then $w = -v - u \in U$; so $w \in U \cap W$. This implies

$$w = \gamma_1 w_1 + \cdots + \gamma_{s-p} w_{s-p} = \lambda_1 v_1 + \cdots + \lambda_p v_p$$

for suitable scalars λ_i: in fact we know that $\{v_1, \ldots, v_p, w_1, \ldots, w_{s-p}\}$ is a free system, so any γ_k must be zero. We need then to prove that, from

$$\alpha_1 v_1 + \cdots + \alpha_p v_p + \beta_1 u_1 + \cdots + \beta_{r-p} u_{r-p} = 0_V$$

it follows that all the coefficients α_i and β_j are zero. This is true, since $(v_1, \ldots, v_p, u_1, \ldots, u_{r-p})$ is a basis for U. Thus I is a free system. □

Corollary 2.5.7 *Let W_1 and W_2 be vector subspaces of V. If $W_1 \oplus W_2$ can be defined, then*

$$\dim(W_1 \oplus W_2) = \dim(W_1) + \dim(W_2).$$

Also, if $\mathcal{B}_1 = (w'_1, \ldots, w'_s)$ and $\mathcal{B}_2 = (w''_1, \ldots, w''_r)$ are basis for W_1 and W_2 respectively, a basis for $W_1 \oplus W_2$ is given by $\mathcal{B} = (w'_1, \ldots, w'_s, w''_1, \ldots, w''_r)$.

Proof By the Grassmann theorem, one has

$$\dim(W_1 + W_2) + \dim(W_1 \cap W_2) = \dim(W_1) + \dim(W_2)$$

and from the Definition 2.2.12 we also have $\dim(W_1 \cap W_2) = 0$, which gives the first claim.

With the basis $\mathcal{B}_1$ and $\mathcal{B}_2$ one considers $\mathcal{B} = \mathcal{B}_1 \cup \mathcal{B}_2$ which obviously generates $W_1 \oplus W_2$. The second claim directly follows from the Corollary 2.5.5. □

The following proposition is a direct generalization.

Proposition 2.5.8 *Let $W_1, \ldots, W_n$ be subspaces of a real vector space V and let the direct sum $W_1 \oplus \cdots \oplus W_n$ be defined. One has that*

$$\dim(W_1 \oplus \cdots \oplus W_n) = \dim(W_1) + \cdots + \dim(W_n).$$

Chapter 3
Euclidean Vector Spaces

When dealing with vectors of $\mathcal{V}_O^3$ in Chap. 1, we have somehow implicitly used the notions of length for a vector and of orthogonality of vectors as well as amplitude of plane angle between vectors. In order to generalise all of this, in the present chapter we introduce the structure of *scalar product* for any vector space, thus coming to the notion of *euclidean* vector space. A scalar product allows one to speak, among other things, of orthogonality of vectors or of the length of a vector in an arbitrary vector space.

3.1 Scalar Product, Norm

We start by recalling, through an example, how the vector space $\mathbb{R}^3$ can be endowed with a euclidean scalar product using the usual scalar product in the space $\mathcal{V}_O^3$.

Example 3.1.1 The usual scalar product in $\mathcal{V}_O^3$, under the isomorphism $\mathbb{R}^3 \simeq \mathcal{V}_O^3$ (see the Proposition 1.3.9), induces a map

$$\cdot : \mathbb{R}^3 \times \mathbb{R}^3 \longrightarrow \mathbb{R}$$

defined as

$$(x_1, x_2, x_3) \cdot (y_1, y_2, y_3) = x_1 y_1 + x_2 y_2 + x_3 y_3.$$

For vectors $(x_1, x_2, x_3), (y_1, y_2, y_3), (z_1, z_2, z_3) \in \mathbb{R}^3$ and scalars $a, b \in \mathbb{R}$, the following properties are easy to verify.

G. Landi and A. Zampini, *Linear Algebra and Analytic Geometry for Physical Sciences*, Undergraduate Lecture Notes in Physics,
https://doi.org/10.1007/978-3-319-78361-1_3

(i) Symmetry, that is:

$$(x_1, x_2, x_3) \cdot (y_1, y_2, y_3) = x_1 y_1 + x_2 y_2 + x_3 y_3$$
$$= y_1 x_1 + y_2 x_2 + y_3 x_3 = (y_1, y_2, y_3) \cdot (x_1, x_2, x_3).$$

(ii) Linearity, that is:

$$\begin{aligned}(a(x_1, x_2, x_3) + b(y_1, y_2, y_3)) \cdot (z_1, z_2, z_3) \\ &= (ax_1 + by_1)z_1 + (ax_2 + by_2)z_2 + (ax_3 + by_3)z_3 \\ &= a(x_1 z_1 + x_2 z_2 + x_3 z_3) + b(y_1 z_1 + y_2 z_2 + by_3 z_3) \\ &= a(x_1, x_2, x_3) \cdot (z_1, z_2, z_3) + b(y_1, y_2, y_3) \cdot (z_1, z_2, z_3).\end{aligned}$$

(iii) Non negativity, that is:

$$(x_1, x_2, x_3) \cdot (x_1, x_2, x_3) = x_1^2 + x_2^2 + x_3^2 \geq 0.$$

(iv) Non degeneracy, that is:

$$(x_1, x_2, x_3) \cdot (x_1, x_2, x_3) = 0 \quad \Leftrightarrow \quad (x_1, x_2, x_3) = (0, 0, 0).$$

These last two properties are summarised by saying that the scalar product in $\mathbb{R}^3$ is positive definite.

The above properties suggest the following definition.

Definition 3.1.2 Let V be a finite dimensional real vector space. A *scalar product* on V is a map

$$\cdot : V \times V \longrightarrow \mathbb{R} \qquad (v, w) \mapsto v \cdot w$$

that fulfils the following properties. For any $v, w, v_1, v_2 \in V$ and $a_1, a_2 \in \mathbb{R}$ it holds that:

(i) $v \cdot w = w \cdot v$,
(ii) $(a_1 v_1 + a_2 v_2) \cdot w = a_1(v_1 \cdot w) + a_2(v_2 \cdot w)$,
(iii) $v \cdot v \geq 0$,
(iv) $v \cdot v = 0 \quad \Leftrightarrow \quad v = 0_V$.

A finite dimensional real vector space V equipped with a scalar product will be denoted $(V, \cdot)$ and will be referred to as a *euclidean vector space*.

Clearly the properties (i) and (ii) in the previous definition allows one to prove that the scalar product map $\cdot$ is linear also with respect to the second argument. A scalar product is then a suitable *bilinear symmetric map*, also called a bilinear symmetric real form since its range is in $\mathbb{R}$.

Exercise 3.1.3 It is clear that the scalar product considered in $\mathcal{V}_O^3$ satisfies the conditions given in the Definition 3.1.2. The map in the Example 3.1.1 is a scalar product on the vector space $\mathbb{R}^3$. This scalar product is not unique. Indeed, consider for instance $p : \mathbb{R}^3 \times \mathbb{R}^3 \longrightarrow \mathbb{R}$ given by

$$p((x_1, x_2, x_3), (y_1, y_2, y_3)) = 2x_1y_1 + 3x_2y_2 + x_3y_3.$$

It is easy to verify that such a map p is bilinear and symmetric. With $v = (v_1, v_2, v_3)$, from $p(v, v) = 2v_1^2 + 3v_2^2 + v_3^2$ one has $p(v, v) \geq 0$ and $p(v, v) = 0 \Leftrightarrow v = 0$. We have then that p is a scalar product on $\mathbb{R}^3$.

Definition 3.1.4 On $\mathbb{R}^n$ there is a *canonical scalar product*

$$\cdot : \mathbb{R}^n \times \mathbb{R}^n \longrightarrow \mathbb{R}$$

defined by

$$(x_1, \ldots, x_n) \cdot (y_1, \ldots, y_n) = x_1y_1 + \cdots + x_ny_n = \sum_{j=1}^{n} x_jy_j.$$

The datum $(\mathbb{R}^n, \cdot)$ is referred to as the *canonical euclidean space* and denoted E^n.

The following lines sketch the proof that the above map satisfies the conditions of Definition 3.1.2.

(i) $(x_1, \ldots, x_n) \cdot (y_1, \ldots, y_n) = \sum_{j=1}^{n} x_jy_j$
$\qquad = \sum_{j=1}^{n} y_jx_j = (y_1, \ldots, y_n) \cdot (x_1, \ldots, x_n)$,

(ii) left to the reader,

(iii) $(x_1, \ldots, x_n) \cdot (x_1, \ldots, x_n) = \sum_{i=1}^{n} x_i^2 \geq 0$,

(iv) $(x_1, \ldots, x_n) \cdot (x_1, \ldots, x_n) = 0 \Leftrightarrow \sum_{i=1}^{n} x_i^2 = 0 \Leftrightarrow x_i = 0, \ \forall i \Leftrightarrow (x_1, \ldots, x_n) = (0, \ldots, 0)$.

Definition 3.1.5 Let $(V, \cdot)$ be a finite dimensional euclidean vector space. The map

$$\| - \| : V \longrightarrow \mathbb{R}, \qquad v \mapsto \|v\| = \sqrt{v \cdot v}$$

is called *norm*. For any $v \in V$, the real number $\|v\|$ is the norm or the length of the vector v.

Exercise 3.1.6 The norm of a vector $v = (x_1, \ldots, x_n)$ in $E^n = (\mathbb{R}^n, \cdot)$ is

$$\|(x_1, \ldots, x_n)\| = \sqrt{\sum_{i=1}^{n} x_i^2}.$$

In particular, for E^3 one has $\|(x_1, x_2, x_3)\| = \sqrt{x_1^2 + x_2^2 + x_3^2}$.

The proof of the following proposition is immediate.

Proposition 3.1.7 *Let $(V, \cdot)$ be a finite dimensional euclidean vector space. For any $v \in V$ and any $a \in \mathbb{R}$, the following properties hold:*

(1) $\|v\| \geq 0$,
(2) $\|v\| = 0 \Leftrightarrow v = 0_V$,
(3) $\|av\| = |a| \, \|v\|$.

Proposition 3.1.8 *Let $(V, \cdot)$ be a finite dimensional euclidean vector space. For any $v, w \in V$ the following inequality holds:*

$$|v \cdot w| \;\leq\; \|v\| \; \|w\|.$$

This is called the Schwarz *inequality.*

Proof If either $v = 0_V$ or $w = 0_V$ the claim is obvious, so we may assume that both vectors $v, w \neq 0_V$. Set $a = \|w\|$ and $b = \|v\|$; from (iii) in the Definition 3.1.2, one can write

$$\begin{aligned} 0 \leq \|av \pm bw\|^2 &= (av \pm bw) \cdot (av \pm bw) \\ &= a^2 \|v\|^2 \pm \; 2ab(v \cdot w) + b^2 \|w\|^2 \\ &= 2ab(\|v\| \|w\| \pm \; v \cdot w). \end{aligned}$$

Since both a, b are real positive scalars, the above expression reads

$$\mp \, v \cdot w \leq \; \|v\| \; \|w\|$$

which is the claim. □

Definition 3.1.9 The Schwarz inequality can be written in the form

$$\frac{|v \cdot w|}{\|v\| \; \|w\|} \leq 1, \qquad \text{that is} \qquad -1 \leq \frac{v \cdot w}{\|v\| \; \|w\|} \leq 1.$$

Then one can define then *angle* α between the vectors v, w, by requiring that

$$\frac{v \cdot w}{\|v\| \; \|w\|} = \cos \alpha$$

with $0 \leq \alpha \leq \pi$. Notice the analogy between such a definition and the one in Definition (1.3.2) for the geometric vectors in $\mathcal{V}_O^3$.

Proposition 3.1.10 *Let $(V, \cdot)$ be a finite dimensional euclidean vector space. For any $v, w \in V$ the following inequality holds:*

$$\|v + w\| \;\leq\; \|v\| + \|w\|.$$

This is called the triangle, *or* Minkowski *inequality.*

Proof From the definition of the norm and the Schwarz inequality in Proposition 3.1.8, one has $v \cdot w \leq |v \cdot w| \leq \|v\| \, \|w\|$. The following relations are immediate,

$$\begin{aligned} \|v + w\|^2 &= (v + w) \cdot (v + w) \\ &= \|v\|^2 + 2(v \cdot w) + \|w\|^2 \\ &\leq \|v\|^2 + 2\|v\| \, \|w\| + \|w\|^2 \\ &= (\|v\| + \|w\|)^2 \end{aligned}$$

and prove the claim. □

3.2 Orthogonality

As mentioned, with a scalar product one generalises the notion of orthogonality between vectors and then between vector subspaces.

Definition 3.2.1 Let $(V, \cdot)$ be a finite dimensional euclidean vector space. Two vectors $v, w \in V$ are said to be *orthogonal* if $v \cdot w = 0$.

Proposition 3.2.2 *Let $(V, \cdot)$ be a finite dimensional euclidean vector space, and let $w_1, \cdots, w_s$ and v be vectors in V. If v is orthogonal to each w_i, then v is orthogonal to any vector in the linear span $\mathcal{L}(w_1, \ldots, w_s)$.*

Proof From the bilinearity of the scalar product, one has

$$v \cdot (\lambda_1 w_1 + \cdots + \lambda_s w_s) = \lambda_1 (v \cdot w_1) + \cdots + \lambda_s (v \cdot w_s).$$

The right hand side of such expression is obviously zero under the hypothesis of orthogonality, that is $v \cdot w_i = 0$ for any i. □

Proposition 3.2.3 *Let $(V, \cdot)$ be a finite dimensional euclidean vector space. If $v_1, \ldots, v_s$ is a collection of non zero vectors which are mutually orthogonal, that is $v_i \cdot v_j = 0$ for $i \neq j$, then the vectors $v_1, \ldots, v_s$ are linearly independent.*

Proof Let us equate to the zero vector a linear combination of the vectors $v_1, \ldots, v_s$, that is, let

$$\lambda_1 v_1 + \cdots + \lambda_s v_s = 0_V.$$

For $v_i \in \{v_1, \ldots, v_s\}$, we have

$$0 = v_i \cdot (\lambda_1 v_1 + \cdots + \lambda_s v_s) = \lambda_1 (v_i \cdot v_1) + \cdots + \lambda_s (v_i \cdot v_s) = \lambda_i \, \|v_i\|^2.$$

Being $v_i \neq 0_V$ it must be $\lambda_i = 0$. One gets $\lambda_1 = \ldots = \lambda_s = 0$, with the same argument for any vector v_i. □

Definition 3.2.4 Let $(V, \cdot)$ be a finite dimensional euclidean vector space. If $W \subseteq V$ is a vector subspace of V, then the set

$$W^{\perp} = \{v \in V \ : s\, v \cdot w = 0, \forall w \in W\}$$

is called the *orthogonal* complement to W.

Proposition 3.2.5 *Let $W \subseteq V$ be a vector subspace of a euclidean vector space $(V, \cdot)$. Then,*

(i) $W^{\perp}$ is a vector subspace of V,
(ii) $W \cap W^{\perp} = \{0_V\}$, and the sum between W and $W^{\perp}$ is direct.

Proof (i) Let $v_1, v_2 \in W^{\perp}$, that is $v_1 \cdot w = 0$ and $v_2 \cdot w = 0$ for any $w \in W$. With arbitrary scalars $\lambda_1, \lambda_2 \in \mathbb{R}$, one has

$$(\lambda_1 v_1 + \lambda_2 v_2) \cdot w = \lambda_1 (v_1 \cdot w) + \lambda_2 (v_2 \cdot w) = 0$$

for any $w \in W$; thus $\lambda_1 v_1 + \lambda_2 v_2 \in W^{\perp}$. The claim follows by recalling the Proposition 2.2.2.
(ii) If $w \in W \cap W^{\perp}$, then $w \cdot w = 0$, which then gives $w = 0_V$. □

Remark 3.2.6 Let $W = \mathcal{L}(w_1, \dots, w_s) \subset V$. One has

$$W^{\perp} = \{v \in V \mid v \cdot w_i = 0, \forall i = 1, \dots, s\}.$$

The inclusion $W^{\perp} \subseteq \mathcal{L}(w_1, \dots, v_s)$ is obvious, while the opposite inclusion $\mathcal{L}(w_1, \dots, w_s) \subseteq W^{\perp}$ follows from the Proposition 3.2.2.

Exercise 3.2.7 Consider the vector subspace $W = \mathcal{L}((1, 0, 1)) \subset E^3$. From the previous remark we have

$$W^{\perp} = \{(x, y, z) \in E^3 \mid (x, y, z) \cdot (1, 0, 1) = 0\} = \{(x, y, z) \in E^3 \mid x + z = 0\},$$

that is $W^{\perp} = \mathcal{L}((1, 0, -1), (0, 1, 0))$.

Exercise 3.2.8 Let $W \subset E^4$ be defined by

$$W = \mathcal{L}((1, -1, 1, 0), (2, 1, 0, 1)).$$

By recalling the Proposition 3.2.3 and the Corollary 2.5.7 we know that the orthogonal subspace $W^{\perp}$ has dimension 2. From the Remark 3.2.6, it is given by

$$W^{\perp} = \left\{ (x, y, z, t) \in E^4 \ : \ \left\{ \begin{array}{l} (x, y, z, t) \cdot (1, -1, 1, 0) = 0 \\ (x, y, z, t) \cdot (2, 1, 0, 1) = 0 \end{array} \right. \right\},$$

that is by the common solutions of the following linear equations,

$$\begin{aligned} x - y + z &= 0 \\ 2x + y + t &= 0\,. \end{aligned}$$

Such solutions can be written as

$$\begin{cases} z = y - x \\ t = -2x - y \end{cases}$$

for arbitrary values of x, y. By choosing, for example, $(x, y) = (1, 0)$ and $(x, y) = (0, 1)$, for the orthogonal subspace $W^{\perp}$ one can show that $W^{\perp} = \mathcal{L}((1, 0, -1, -2), (0, 1, 1, -1))$ (this kind of examples and exercises will be clearer after studying homogeneous linear systems of equations).

3.3 Orthonormal Basis

We have seen in Chap. 2 that the orthogonal cartesian coordinate system $(O, \mathbf{i}, \mathbf{j}, \mathbf{k})$ for the vector space $\mathcal{V}_O^3$ can be seen as having a basis whose vectors are mutually orthogonal and have norm one.

In this section we analyse how to select in a finite dimensional euclidean vector space $(V, \cdot)$, a basis whose vectors are mutually orthogonal and have norm one.

Definition 3.3.1 Let $I = \{v_1, \ldots, v_r\}$ be a system of vectors of a vector space V. If V is endowed with a scalar product, I is called *orthonormal* if

$$v_i \cdot v_j = \delta_{ij} = \begin{cases} 1 \text{ if } & i = j \\ 0 \text{ if } & i \neq j \end{cases}.$$

Remark 3.3.2 From the Proposition 3.2.3 one has that any orthonormal system of vectors if free, that is its vectors are linearly independent.

Definition 3.3.3 A basis $\mathcal{B}$ for $(V, \cdot)$ is called *orthonormal* if it is an orthonormal system.

By such a definition, the basis $(\mathbf{i}, \mathbf{j}, \mathbf{k})$ of $\mathcal{V}_O^3$ as well as the canonical basis for E^n are orthonormal.

Remark 3.3.4 Let $\mathcal{B} = (e_1, \ldots, e_n)$ be an orthonormal basis for $(V, \cdot)$ and let $v \in V$. The vector v can be written with respect to $\mathcal{B}$ as

$$v = (v \cdot e_1)e_1 + \cdots + (v \cdot e_n)e_n.$$

Indeed, from

$$v = a_1 e_1 + \cdots + a_n e_n$$

one can consider the scalar products of v with each e_i, and the orthogonality of these yields

$$a_1 = v \cdot e_1, \quad \dots \quad , a_n = v \cdot e_n.$$

Thus the components of a vector with respect to an orthonormal basis are given by the scalar products of the vector with the corresponding basis elements.

Definition 3.3.5 Let $\mathcal{B} = (e_1, \dots, e_n)$ be an orthonormal basis for $(V, \cdot)$. With $v \in V$, the vectors

$$(v \cdot e_1)e_1, \ \dots \ , (v \cdot e_n)e_n,$$

which give a linear decomposition of v, are called the *orthogonal projections* of v along $e_1, \dots, e_n$.

The next proposition shows that in an any finite dimensional real vector space $(V, \cdot)$, with respect to an orthonormal basis for V the scalar product has the same form than the canonical scalar product in E^n.

Proposition 3.3.6 *Let* $\mathcal{B} = (e_1, \dots, e_n)$ *be an orthonormal basis for* $(V, \cdot)$. *With* $v, w \in V$, *let it be* $v = (a_1, \dots, a_n)_{\mathcal{B}}$ *and* $w = (b_1, \dots, b_n)_{\mathcal{B}}$. *Then one has*

$$v \cdot w = a_1 b_1 + \cdots + a_n b_n.$$

Proof This follows by using the bilinearity of the scalar product and the relations $e_i \cdot e_j = \delta_{ij}$. □

Any finite dimensional real vector space can be shown to admit an orthonormal basis. This is done via the so called *Gram-Schmidt orthonormalisation method*. Its proof is constructive since, out of any given basis, the method provides an explicit orthonormal basis via linear algebra computations.

Proposition 3.3.7 (Gram-Schmidt method) *Let* $\mathcal{B} = (v_1, \dots, v_n)$ *be a basis for the finite dimensional euclidean space* $(V, \cdot)$. *The vectors*

$$\begin{aligned} e_1 &= \frac{v_1}{\|v_1\|}, \\ e_2 &= \frac{v_2 - (v_2 \cdot e_1)e_1}{\|v_2 - (v_2 \cdot e_1)e_1\|}, \\ &\vdots \\ e_n &= \frac{v_n - \sum_{i=1}^{n-1}(v_n \cdot e_i)e_i}{\|v_n - \sum_{i=1}^{n-1}(v_n \cdot e_i)e_i\|} \end{aligned}$$

form an orthonormal basis $(e_1, \dots, e_n)$ *for* V.

Proof We start by noticing that $\|e_j\| = 1$, for $j = 1, \ldots, n$, from the way these vectors are defined. The proof of orthogonality is done by induction. As induction basis we prove explicitly that $e_1 \cdot e_2 = 0$. Being $e_1 \cdot e_1 = 1$, one has

$$e_1 \cdot e_2 = \frac{e_1 \cdot v_2 - (v_2 \cdot e_1)e_1 \cdot e_1}{\|v_1\| \, \|v_2 - (v_2 \cdot e_1)e_1\|} = 0 \,.$$

We then assume that $e_1, \ldots, e_h$ are pairwise orthogonal (this is the inductive hypothesis) and show that $e_1, \ldots, e_{h+1}$ are pairwise orthogonal. Consider an integer k such that $1 \leq k \leq h$. Then,

$$\begin{aligned} e_{h+1} \cdot e_k &= \frac{v_{h+1} - \sum_{i=1}^{h}(v_{h+1} \cdot e_i)e_i}{\|v_{h+1} - \sum_{i=1}^{h}(v_{h+1} \cdot e_i)e_i\|} \cdot e_k \\ &= \frac{v_{h+1} \cdot e_k - \sum_{i=1}^{h}((v_{h+1} \cdot e_i)(e_i \cdot e_k))}{\|v_{h+1} - \sum_{i=1}^{h}(v_{h+1} \cdot e_i)e_i\|} \\ &= \frac{v_{h+1} \cdot e_k - v_{h+1} \cdot e_k}{\|v_{h+1} - \sum_{i=1}^{h}(v_{h+1} \cdot e_i)e_i\|} = 0 \end{aligned}$$

where the last equality follows from the inductive hypothesis $e_i \cdot e_k = 0$. The system $(e_1, \ldots, e_n)$ is free by Remark 3.3.2, thus giving an orthonormal basis for V. □

Exercise 3.3.8 Let $V = \mathcal{L}(v_1, v_2) \subset E^4$, with $v_1 = (1, 1, 0, 0)$, and $v_2 = (0, 2, 1, 1)$. With the Gram-Schmidt orthogonalization method, we obtain an orthonormal basis for V. Firstly, we have

$$e_1 = \frac{v_1}{\|v_1\|} = \frac{1}{\sqrt{2}}\,(1, 1, 0, 0)\,.$$

Set $f_2 = v_2 - (v_2 \cdot e_1)e_1$. We have then

$$\begin{aligned} f_2 &= (0, 2, 1, 1) - \left((0, 2, 1, 1) \cdot \frac{1}{\sqrt{2}}\,(1, 1, 0, 0)\right) \frac{1}{\sqrt{2}}\,(1, 1, 0, 0) \\ &= (0, 2, 1, 1) - (1, 1, 0, 0) \\ &= (-1, 1, 1, 1). \end{aligned}$$

Then, the second vector $e_2 = \frac{f_2}{\|f_2\|}$ is

$$e_2 = \frac{1}{2}\,(-1, 1, 1, 1)\,.$$

Theorem 3.3.9 *Any finite dimensional euclidean vector space $(V, \cdot)$ admits an orthonormal basis.*

Proof Since V is finite dimensional, by the Corollary 2.4.4 it has a basis, which can be orthonormalised using the Gram-Schmidt method. □

Theorem 3.3.10 *Let $(V, \cdot)$ be finite dimensional with $\{e_1, \ldots, e_r\}$ an orthonormal system of vectors of V. The system can be completed to an orthonormal basis $(e_1, \ldots, e_r, e_{r+1}, \ldots, e_n)$ for V.*

Proof From the Theorem 2.4.19 the free system $\{e_1, \ldots, e_r\}$ can be completed to a basis for V, say

$$\mathcal{B} = (e_1, \ldots, e_r, v_{r+1}, \ldots, v_n).$$

The Gram-Schmidt method for the system $\mathcal{B}$ does not alter the first r vectors, and provides an orthonormal basis for V. □

Corollary 3.3.11 *Let $(V, \cdot)$ have finite dimension n and let W be a vector subspace of V. Then,*

(1) $\dim(W) + \dim(W^{\perp}) = n$,
(2) $V = W \oplus W^{\perp}$,
(3) $(W^{\perp})^{\perp} = W$.

Proof

(1) Let $(e_1, \ldots, e_r)$ be an orthonormal basis for W completed (by the theorem above) to an orthonormal basis $(e_1, \ldots, e_r, e_{r+1}, \ldots, e_n)$ for V. Since the vectors $e_{r+1}, \ldots, e_n$ are then orthogonal to the vectors $e_1, \ldots, e_r$, they are (see the Definition 3.2.1) orthogonal to any vector in W, so $e_{r+1}, \ldots, e_n \in W^{\perp}$. This gives $\dim(W^{\perp}) \geq n - r$, that is $\dim(W) + \dim(W^{\perp}) \geq n$. From the Definition 3.2.4 the sum of W and $W^{\perp}$ is direct, so, recalling the Corollary 2.5.7, one has $\dim(W) + \dim(W^{\perp}) = \dim(W \oplus W^{\perp}) \leq n$, thus proving the claim.
(2) From (1) we have $\dim(W \oplus W^{\perp}) = \dim(W) + \dim(W^{\perp}) = n = \dim(V)$; thus $W \oplus W^{\perp} = V$.
(3) We start by proving the inclusion $(W^{\perp})^{\perp} \supseteq W$.
By definition, it is $(W^{\perp})^{\perp} = \{v \in V \mid v \cdot w = 0, \ \forall w \in W^{\perp}\}$. If $v \in W$, then $v \cdot w = 0$ for any $w \in W^{\perp}$, thus $W \subseteq (W^{\perp})^{\perp}$. Apply now the result in point 1) to $W^{\perp}$: one has

$$\dim(W^{\perp}) + \dim((W^{\perp})^{\perp}) = n.$$

This inequality, together with the point 1) gives $\dim((W^{\perp})^{\perp}) = \dim(W)$; the spaces W and $(W^{\perp})^{\perp}$ are each other subspace with the same dimension, thus they coincide. □

It is worth stressing that for the identity $(W^{\perp})^{\perp} = W$ it is crucial that the vector space V be finite dimensional. For infinite dimensional vector spaces in general only the inclusion $(W^{\perp})^{\perp} \supseteq W$ holds.

Exercise 3.3.12 In Exercise 3.2.7 we considered the subspace of E^3 given by $W = \mathcal{L}((1, 0, 1))$, and computed $W^{\perp} = \mathcal{L}((1, 0, -1), (0, 1, 0))$. It is immediate to verify that

$$\dim(W) + \dim(W^{\perp}) = 1 + 2 = 3 = \dim(E^3).$$

3.4 Hermitian Products

The canonical scalar product in $\mathbb{R}^n$ can be naturally extended to the complex vector space $\mathbb{C}^n$ with a minor modification.

Definition 3.4.1 The *canonical hermitian product* on $\mathbb{C}^n$ is the map

$$\cdot : \mathbb{C}^n \times \mathbb{C}^n \longrightarrow \mathbb{C}$$

defined by

$$(z_1, \ldots, z_n) \cdot (w_1, \ldots, w_n) = \bar{z}_1 w_1 + \cdots + \bar{z}_n w_n$$

where $\bar{z}$ denotes the complex conjugate of z (see the Sect. A.5). The datum $(\mathbb{C}^n, \cdot)$ is called the *canonical hermitian vector space* of dimension n.

The following proposition—whose straightforward proof we omit—generalises to the complex case the properties of the canonical scalar product on $\mathbb{R}^n$ shown after Definition 3.1.4. For easy of notation, we shall denote the vectors in $\mathbb{C}^n$ by $z = (z_1, \ldots, z_n)$.

Proposition 3.4.2 *For any $z, w, v \in \mathbb{C}^n$ and $a, b \in \mathbb{C}$, the following properties hold:*

(i) $w \cdot z = \overline{z \cdot w}$,
(ii) $(az + bw) \cdot v = \bar{a}(z \cdot v) + \bar{b}(w \cdot v)$
while $v \cdot (az + bw) = a(v \cdot z) + b(v \cdot w)$,
(iii) $z \cdot z = \sum_{i=1}^{n} |z_i|^2 \geq 0$,
(iv) $z \cdot z = 0 \quad \Leftrightarrow \quad z = (0, \ldots, 0) \in \mathbb{C}^n$.

Notice that the complex conjugation for the first entry of the hermitian scalar product implies that the hermitian product of a vector with itself is a real positive number. It is this number that gives the *real* norm of a *complex* vector $z = (z_1, \ldots, z_n)$, defined as

$$\|z\| = \sqrt{(z_1, \ldots, z_n) \cdot (z_1, \ldots, z_n)} = \sqrt{\sum_{i=1}^{n} |z_i|^2}\,.$$

Chapter 4
Matrices

Matrices are an important tool when dealing with many problems, notably the theory of systems of linear equations and the study of maps (operators) between vector spaces. This chapter is devoted to their basic notions and properties.

4.1 Basic Notions

Definition 4.1.1 A *matrix* M with entries in $\mathbb{R}$ (or a *real matrix*) is a collection of elements $a_{ij} \in \mathbb{R}$, with $i = 1, \dots, m$; $j = 1, \dots, n$ and $m, n \in \mathbb{N}$, displayed as follows

$$M = \begin{pmatrix} a_{11} & a_{12} & \dots & a_{1n} \\ a_{21} & a_{22} & \dots & a_{2n} \\ \vdots & \vdots & & \vdots \\ a_{m1} & a_{m2} & \dots & a_{mn} \end{pmatrix}.$$

The matrix M above is said to be made of m *row* vectors in $\mathbb{R}^n$, that is

$$R_1 = (a_{11}, \dots, a_{1n}) \,, \quad \dots \,, \quad R_m = (a_{m1}, \dots, a_{mn})$$

or by n *column* vectors in $\mathbb{R}^m$, that is

$$C_1 = (a_{11}, \dots, a_{m1}) \,, \quad \dots \,, \quad C_n = (a_{1n}, \dots, a_{mn}).$$

Thus the matrix M above is a $m \times n$-matrix (m rows $R_i \in \mathbb{R}^n$ and n columns $R_i \in \mathbb{R}^n$). As a shorthand, by $M = (a_{ij})$ we shall denote a matrix M with entry a_{ij} at the i-th row and j-th column. We denote by $\mathbb{R}^{m,n}$ the collection of all $m \times n$-matrices whose entries are in $\mathbb{R}$.

G. Landi and A. Zampini, *Linear Algebra and Analytic Geometry for Physical Sciences*, Undergraduate Lecture Notes in Physics,
https://doi.org/10.1007/978-3-319-78361-1_4

Remark 4.1.2 It is sometime useful to consider a matrix $M \in \mathbb{R}^{m,n}$ as the collection of its n columns, or as the collection of its m rows, that is to write

$$M = (C_1 \ C_2 \ \dots \ C_n) = \begin{pmatrix} R_1 \\ R_2 \\ \vdots \\ R_m \end{pmatrix}.$$

An element $M \in \mathbb{R}^{1,n}$ is called a n-dimensional *row matrix*,

$$M = (a_{11} \ a_{12} \ \dots \ a_{1n})$$

while an element $M \in \mathbb{R}^{m,1}$ is called a m-dimensional *column matrix*,

$$M = \begin{pmatrix} a_{11} \\ a_{21} \\ \vdots \\ a_{m1} \end{pmatrix}.$$

A *square* matrices of order n is a $n \times n$ matrix, that is an element in $\mathbb{R}^{n,n}$. An element $M \in \mathbb{R}^{1,1}$ is a scalar, that is a single element in $\mathbb{R}$. If $A = (a_{ij}) \in \mathbb{R}^{n,n}$ is a square matrix, the entries $(a_{11}, a_{22}, \dots, a_{nn})$ give the *(principal) diagonal* of A.

Example 4.1.3 The bold typeset entries in

$$A = \begin{pmatrix} \mathbf{1} & 2 & 2 \\ -1 & \mathbf{0} & 3 \\ 2 & 4 & \mathbf{7} \end{pmatrix}$$

give the diagonal of A.

Proposition 4.1.4 *The set $\mathbb{R}^{m,n}$ is a real vector space whose dimension is mn. With $A = (a_{ij})$, $B = (b_{ij}) \in \mathbb{R}^{m,n}$ and $\lambda \in \mathbb{R}$, the vector space operations are defined by*

$$A + B = (a_{ij} + b_{ij}), \quad \lambda A = (\lambda a_{ij}).$$

Proof We task the reader to show that $\mathbb{R}^{m,n}$ equipped with the above defined operations is a vector space. We only remark that the zero element in $\mathbb{R}^{m,n}$ is given by the matrix A with entries $a_{ij} = 0_{\mathbb{R}}$; such a matrix is also called the *null* matrix and denoted $0_{\mathbb{R}^{m,n}}$.

In order to show that the dimension of $\mathbb{R}^{m,n}$ is mn, consider the *elementary* $m \times n$-matrices

$$E_{rs} = (e_{jk}^{(rs)}), \quad \text{with} \quad e_{jk}^{(rs)} = \begin{cases} 1 \text{ if } (j,k) = (r,s) \\ 0 \text{ if } (j,k) \neq (r,s) \end{cases}.$$

Thus the matrix E_{rs} has entries all zero but for the entry (r, s) which is 1. Clearly there are mn of them and it is immediate to show that they form a basis for $\mathbb{R}^{m,n}$. □

Exercise 4.1.5 Let $A = \begin{pmatrix} 1 & 2 & -1 \\ 0 & -1 & 1 \end{pmatrix} \in \mathbb{R}^{2,3}$. One computes

$$\begin{aligned} \begin{pmatrix} 1 & 2 & -1 \\ 0 & -1 & 1 \end{pmatrix} &= \begin{pmatrix} 1 & 0 & 0 \\ 0 & 0 & 0 \end{pmatrix} + 2\begin{pmatrix} 0 & 1 & 0 \\ 0 & 0 & 0 \end{pmatrix} - \begin{pmatrix} 0 & 0 & 1 \\ 0 & 0 & 0 \end{pmatrix} \\ &\quad + 0\begin{pmatrix} 0 & 0 & 0 \\ 1 & 0 & 0 \end{pmatrix} - \begin{pmatrix} 0 & 0 & 0 \\ 0 & 1 & 0 \end{pmatrix} + \begin{pmatrix} 0 & 0 & 0 \\ 0 & 0 & 1 \end{pmatrix} \\ &= E_{11} + 2E_{12} - E_{13} - E_{22} + E_{23}. \end{aligned}$$

In addition to forming a vector space, matrices of ‘matching size’ can be multiplied.

Definition 4.1.6 If $A = (a_{ij}) \in \mathbb{R}^{m,n}$ and $B = (b_{jk}) \in \mathbb{R}^{n,p}$ the *product* between A and B is the matrix in $\mathbb{R}^{m,p}$ defined by

$$C = (c_{ik}) = AB \in \mathbb{R}^{m,p}, \quad \text{where} \quad c_{ik} = R_i^{(A)} \cdot C_k^{(B)} = \sum_{j=1}^{n} a_{ij} b_{jk},$$

with $i = 1, \dots, m$ and $k = 1, \dots, p$. Here $R_i^{(A)} \cdot C_k^{(B)}$ denotes the scalar product in $\mathbb{R}^n$ between the i-th row vector $R_i^{(A)}$ of A with the k-th column vector $C_k^{(B)}$ of B.

Remark 4.1.7 Notice that the product AB—also called the *row by column* product—is defined only if the number of columns of A equals the number of rows of B.

Exercise 4.1.8 Consider the matrices

$$A = \begin{pmatrix} 1 & 2 & -1 \\ 3 & 0 & 1 \end{pmatrix} \in \mathbb{R}^{2,3}, \qquad B = \begin{pmatrix} 1 & 2 \\ 2 & 1 \\ 3 & 4 \end{pmatrix} \in \mathbb{R}^{3,2}.$$

One has $AB = C = (c_{ik}) \in \mathbb{R}^{2,2}$ with

$$C = \begin{pmatrix} 2 & 0 \\ 6 & 10 \end{pmatrix}.$$

On the other hand, $BA = C' = (c'_{st}) \in \mathbb{R}^{3,3}$ with

$$C' = \begin{pmatrix} 7 & 2 & 1 \\ 5 & 4 & -1 \\ 15 & 6 & 1 \end{pmatrix}.$$

Clearly, comparing C with C' is meaningless, since they are in different spaces.

Remark 4.1.9 With $A \in \mathbb{R}^{m,n}$ and $B \in \mathbb{R}^{p,q}$, the product AB is defined only if $n = p$, giving a matrix $AB \in \mathbb{R}^{m,q}$. It is clear that the product BA is defined only if $m = q$ and in such a case one has $BA \in \mathbb{R}^{p,n}$. When both the conditions $m = q$ and $n = p$ are satisfied both products are defined. This is the case of the matrices A and B in the previous exercise.
Let us consider the space $\mathbb{R}^{n,n}$ of square matrices of order n. If A, B are in $\mathbb{R}^{n,n}$ then both AB and BA are square matrices of order n. An example shows that in general one has $AB \neq BA$. If

$$A = \begin{pmatrix} 1 & 2 \\ 1 & -1 \end{pmatrix}, \qquad B = \begin{pmatrix} 1 & -1 \\ 1 & 0 \end{pmatrix},$$

one computes that

$$AB = \begin{pmatrix} 3 & -1 \\ 0 & -1 \end{pmatrix} \neq BA = \begin{pmatrix} 0 & 3 \\ 1 & 2 \end{pmatrix}.$$

Thus the product of matrices is *non commutative*. We shall say that two matrices $A, B \in \mathbb{R}^{n,n}$ commute if $AB = BA$. On the other hand, the associativity of the product in $\mathbb{R}$ and its distributivity with respect to the sum, allow one to prove analogous properties for the space of matrices.

Proposition 4.1.10 *The following identities hold:*

(i) $A(BC) = (AB)C$, *for any* $A \in \mathbb{R}^{m,n}$, $B \in \mathbb{R}^{n,p}$, $C \in \mathbb{R}^{p,q}$,
(ii) $A(B + C) = AB + AC$, *for any* $A \in \mathbb{R}^{m,n}$, $B, C \in \mathbb{R}^{n,p}$,
(iii) $\lambda(AB) = (\lambda A)B = A(\lambda B)$, *for any* $A \in \mathbb{R}^{m,n}$, $B \in \mathbb{R}^{n,p}$, $\lambda \in \mathbb{R}$.

Proof (i) Consider three matrices $A = (a_{ih}) \in \mathbb{R}^{m,n}$, $B = (b_{hk}) \in \mathbb{R}^{n,p}$ and $C = (c_{kj}) \in \mathbb{R}^{p,q}$. From the definition of row by column product one has $AB = (d_{ik})$ with $d_{ik} = \sum_{h=1}^{n} a_{ih}b_{hk}$, while $BC = (e_{hj})$ with $e_{hj} = \sum_{k=1}^{p} b_{hk}c_{kj}$. The ij-entries of $(AB)C$ and $A(BC)$ are

$$\sum_{k=1}^{p} d_{ik}c_{kj} = \sum_{k=1}^{p} \left(\sum_{h=1}^{n} a_{ih}b_{hk} \right) c_{kj} = \sum_{k=1}^{p} \sum_{h=1}^{n} (a_{ih}b_{hk}c_{kj}),$$

$$\sum_{h=1}^{n} a_{ih}e_{hj} = \sum_{h=1}^{n} a_{ih} \left(\sum_{k=1}^{p} b_{hk}c_{kj} \right) = \sum_{h=1}^{n} \sum_{k=1}^{p} (a_{ih}b_{hk}c_{kj}).$$

These two expressions coincide (the last equality on both lines follows from the distributivity in $\mathbb{R}$ of the product with respect to the sum).

(ii) Take matrices $A = (a_{ih}) \in \mathbb{R}^{m,n}$, $B = (b_{hj}) \in \mathbb{R}^{n,p}$ and $C = (c_{hj}) \in \mathbb{R}^{n,p}$. The equality $A(B + C) = AB + AC$ is proven again by a direct computation of the ij-entry for both sides:

$$
\begin{aligned}
[A(B+C)]_{ij} &= \sum_{h=1}^{n} a_{ih}(b_{hj}+c_{hj}) \\
&= \sum_{h=1}^{n} a_{ih}b_{hj} + \sum_{h=1}^{n} a_{ih}c_{hj} \\
&= [AB]_{ij} + [AC]_{ij} \\
&= [AB+AC]_{ij}.
\end{aligned}
$$

(iii) This is immediate. □

The matrix product in $\mathbb{R}^{n,n}$ is inner and it has a neutral element, a multiplication unit.

Definition 4.1.11 The *unit* matrix of order n, denoted by I_n, is the element in $\mathbb{R}^{n,n}$ given by

$$I_n = (\delta_{ij}), \quad \text{with} \quad \delta_{ij} = \begin{cases} 1 & \text{if } i = j \\ 0 & \text{if } i \neq j \end{cases}.$$

The diagonal entries of I_n are all 1, while its off-diagonal entries are all zero.

Remark 4.1.12 It is easy to prove that, with $A \in \mathbb{R}^{m,n}$, one has

$$AI_n = A \quad \text{and} \quad I_m A = A.$$

Proposition 4.1.13 *The space $\mathbb{R}^{n,n}$ of square matrices of order n, endowed with the sum and the product as defined above, is a non abelian ring.*

Proof Recall the definition of a ring given in A.1.6. The matrix product is an inner operation in $\mathbb{R}^{n,n}$, so the claim follows from the fact that $(\mathbb{R}^{n,n}, +, 0_{\mathbb{R}^{n,n}})$ is an abelian group and the results of the Proposition 4.1.10. □

Definition 4.1.14 A matrix $A \in \mathbb{R}^{n,n}$ is said to be *invertible* (also *non-singular* or *non-degenerate*) if there exists a matrix $B \in \mathbb{R}^{n,n}$, such that $AB = BA = I_n$. Such a matrix B is denoted by A^{-1} and is called the *inverse* of A.

Definition 4.1.15 If a matrix is non invertible, then it is called *singular* or *degenerate*.

Exercise 4.1.16 An element of the ring $\mathbb{R}^{n,n}$ needs not be invertible. The matrix

$$A = \begin{pmatrix} 1 & 1 \\ 0 & 1 \end{pmatrix} \in \mathbb{R}^{2,2}$$

is invertible, with inverse

$$A^{-1} = \begin{pmatrix} 1 & -1 \\ 0 & 1 \end{pmatrix}$$

as it can be easily checked. On the other hand, the matrix

$$A' = \begin{pmatrix} 1 & 0 \\ k & 0 \end{pmatrix} \in \mathbb{R}^{2,2}$$

is non invertible, for any value of the parameter $k \in \mathbb{R}$. It is easy to verify that the matrix equation

$$\begin{pmatrix} 1 & 0 \\ k & 0 \end{pmatrix} \begin{pmatrix} x & y \\ z & t \end{pmatrix} = \begin{pmatrix} 1 & 0 \\ 0 & 1 \end{pmatrix}$$

has no solutions.

Proposition 4.1.17 *The subset of invertible elements in* $\mathbb{R}^{n,n}$ *is a group with respect to the matrix product. It is called the* general linear group *of order n and is denoted by* $\mathrm{GL}(n, \mathbb{R})$ *or simply by* $\mathrm{GL}(n)$.

Proof Recall the definition of a group in A.2.7. We observe first that if A and B are both invertible then AB is invertible since $AB(B^{-1}A^{-1}) = (B^{-1}A^{-1})AB = I_n$; this means that $(AB)^{-1} = B^{-1}A^{-1}$ so $\mathrm{GL}(n)$ is closed under the matrix product. It is evident that $I_n^{-1} = I_n$ and that if A is invertible, then A is the inverse of A^{-1}, thus the latter is invertible. □

Notice that since AB is in general different from BA the group $\mathrm{GL}(n)$ is non abelian. As an example, the non commuting matrices A and B considered in the Remark 4.1.9 are both invertible.

Definition 4.1.18 Given $A = (a_{ij}) \in \mathbb{R}^{m,n}$ its *transpose*, denoted by ${}^tA$, is the matrix obtained from A when exchanging its rows with its columns, that is ${}^tA = (b_{ij}) \in \mathbb{R}^{n,m}$ with $b_{ij} = a_{ji}$.

Exercise 4.1.19 The matrix

$$A = \begin{pmatrix} 1 & 2 & -1 \\ 3 & 0 & 1 \end{pmatrix} \in \mathbb{R}^{2,3}$$

has transpose ${}^tA \in \mathbb{R}^{3,2}$ given by

$${}^tA = \begin{pmatrix} 1 & 3 \\ 2 & 0 \\ -1 & 1 \end{pmatrix}.$$

Proposition 4.1.20 *The following identities hold:*

(i) ${}^t(A + B) = {}^tA + {}^tB$, *for any* $A, B \in \mathbb{R}^{m,n}$,
(ii) ${}^t(AB) = {}^tB\,{}^tA$, *for* $A \in \mathbb{R}^{m,n}$ *and* $B \in \mathbb{R}^{n,p}$,
(iii) if $A \in \mathrm{GL}(n)$ *then* ${}^tA \in \mathrm{GL}(n)$ *that is, if A is invertible its transpose is invertible as well with* $({}^tA)^{-1} = {}^t(A^{-1})$.

Proof (i) It is immediate.

(ii) Given $A = (a_{ij})$ and $B = (b_{ij})$, denote ${}^tA = (a'_{ij})$ and ${}^tB = (b'_{ij})$ with $a'_{ij} = a_{ji}$ and $b'_{ij} = b_{ji}$. If $AB = (c_{ij})$, then $c_{ij} = \sum_{h=1}^{n} a_{ih}b_{hj}$. The ij-element in ${}^t(AB)$ is then $\sum_{h=1}^{n} a_{jh}b_{hi}$; the ij-element in ${}^tB{}^tA$ is $\sum_{h=1}^{n} b'_{ih}a'_{hj}$ and these elements clearly coincide, for any i and j.

(iii) It is enough to show that ${}^t(A^{-1})\,{}^tA = I_n$. From (ii) one has indeed

$${}^t(A^{-1})\,{}^tA = {}^t(AA^{-1}) = {}^tI_n = I_n.$$

This finishes the proof. □

Definition 4.1.21 A square matrix of order n, $A = (a_{ij}) \in \mathbb{R}^{n,n}$, is said to be *symmetric* if ${}^tA = A$ that is, if for any i, j it holds that $a_{ij} = a_{ji}$.

Exercise 4.1.22 The matrix $A = \begin{pmatrix} 1 & 2 & -1 \\ 2 & 0 & 1 \\ -1 & 1 & -1 \end{pmatrix}$ is symmetric.

4.2 The Rank of a Matrix

Definition 4.2.1 Let $A = (a_{ij})$ be a matrix in $\mathbb{R}^{m,n}$. We have seen that the m rows of A,

$$R_1 = (a_{11}, \dots, a_{1n}),$$
$$\vdots$$
$$R_m = (a_{m1}, \dots, a_{mn})$$

are elements (vectors, indeed) in $\mathbb{R}^n$. By $R(A)$ we denote the vector subspace of $\mathbb{R}^n$ generated by the vectors $R_1, \dots, R_m$ that is,

$$R(A) = \mathcal{L}(R_1, \dots, R_m).$$

We call $R(A)$ the *row space* of A. Analogously, given the columns

$$C_1 = \begin{pmatrix} a_{11} \\ a_{21} \\ \vdots \\ a_{m1} \end{pmatrix}, \quad \cdots \quad , C_n = \begin{pmatrix} a_{1n} \\ a_{2n} \\ \vdots \\ a_{mn} \end{pmatrix}$$

of A, we define the vector subspace $C(A)$ in $\mathbb{R}^m$,

$$C(A) = \mathcal{L}(C_1, \dots, C_n)$$

as the *column space* of A.

Remark 4.2.2 Clearly $C({}^tA) = R(A)$ since the columns of ${}^tA$ are the rows of A.

Theorem 4.2.3 *Given* $A = (a_{ij}) \in \mathbb{R}^{m,n}$ *one has that* $\dim(R(A)) = \dim(C(A))$.

Proof Since A is fixed, to simplify notations we set $R = R(A)$ and $C = C(A)$. The first step is to show that $\dim(C) \leq \dim(R)$. Let $\dim R = r$; up to a permutation, we can take the first r rows in A as linearly independent. The remaining rows $R_{r+1}, \ldots, R_m$ are elements in $R = \mathcal{L}(R_1, \ldots, R_r)$ and we can write

$$A = \begin{pmatrix} R_1 \\ \vdots \\ R_r \\ \sum_{i=1}^r \lambda_i^{r+1} R_i \\ \vdots \\ \sum_{i=1}^r \lambda_i^m R_i \end{pmatrix} = \begin{pmatrix} a_{11} & \cdots & a_{1n} \\ \vdots & & \vdots \\ a_{r1} & \cdots & a_{rn} \\ \sum_{i=1}^r \lambda_i^{r+1} a_{i1} & \cdots & \sum_{i=1}^r \lambda_i^{r+1} a_{in} \\ \vdots & & \vdots \\ \sum_{i=1}^r \lambda_i^m a_{i1} & \cdots & \sum_{i=1}^r \lambda_i^m a_{in} \end{pmatrix}$$

for suitable scalars λ_i^j (with $i \in 1, \ldots, r$, and $j \in r+1, \ldots, m$). Given $h = 1, \ldots, n$, consider the h-th column,

$$C_h = \begin{pmatrix} a_{1h} \\ a_{2h} \\ \vdots \\ a_{rh} \\ \sum_{i=1}^r \lambda_i^{r+1} a_{ih} \\ \vdots \\ \sum_{i=1}^r \lambda_i^m a_{ih} \end{pmatrix} = a_{1h} \begin{pmatrix} 1 \\ 0 \\ \vdots \\ 0 \\ \lambda_1^{r+1} \\ \vdots \\ \lambda_1^m \end{pmatrix} + \cdots + a_{rh} \begin{pmatrix} 0 \\ 0 \\ \vdots \\ 1 \\ \lambda_r^{r+1} \\ \vdots \\ \lambda_r^m \end{pmatrix}.$$

This means that C is generated by the r columns

$$\begin{pmatrix} 1 \\ 0 \\ \vdots \\ 0 \\ \lambda_1^{r+1} \\ \vdots \\ \lambda_1^m \end{pmatrix}, \ldots, \begin{pmatrix} 0 \\ 0 \\ \vdots \\ 1 \\ \lambda_r^{r+1} \\ \vdots \\ \lambda_r^m \end{pmatrix},$$

so we have $\dim(C) \leq r = \dim R$. By exchanging the rows with columns, a similar argument shows also that $\dim(C) \geq \dim(R)$ thus the claim. □

This theorem shows that $\dim(R(A)) = \dim(C(A))$ is an integer number that characterises A.

Definition 4.2.4 Given a matrix $A \in \mathbb{R}^{m,n}$, its *rank* is the number

$$\mathrm{rk}(A) = \dim(C(A)) = \dim(R(A))$$

that is the common dimension of its space of rows, or columns.

Corollary 4.2.5 *For any* $A \in \mathbb{R}^{m,n}$ *one has* $\mathrm{rk}(A) = \mathrm{rk}({}^tA)$.

Proof This follows from Remark 4.2.2 since $C({}^tA) = R(A)$. □

It is clear that $\mathrm{rk}(A) \leq \min(m, n)$.

Definition 4.2.6 A matrix $A \in \mathbb{R}^{m,n}$ has *maximal rank* if $\mathrm{rk}(A) = \min(m, n)$.

Our task next is to give methods to compute the rank of a given matrix. We first identify a class of matrices whose rank is easy to determine.

Remark 4.2.7 It is immediate to convince one-self that the rank of a matrix A does not change by enlarging it with an arbitrary number of zero rows or columns. Moreover, if a matrix B is obtained from a matrix A by a permutation of either its rows or columns, that is, if it is

$$A = \begin{pmatrix} R_1 \\ R_2 \\ \vdots \\ R_m \end{pmatrix} \quad \text{and} \quad B = \begin{pmatrix} R_{\sigma(1)} \\ R_{\sigma(2)} \\ \vdots \\ R_{\sigma(m)} \end{pmatrix}$$

(where σ denotes a permutation of m objects) or if

$$A = (C_1, \dots, C_n) \quad \text{and} \quad B' = (C_{\sigma'(1)}, \dots, C_{\sigma'(n)})$$

(where σ' denotes a permutation of n objects), then $\mathrm{rk}(A) = \mathrm{rk}(B) = \mathrm{rk}(B')$. These equalities are true since the dimension of a vector space does not depend on the ordering of its basis.

Definition 4.2.8 A square matrix $A = (a_{ij}) \in \mathbb{R}^{n,n}$ is called *diagonal* if $a_{ij} = 0$ for $i \neq j$.

Exercise 4.2.9 The following matrix is diagonal,

$$A = \begin{pmatrix} 1 & 0 & 0 & 0 \\ 0 & 2 & 0 & 0 \\ 0 & 0 & 0 & 0 \\ 0 & 0 & 0 & -3 \end{pmatrix}.$$

Its rows and columns are vectors in $\mathbb{R}^4$, with $R_1 = e_1$, $R_2 = 2e_2$, $R_3 = 0$, $R_4 = -3e_4$ with respect to the canonical basis. As a consequence $R(A) = \mathcal{L}(R_1, R_2, R_3, R_4) = \mathcal{L}(e_1, e_2, e_4)$ so that $\mathrm{rk}(A) = 3$.

The rank of a diagonal matrix of order n coincides with the number of its non zero diagonal elements, since, as the previous exercise shows, its non zero rows or columns correspond to multiples of vectors of the canonical basis of $\mathbb{R}^n$. Beside the diagonal ones, a larger class of matrices for which the rank is easy to compute is given in the following definition.

Definition 4.2.10 Let $A = (a_{ij})$ be a square matrix in $\mathbb{R}^{n,n}$. The matrix A is called *upper triangular* if $a_{ij} = 0$ for $i > j$. An upper triangular matrix for which $a_{ii} \neq 0$ for any i, is called a *complete upper triangular* matrix.

Exercise 4.2.11 Given

$$A = \begin{pmatrix} 1 & 0 & 3 \\ 0 & 0 & 2 \\ 0 & 0 & -1 \end{pmatrix}, \qquad B = \begin{pmatrix} 1 & 0 & 3 \\ 0 & 2 & 2 \\ 0 & 0 & -1 \end{pmatrix},$$

then A is upper triangular and B is complete upper triangular.

Theorem 4.2.12 *Let $A \in \mathbb{R}^{n,n}$ be a complete upper triangular matrix. Then,*

$$\text{rk}(A) = n.$$

Proof Let

$$A = \begin{pmatrix} a_{11} & a_{12} & \cdots & a_{1n} \\ 0 & a_{22} & \cdots & a_{2n} \\ \vdots & \vdots & & \vdots \\ 0 & 0 & \cdots & a_{nn} \end{pmatrix}.$$

To prove the claim we show that the n columns $C_1, \dots, C_n$ of A are linearly independent. The equation $\lambda_1 C_1 + \cdots + \lambda_n C_n = 0$ can be written in the form

$$\begin{pmatrix} \lambda_1 a_{11} + \cdots + \lambda_{n-1} a_{1n-1} + \lambda_n a_{1n} \\ \vdots \\ \lambda_{n-1} a_{n-1\,n-1} + \lambda_n a_{n-1\,n} \\ \lambda_n a_{nn} \end{pmatrix} = \begin{pmatrix} 0 \\ \vdots \\ 0 \\ 0 \end{pmatrix}.$$

Equating term by term, one has for the n-th component $\lambda_n a_{nn} = 0$, which gives $\lambda_n = 0$ since $a_{nn} \neq 0$. For the $(n-1)$-th component, one has

$$\lambda_{n-1} a_{n-1,n-1} + \lambda_n a_{n-1,n} = 0$$

which gives, from $\lambda_n = 0$ and $a_{n-1,n-1} \neq 0$, that $\lambda_{n-1} = 0$. This can be extended step by step to all components, thus getting $\lambda_n = \lambda_{n-1} = \cdots = \lambda_1 = 0$. □

The notion of upper triangularity can be extended to non square matrices.

Definition 4.2.13 A matrix $A = (a_{ij}) \in \mathbb{R}^{m,n}$ is called *upper triangular* if it satisfies $a_{ij} = 0$ for $i > j$ and *complete upper triangular* if it is upper triangular with $a_{ii} \neq 0$ for any i.

Remark 4.2.14 Given a matrix $A \in \mathbb{R}^{m,n}$ set $p = \min(m, n)$. If A is a complete upper triangular matrix, the submatrix B made by the first p rows of A when $m > n$, or the first p columns of A when $m < n$, is a square complete upper triangular matrix of order p.

Exercise 4.2.15 The following matrices are complete upper triangular:

$$A = \begin{pmatrix} 1 & 0 & -3 \\ 0 & 2 & 0 \\ 0 & 0 & -1 \\ 0 & 0 & 0 \end{pmatrix}, \qquad A' = \begin{pmatrix} 1 & 2 & 3 & 9 \\ 0 & 2 & 0 & 7 \\ 0 & 0 & 4 & -3 \end{pmatrix}.$$

The submatrices

$$B = \begin{pmatrix} 1 & 0 & -3 \\ 0 & 2 & 0 \\ 0 & 0 & -1 \end{pmatrix}, \qquad B' = \begin{pmatrix} 1 & 2 & 3 \\ 0 & 2 & 0 \\ 0 & 0 & 4 \end{pmatrix}$$

are (square) complete upper triangular as mentioned in the previous remark.

Corollary 4.2.16 *If $A \in \mathbb{R}^{m,n}$ is a complete upper triangular matrix then* $\mathrm{rk}(A) = \min(m, n)$.

Proof We consider two cases.

- $n \geq m$. One has $\mathrm{rk}(A) \leq \min(m, n) = m$, with

$$A = \begin{pmatrix} a_{11} & a_{12} & a_{13} & \dots & a_{1m-1} & a_{1m} & * & \dots & * \\ 0 & a_{22} & a_{23} & \dots & a_{2m-1} & a_{2m} & * & \dots & * \\ 0 & 0 & a_{33} & \dots & a_{3m-1} & a_{3m} & * & \dots & * \\ \vdots & \vdots & \vdots & & \vdots & \vdots & \vdots & & \vdots \\ 0 & 0 & 0 & \dots & 0 & a_{mm} & * & \dots & * \end{pmatrix}.$$

  Let B be the submatrix of A given by the its first m columns. Since B is (Remark 4.2.14) a complete upper triangular square matrix of order m, the columns $C_1, \dots, C_m$ are linearly independent. This means that $\mathrm{rk}(A) \geq m$ and the claim follows.
- $n < m$. One has $\mathrm{rk}(A) \leq \min(m, n) = n$, with

$$A = \begin{pmatrix} a_{11} & a_{12} & a_{13} & \dots & a_{1n} \\ 0 & a_{22} & a_{23} & \dots & a_{2n} \\ 0 & 0 & a_{33} & \dots & a_{3n} \\ \vdots & \vdots & \vdots & & \vdots \\ 0 & 0 & 0 & \dots & a_{nn} \\ 0 & 0 & 0 & \dots & 0 \\ \vdots & \vdots & \vdots & & \vdots \\ 0 & 0 & 0 & \dots & 0 \end{pmatrix}.$$

By deleting all zero rows, one gets a matrix of the previous type, thus $\mathrm{rk}(A) = n$. □

The matrices A and A' in the Exercise 4.2.15 are both complete upper triangular. Their rank is 3.

Remark 4.2.17 The notions introduced in the present section can be formulated by considering columns instead of rows. One has:

- A matrix $A \in \mathbb{R}^{m,n}$ is called *lower triangular* if $a_{ij} = 0$ for $i < j$. A lower triangular matrix is called *complete* if $a_{ii} \neq 0$ for any i.
- Given $A \in \mathbb{R}^{m,n}$, one has that A is (complete) upper triangular if and only if ${}^tA$ is (complete) lower triangular.
- If $A \in \mathbb{R}^{m,n}$ is a complete lower triangular matrix then $\mathrm{rk}(A) = \min(m, n)$.

4.3 Reduced Matrices

Definition 4.3.1 A matrix $A \in \mathbb{R}^{m,n}$ is said to be *reduced by rows* if any non zero row has a non zero element such that the entries below it are all zero. Such an element, which is not necessarily unique if $m \leq n$, is called the *pivot* of its row.

Exercise 4.3.2 The matrix

$$A = \begin{pmatrix} 0 & 1 & 3 \\ 0 & 0 & 0 \\ 2 & 0 & 0 \\ 0 & 0 & -1 \end{pmatrix}$$

is reduced by row. The pivot element for the first row is 1, the pivot element for the third row is 2, the pivot element for the fourth row is -1. Note that $\mathrm{rk}(A) = 3$ since the three non zero rows are linearly independent.

Exercise 4.3.3 Any complete upper triangular matrix is reduced by rows.

Theorem 4.3.4 *The rank of a matrix A which is reduced by row coincides with the number of its non zero rows. Indeed, the non zero rows of a reduced by rows matrix are linearly independent.*

Proof Let A be a reduced by rows matrix and let A' be the submatrix of A obtained by deleting the zero rows of A. From the Remark 4.2.7, $\mathrm{rk}(A') = \mathrm{rk}(A)$. Let A'' be the matrix obtained by A' by the following permutation of its columns: the first column of A'' is the column of A' containing the pivot element for the first row of A', the second column of A'' is the column of A' containing the pivot element for the second row of A' and so on. By such a permutation A'' turns out to be a complete upper triangular matrix and again from the Remark 4.2.7 it is $\mathrm{rk}(A') = \mathrm{rk}(A'')$. Since A'' is complete upper triangular its rank is given by the number of its rows, the rank of A is given by the number of non zero rows of A. □

Since the proof of such a theorem is constructive, an example clarifies it.

Example 4.3.5 Let us consider the following matrix A which is reduced by rows (its pivot elements are bold typed):

$$A = \begin{pmatrix} 1 & \mathbf{-1} & 1 & 1 \\ 0 & 0 & \mathbf{2} & -1 \\ \mathbf{2} & 0 & 0 & 0 \\ 0 & 0 & 0 & \mathbf{1} \end{pmatrix}.$$

The first column of A' is the column containing the pivot element for the first row of A, the second column of A' is the column containing the pivot element for the second row of A and so on. The matrix A' is then

$$A' = \begin{pmatrix} -1 & 1 & 1 & 1 \\ 0 & 2 & 0 & -1 \\ 0 & 0 & 2 & 0 \\ 0 & 0 & 0 & 1 \end{pmatrix}$$

and A' is complete upper triangular; so $\mathrm{rk}(A) = \mathrm{rk}(A') = 4$.

Remark 4.3.6 As we noticed in Remark 4.2.17, the notions introduced above can be formulated by exchanging the role of the columns with that of the rows of a matrix.

- A matrix $A \in \mathbb{R}^{m,n}$ is said to be *reduced by columns* if any non zero column has a non zero element such that the entries at its right are all zero. Such an element, which is not necessarily unique, is called the *pivot* of its column.
- If A is a reduced by columns matrix its rank coincides with the number of its non zero columns. The non zero columns are linearly independent.
- By mimicking the proof of the Theorem 4.3.4 it is clear that a matrix A is reduced by rows if and only if ${}^tA$ is reduced by column.

4.4 Reduction of Matrices

In the previous section we have learnt how to compute the rank of a reduced matrix. In this section we outline a procedure that associates to any given matrix a reduced matrix having the same rank.
We shall consider the following set of transformations acting on the rows of a matrix. They are called *elementary transformations of rows* and their action preserves the vector space structure of the space of rows.

- (λ) The transformation $R_i \mapsto \lambda R_i$ that replace the row R_i with its multiple λR_i, with $\mathbb{R} \ni \lambda \neq 0$,
- (e) The transformation $R_i \leftrightarrow R_j$, that exchanges the rows R_i and R_j,
- (D) The transformation $R_i \mapsto R_i + aR_j$ that replace the row R_i with the linear combination $R_i + aR_j$, with $a \in \mathbb{R}$ and $i \neq j$.

Given a matrix $A \in \mathbb{R}^{m,n}$ the matrix $A' \in \mathbb{R}^{m,n}$ is said to be *row-transformed* from A if A' is obtained from A by the action of a finite number of the elementary transformations (λ), (e) and (D) listed above.

Proposition 4.4.1 *Let $A \in \mathbb{R}^{m,n}$ and $A' \in \mathbb{R}^{m,n}$ be row-transformed form A. Then $R(A) = R(A')$ as vector spaces and* $\mathrm{rk}(A) = \mathrm{rk}(A')$.

Proof It is obvious that for an elementary transformation (e) or (λ) the vector spaces $R(A)$ and $R(A')$ coincide. Let us take A' to be row-transformed from A by a transformation (D). Since

$$R(A) = \mathcal{L}(R_1, \ldots, R_{i-1}, R_i, R_{i+1}, \ldots, R_m)$$

and

$$R(A') = \mathcal{L}(R_1, \ldots, R_{i-1}, R_i + aR_j, R_{i+1}, \ldots, R_m)$$

it is clear that $R(A') \subseteq R(A)$. To prove the opposite inclusion, $R(A) \subseteq R(A')$, it is enough to show that the row R_i in A is in the linear span of the rows of A'. Indeed $R_i = (R_i + aR_j) - aR_j$, thus the claim. □

Exercise 4.4.2 Let

$$A = \begin{pmatrix} 1 & 0 & 1 \\ 2 & 1 & -1 \\ -1 & 1 & 0 \end{pmatrix}.$$

We act on A with the following (D) elementary transformations:

$$A \xrightarrow{R_2 \mapsto R_2 - 2R_1} A' = \begin{pmatrix} 1 & 0 & 1 \\ 0 & 1 & -3 \\ -1 & 1 & 0 \end{pmatrix}$$

$$A' \xrightarrow{R'_3 \mapsto R'_3 + R'_1} A'' = \begin{pmatrix} 1 & 0 & 1 \\ 0 & 1 & -3 \\ 0 & 1 & 1 \end{pmatrix}$$

$$A'' \xrightarrow{R''_3 \mapsto R''_3 - R''_2} A''' = \begin{pmatrix} 1 & 0 & 1 \\ 0 & 1 & -3 \\ 0 & 0 & 4 \end{pmatrix}.$$

The matrix A''' is reduced by rows with $\mathrm{rk}(A''') = 3$. From the proposition above, we conclude that $\mathrm{rk}(A) = \mathrm{rk}(A''') = 3$. This exercise shows how the so called *Gauss' algorithm* works.

Proposition 4.4.3 *Given any matrix A it is always possible to find a finite sequence of type (D) elementary transformations whose action results in a matrix (say B) which is reduced by rows.*

Proof Let $A = (a_{ij}) \in \mathbb{R}^{m,n}$. We denote by R_i the first non zero row in A and by a_{ij} the first non zero element in R_i. In order to obtain a matrix A' such that the elements under a_{ij} are zero one acts with the following (D) transformation

$$R_k \mapsto R_k - a_{kj} a_{ij}{}^{-1} R_i, \quad \text{for any } k > i.$$

We denote such a transformed matrix by $A' = (a'_{ij})$. Notice that the first i rows in A' coincide with the first i rows in A, with all the elements in the column j below the element $a'_{ij} = a_{ij}$ being null. Next, let R'_h be the first non zero row in A' with $h > i$ and let a'_{hp} be the first non zero element in R'_h. As before we now act with the following (D) elementary transformation

$$R'_k \longrightarrow R'_k - a'_{kp} a'_{hp}{}^{-1} R'_h, \quad \text{for any } k > h.$$

Let A'' the matrix obtained with this transformation and iterate. It is clear that a finite number of iterations of this procedure yield a matrix B which is—by construction—reduced by row. □

With the expression of *reduction by rows* of a matrix A we mean a finite sequence of elementary transformations on the rows of A whose final image is a matrix A' which is reduced by rows.

Remark 4.4.4 The proof of the Proposition 4.4.3 made use only of type (D) transformations. It is clear that, depending on the specific elements of the matrix one is considering, it can be easier to use also type (e) and (λ) transformations. The claim of the Proposition 4.4.1 does not change.

Exercise 4.4.5 Let us reduce by rows the following matrix

$$A = \begin{pmatrix} 0 & 1 & 0 & 0 \\ 0 & 1 & 2 & -1 \\ 0 & 0 & 0 & 9 \\ 1 & 3 & 1 & 5 \end{pmatrix}.$$

This matrix can be reduced as in the proof of the Proposition 4.4.3 by type (D) transformations alone. A look at it shows that it is convenient to swap the first row with the fourth. We have

$$A \xrightarrow{R_1 \leftrightarrow R_4} \begin{pmatrix} \mathbf{1} & 3 & 1 & 5 \\ 0 & 1 & \mathbf{2} & -1 \\ 0 & 0 & 0 & \mathbf{9} \\ 0 & \mathbf{1} & 0 & 0 \end{pmatrix} = B.$$

It is evident that the matrix B is *already* reduced by row so we can write $\mathrm{rk}(A) = \mathrm{rk}(B) = 4$.

Exercise 4.4.6 Let us consider the matrix

$$A = \begin{pmatrix} 2 & 1 & -1 & 1 \\ 3 & 1 & 1 & -1 \\ 0 & 1 & 1 & 9 \end{pmatrix}.$$

To reduce A we start with the type (D) transformation $R_2 \mapsto R_2 - 3/2R_1$, that leads to

$$A' = \begin{pmatrix} 2 & 1 & -1 & 1 \\ 0 & -1/2 & 5/2 & -5/2 \\ 0 & 1 & 1 & 9 \end{pmatrix}.$$

Since we are interested in computing the rank of the matrix A in order to avoid non integers matrix entries (which would give heavier computations) we can instead reduce by rows the matrix A' as

$$A' \xrightarrow{R_2 \mapsto 2R_2} A'' = \begin{pmatrix} 2 & 1 & -1 & 1 \\ 0 & -1 & 5 & -5 \\ 0 & 1 & 1 & 9 \end{pmatrix}$$

$$A'' \xrightarrow{R_3' \mapsto R_2' + R_3'} A''' = \begin{pmatrix} 2 & 1 & -1 & 1 \\ 0 & -1 & 5 & -5 \\ 0 & 0 & 6 & 4 \end{pmatrix}.$$

The matrix A''' is upper triangular so we have $\mathrm{rk}(A) = 3$.

The method of reducing by rows a matrix can be used to select a basis for a vector space V given as a linear span of a system of vectors in some $\mathbb{R}^n$, that is

$V = \mathcal{L}(v_1, \ldots, v_r)$. To this end, given the vectors $v_1, \ldots, v_r$ spanning V, one considers the matrix A with rows $v_1, \ldots, v_r$ or alternatively a matrix B with columns $v_1, \ldots, v_r$:

$$A = \begin{pmatrix} v_1 \\ \vdots \\ v_r \end{pmatrix}, \qquad B = (v_1 \cdots v_r).$$

One then has $R(A) = V$ using A, which is reduced by rows to a matrix

$$A' = \begin{pmatrix} w_1 \\ \vdots \\ w_r \end{pmatrix}.$$

Clearly $V = R(A) = R(A')$ and $\dim(V) = \dim(R(A)) = \mathrm{rk}(A) = \mathrm{rk}(A')$. That is $\dim(V)$ is the number of non zero rows in A' and these non zero rows in A' are a basis for V.

Exercise 4.4.7 In $\mathbb{R}^4$ consider the system of vectors $I = \{v_1, v_2, v_3, v_4, v_5\}$ with $v_1 = (1, -1, 2, 1), v_2 = (-2, 2, -4, -2), v_3 = (1, 1, 1, -1), v_4 = (-1, 3, -3, -3)$, $v_5 = (1, 2, 1, 2)$. We would like to

(a) exhibit a basis $\mathcal{B}$ for $V = \mathcal{L}(I) \subset \mathbb{R}^4$, with $\mathcal{B} \subset I$,
(b) complete $\mathcal{B}$ to a basis $\mathcal{C}$ for $\mathbb{R}^4$.

For point (a) we let A be the matrix whose rows are the vectors in I that is,

$$A = \begin{pmatrix} v_1 \\ v_2 \\ v_3 \\ v_4 \\ v_5 \end{pmatrix} = \begin{pmatrix} 1 & -1 & 2 & 1 \\ -2 & 2 & -4 & -2 \\ 1 & 1 & 1 & -1 \\ -1 & 3 & -3 & -3 \\ 1 & 2 & 1 & 2 \end{pmatrix}.$$

We reduce the matrix A by rows using the following transformations:

$$A \xrightarrow[\substack{R_4 \mapsto R_4 + R_1 \\ R_5 \mapsto R_5 - R_1}]{\substack{R_2 \mapsto R_2 + 2R_1 \\ R_3 \mapsto R_3 - R_1}} A' = \begin{pmatrix} 1 & -1 & 2 & 1 \\ 0 & 0 & 0 & 0 \\ 0 & 2 & -1 & -2 \\ 0 & 2 & -1 & -2 \\ 0 & 3 & -1 & 1 \end{pmatrix}$$

$$A' \xrightarrow[R'_5 \mapsto 2R'_5 - 3R'_3]{R'_4 \mapsto R'_4 - R'_3} A'' = \begin{pmatrix} 1 & -1 & 2 & 1 \\ 0 & 0 & 0 & 0 \\ 0 & 2 & -1 & -2 \\ 0 & 0 & 0 & 0 \\ 0 & 0 & 1 & 8 \end{pmatrix}.$$

As a result we have $\text{rk}(A) = 3$ and then $\dim(V) = 3$. A basis for V is for example given by the three non zero rows in A'' since $R(A) = R(A'')$. The basis $\mathcal{B}$ is made by the vectors in I corresponding to the three non zero rows in A'' that is $\mathcal{B} = (v_1, v_3, v_5)$. Cleary, with the transformations given above one has also that

$$B = \begin{pmatrix} v_1 \\ v_3 \\ v_5 \end{pmatrix} \quad \mapsto \quad B' = \begin{pmatrix} 1 & -1 & 2 & 1 \\ 0 & 2 & -1 & -2 \\ 0 & 0 & 1 & 8 \end{pmatrix}.$$

To complete the basis $\mathcal{B}$ to a basis for $\mathbb{R}^4$ one can use the vectors of the canonical basis. From the form of the matrix B' it is clear that it suffices to add the vector e_4 to the three row vectors in B to meet the requirement:

$$\begin{pmatrix} v_1 \\ v_3 \\ v_5 \\ e_4 \end{pmatrix} \quad \mapsto \quad \begin{pmatrix} 1 & -1 & 2 & 1 \\ 0 & 2 & -1 & -2 \\ 0 & 0 & 1 & 8 \\ 0 & 0 & 0 & 1 \end{pmatrix}.$$

We can conclude that $\mathcal{C} = (v_1, v_3, v_5, e_4)$.

Exercise 4.4.8 Let $I = \{v_1, v_2, v_3, v_4\} \subset \mathbb{R}^4$ be given by $v_1 = (0, 1, 2, 1)$, $v_2 = (0, 1, 1, 1)$, $v_3 = (0, 2, 3, 2)$, $v_4 = (1, 2, 2, 1)$. With $V = \mathcal{L}(I) \subset \mathbb{R}^4$:

(a) determine a basis $\mathcal{B}$ for V, with $\mathcal{B} \subset I$,
(b) complete $\mathcal{B}$ to a basis $\mathcal{C}$ for $\mathbb{R}^4$.

Let A be the matrix whose rows are given by the vectors in I that is,

$$A = \begin{pmatrix} v_1 \\ v_2 \\ v_3 \\ v_4 \end{pmatrix} = \begin{pmatrix} 0 & 1 & 2 & 1 \\ 0 & 1 & 1 & 1 \\ 0 & 2 & 3 & 2 \\ 1 & 2 & 2 & 1 \end{pmatrix}.$$

After swapping $R_1 \leftrightarrow R_4$, the matrix can be reduced following the lines above, leading to

$$\begin{pmatrix} 1 & 2 & 2 & 1 \\ 0 & 1 & 1 & 1 \\ 0 & 2 & 3 & 2 \\ 0 & 1 & 2 & 1 \end{pmatrix} \xrightarrow[R_4 \mapsto R_4 - R_2]{R_3 \mapsto R_3 - 2R_2} \begin{pmatrix} 1 & 2 & 2 & 1 \\ 0 & 1 & 1 & 1 \\ 0 & 0 & 1 & 0 \\ 0 & 0 & 1 & 0 \end{pmatrix}$$

$$\xrightarrow[R_4 \mapsto R_4 - R_3]{} \begin{pmatrix} 1 & 2 & 2 & 1 \\ 0 & 1 & 1 & 1 \\ 0 & 0 & 1 & 0 \\ 0 & 0 & 0 & 0 \end{pmatrix}.$$

We can then take $\mathcal{B} = (v_4, v_2, v_3)$. Analogously to what we did in the Exercise 4.4.7, we have

$$\begin{pmatrix} v_4 \\ v_2 \\ v_3 \\ e_4 \end{pmatrix} \quad \mapsto \quad \begin{pmatrix} 1 & 2 & 2 & 1 \\ 0 & 1 & 1 & 1 \\ 0 & 0 & 1 & 0 \\ 0 & 0 & 0 & 1 \end{pmatrix}$$

and such a matrix shows that we can tale $\mathcal{C} = (v_4, v_2, v_3, e_4)$ as a basis for $\mathbb{R}^4$.

Exercise 4.4.9 Consider again the set I given in the previous exercise. We now look for a basis $\mathcal{B} \subset I$ via the constructive proof of the Theorem 2.4.2. The reduction by rows procedure can be used in this case as well. Start again with

$$A = \begin{pmatrix} v_1 \\ v_2 \\ v_3 \\ v_4 \end{pmatrix} = \begin{pmatrix} 0 & 1 & 2 & 1 \\ 0 & 1 & 1 & 1 \\ 0 & 2 & 3 & 2 \\ 1 & 2 & 2 & 1 \end{pmatrix}.$$

The swap operated in Exercise 4.4.8 is not admissible with the procedure in the Theorem 2.4.2 so we use type (D) transformations:

$$A \xrightarrow[\substack{R_3 \mapsto R_3 - 2R_2 \\ R_4 - R_4 - R_1}]{R_2 \mapsto R_2 - R_1} \begin{pmatrix} 0 & 1 & 2 & 1 \\ 0 & 0 & -1 & 0 \\ 0 & 0 & -1 & 0 \\ 1 & 1 & 0 & 0 \end{pmatrix} \xrightarrow[R'_3 \mapsto R'_3 - R'_2]{} \begin{pmatrix} 0 & 1 & 2 & 1 \\ 0 & 0 & -1 & 0 \\ 0 & 0 & 0 & 0 \\ 1 & 1 & 0 & 0 \end{pmatrix}.$$

These computations show that $R'_3 - R'_2 = 0$, $R'_3 = R_3 - 2R_1$ and $R'_2 = R_2 - R_1$. From these relations we have that $R_3 - R_2 - R_1 = 0$ which is equivalent to $v_3 = v_1 + v_2$: this shows that v_3 is a linear combination of v_1 and v_2, so we recover the set $\{v_1, v_2, v_4\}$ as a basis for $\mathcal{L}(I)$.

The method we just illustrated in order to exhibit the basis of a vector subspace of $\mathbb{R}^n$ can be used with any vector space: the entries of the relevant matrix will be given by the components of a system of vectors with respect to a fixed basis.

Exercise 4.4.10 Let $V = \mathcal{L}(I) \subset \mathbb{R}^{2,3}$ with $I = \{M_1, M_2, M_3, M_4\}$ given by

$$M_1 = \begin{pmatrix} 1 & 1 & 1 \\ 0 & 1 & 0 \end{pmatrix}, \quad M_2 = \begin{pmatrix} 1 & 2 & 1 \\ 0 & 1 & 1 \end{pmatrix},$$
$$M_3 = \begin{pmatrix} 2 & 3 & 2 \\ 0 & 2 & 1 \end{pmatrix}, \quad M_4 = \begin{pmatrix} 0 & 1 & 1 \\ 0 & 1 & -1 \end{pmatrix};$$

(a) exhibit a basis $\mathcal{B}$ for V, with $\mathcal{B} \subset I$,
(b) complete $\mathcal{B}$ to a basis $\mathcal{C}$ for $\mathbb{R}^{2,3}$.

In order to use the reduction method we need to represent the matrices M_1, M_2, M_3, M_4 as row vectors. The components of these vectors will be given by the components of the matrices in a basis. This we may take to be the basis $\mathcal{E} = (E_{ij} \mid i = 1, 2; j = 1, 2, 3)$ of $\mathbb{R}^{2,3}$ made of elementary matrices as introduced in the proof of the Proposition 4.1.4. One has, for example,

$$M_1 = E_{11} + E_{12} + E_{13} + E_{22} = (1, 1, 1, 0, 1, 0)_{\mathcal{E}}.$$

Proceeding analogously we write the matrix

$$A = \begin{pmatrix} M_1 \\ M_2 \\ M_3 \\ M_4 \end{pmatrix} = \begin{pmatrix} 1 & 1 & 1 & 0 & 1 & 0 \\ 1 & 2 & 1 & 0 & 1 & 1 \\ 2 & 3 & 2 & 0 & 2 & 1 \\ 0 & 1 & 1 & 0 & 1 & -1 \end{pmatrix}.$$

With a suitable reduction we have

$$A \mapsto \begin{pmatrix} 1 & 1 & 1 & 0 & 1 & 0 \\ 0 & 1 & 0 & 0 & 0 & 1 \\ 0 & 1 & 0 & 0 & 0 & 1 \\ 0 & 1 & 1 & 0 & 1 & -1 \end{pmatrix} \mapsto \begin{pmatrix} 1 & 1 & 1 & 0 & 1 & 0 \\ 0 & 1 & 0 & 0 & 0 & 1 \\ 0 & 0 & 0 & 0 & 0 & 0 \\ 0 & 2 & 1 & 0 & 1 & 0 \end{pmatrix},$$

from which we have $\mathcal{B} = (M_1, M_2, M_4)$.
We complete $\mathcal{B}$ to a basis $\mathcal{C}$ for $\mathbb{R}^{2,3}$ by considering 3 elements in $\mathcal{E}$ and the same reduction:

$$\begin{pmatrix} M_1 \\ M_2 \\ M_4 \\ E_{13} \\ E_{21} \\ E_{22} \end{pmatrix} \mapsto \begin{pmatrix} 1 & 1 & 1 & 0 & 1 & 0 \\ 0 & 1 & 0 & 0 & 0 & 1 \\ 0 & 2 & 1 & 0 & 1 & 0 \\ 0 & 0 & 1 & 0 & 0 & 0 \\ 0 & 0 & 0 & 1 & 0 & 0 \\ 0 & 0 & 0 & 0 & 1 & 0 \end{pmatrix}.$$

Since this matrix is reduced by row, the vectors $\{M_1, M_2, M_4, E_{13}, E_{21}, E_{22}\}$ are 6 linearly independent vectors in $\mathbb{R}^{2,3}$ (whose dimension is 6). This is enough to say that they give a basis $\mathcal{C}$ for $\mathbb{R}^{2,3}$ completing $\mathcal{B}$.

4.5 The Trace of a Matrix

We end this chapter with another useful notion for square matrices.

Definition 4.5.1 The *trace* of a square matrix is the function $\text{tr} : \mathbb{R}^{n,n} \to \mathbb{R}$ defined as follows. If $A = (a_{ij})$ its *trace* is given by

$$\operatorname{tr}(A) = a_{11} + a_{22} + \cdots + a_{nn} = \sum_{j=1}^{n} a_{jj}.$$

That is, the trace of a matrix is the sum of its diagonal elements.

The following proposition proves an important property of the trace function for a matrix.

Proposition 4.5.2 *With $A = (a_{ij})$ and $B = (b_{ij}) \in \mathbb{R}^{n,n}$ it holds that*

$$\operatorname{tr}(AB) = \operatorname{tr}(BA).$$

Proof The entry (i, j) in AB is $(AB)_{ij} = \sum_{k=1}^{n} a_{ik} b_{kj}$, while the entry (i, j) in BA is $(BA)_{ij} = \sum_{k=1}^{n} b_{ik} a_{kj}$. From the row by column product of square matrices one obtaines

$$\begin{aligned}
\operatorname{tr}(AB) = \sum_{j=1}^{n} (AB)_{jj} &= \sum_{j=1}^{n} \sum_{k=1}^{n} a_{jk} b_{kj} \\
&= \sum_{k=1}^{n} \sum_{j=1}^{n} b_{kj} a_{jk} \\
&= \sum_{k=1}^{n} (BA)_{kk} = \operatorname{tr}(BA),
\end{aligned}$$

which is the claim. □

Because of the above property one says that the trace is cyclic.

Chapter 5
The Determinant

The notion of *determinant* of a matrix plays an important role in linear algebra. While the rank measures the linear independence of the row (or column) vectors of a matrix, the determinant (which is defined only for square matrices) is used to control the invertibility of a matrix and in explicitly constructing the inverse of an invertible matrix.

5.1 A Multilinear Alternating Mapping

The determinant can be defined as an abstract function by using multilinear algebra. We shall define it constructively and using a recursive procedure.

Definition 5.1.1 The *determinant* of a 2×2 matrix is the map

$$\det : \mathbb{R}^{2,2} \to \mathbb{R}, \qquad A \mapsto \det(A) = |A|$$

defined as

$$A = \begin{pmatrix} a_{11} & a_{12} \\ a_{21} & a_{22} \end{pmatrix} \quad \mapsto \quad \det(A) = \begin{vmatrix} a_{11} & a_{12} \\ a_{21} & a_{22} \end{vmatrix} = a_{11}a_{22} - a_{12}a_{21}.$$

The above definition shows that the determinant can be though of as a function of the column vectors of $A = (C_1, C_2)$, that is

$$\det : \mathbb{R}^2 \times \mathbb{R}^2 \to \mathbb{R}, \qquad (C_1, C_2) \mapsto a_{11}a_{22} - a_{12}a_{21}.$$

It is immediate to see that the map det is bilinear on the column of A, that is

G. Landi and A. Zampini, *Linear Algebra and Analytic Geometry for Physical Sciences*, Undergraduate Lecture Notes in Physics,
https://doi.org/10.1007/978-3-319-78361-1_5

$$\det(\lambda C_1 + \lambda' C_1', C_2) = \lambda \det(C_1, C_2) + \lambda' \det(C_1', C_2)$$
$$\det(C_1, \lambda C_2 + \lambda' C_2') = \lambda \det(C_1, C_2) + \lambda' \det(C_1, C_2') \quad (5.1)$$

for any $C_1, C_1', C_2, C_2' \in \mathbb{R}^2$ and any $\lambda, \lambda' \in \mathbb{R}$.

The map det is indeed alternating (or skew-symmetric), that is

$$\det(C_2, C_1) = -\det(C_1, C_2). \quad (5.2)$$

From (5.2) the determinant of A vanishes if the columns C_1 and C_2 coincide. More generally, $\det(A) = 0$ if $C_2 = \lambda C_1$ for $\lambda \in \mathbb{R}$, since, from (5.1)

$$\det(C_1, C_2) = \det(C_1, \lambda C_1) = \lambda \det(C_1, C_1) = 0.$$

Since the determinant map is bilinear and alternating, one also has

$$\det(C_1 + \lambda C_2, C_2) = \det(C_1, C_2) + \det(\lambda C_2, C_2) = \det(C_1, C_2).$$

Exercise 5.1.2 Given the canonical basis (e_1, e_2) for $\mathbb{R}^2$, we compute

$$\det(e_1, e_1) = \begin{vmatrix} 1 & 1 \\ 0 & 0 \end{vmatrix} = 0, \qquad \det(e_1, e_2) = \begin{vmatrix} 1 & 0 \\ 0 & 1 \end{vmatrix} = 1$$

$$\det(e_2, e_1) = \begin{vmatrix} 0 & 1 \\ 1 & 0 \end{vmatrix} = -1, \qquad \det(e_2, e_2) = \begin{vmatrix} 0 & 0 \\ 1 & 1 \end{vmatrix} = 0.$$

We generalise the definition of determinant to 3×3 and further to $n \times n$ matrices.

Definition 5.1.3 Given a 3×3 matrix

$$A = \begin{pmatrix} a_{11} & a_{12} & a_{13} \\ a_{21} & a_{22} & a_{23} \\ a_{31} & a_{32} & a_{33} \end{pmatrix}$$

one defines $\det : \mathbb{R}^{3,3} \to \mathbb{R}$ as

$$\det(A) = |A| = \begin{vmatrix} a_{11} & a_{12} & a_{13} \\ a_{21} & a_{22} & a_{23} \\ a_{31} & a_{32} & a_{33} \end{vmatrix} \quad (5.3)$$
$$= a_{11} \begin{vmatrix} a_{22} & a_{23} \\ a_{32} & a_{33} \end{vmatrix} - a_{12} \begin{vmatrix} a_{21} & a_{23} \\ a_{31} & a_{33} \end{vmatrix} + a_{13} \begin{vmatrix} a_{21} & a_{22} \\ a_{31} & a_{32} \end{vmatrix}$$
$$= a_{11}a_{22}a_{33} - a_{11}a_{23}a_{32} - a_{12}a_{21}a_{33} + a_{12}a_{23}a_{31} + a_{13}a_{21}a_{32} - a_{13}a_{22}a_{31}.$$

Exercise 5.1.4 Let us compute the determinant of the following matrix,

$$A = \begin{pmatrix} 1 & 0 & -1 \\ 1 & 1 & -1 \\ 2 & 1 & 0 \end{pmatrix}.$$

Using the first row as above one gets:

$$\det(A) = \begin{vmatrix} 1 & -1 \\ 1 & 0 \end{vmatrix} - \begin{vmatrix} 1 & 1 \\ 2 & 1 \end{vmatrix} = 2.$$

It is evident that the map det can be read, as we showed above, as defined on the column vectors of $A = (C_1, C_2, C_3)$, that is

$$\det : \mathbb{R}^3 \times \mathbb{R}^3 \times \mathbb{R}^3 \to \mathbb{R}, \qquad (C_1, C_2, C_3) \mapsto \det(A).$$

Remark 5.1.5 It is easy to see that the map det defined in (5.3) is multilinear, that is it is linear in each column argument. Also, for any swap of the columns of A, $\det(A)$ changes its sign. This means that (5.3) is an alternating map (this property generalises the skew-symmetry of the det map on 2×2 matrices). For example,

$$\det(C_2, C_1, C_3) = -\det(C_1, C_2, C_3),$$

with analogous relations holding for any swap of the columns of A. Then $\det(A) = 0$ if one of the columns of A is a multiple of the others, like in

$$\det(C_1, C_2, \lambda C_2) = \lambda \det(C_1, C_2, C_2) = -\lambda \det(C_1, C_2, C_2) = 0.$$

More generally $\det(A) = 0$ if one of the columns of A is a linear combination of the others as in

$$\det(\lambda C_2 + \mu C_3, C_2, C_3) = \lambda \det(C_2, C_2, C_3) + \mu \det(C_3, C_2, C_3) = 0.$$

Exercise 5.1.6 If (e_1, e_2, e_3) is the canonical basis for $\mathbb{R}^3$, generalising Exercise 5.1.2 one finds $\det(e_i, e_i, e_i) = 0$, $\det(e_i, e_i, e_j) = 0$ and $\det(e_1, e_2, e_3) = \det(I_3) = 1$, with I_3 the 3×3 unit matrix.

We have seen that the determinant of a 3×3 matrix A makes use of the determinant of a 2×2 matrix: such a determinant is given as the alternating sum of the elements in the first row of A, times the determinant of suitable 2×2 submatrices in A. This procedure is generalised to define the determinant of $n \times n$ matrices.

Definition 5.1.7 Consider the matrix $A = (a_{ij}) \in \mathbb{R}^{n,n}$, or

$$A = \begin{pmatrix} a_{11} & a_{12} & \dots & a_{1n} \\ a_{21} & a_{22} & \dots & a_{2n} \\ \vdots & \vdots & & \vdots \\ a_{n1} & a_{n2} & \dots & a_{nn} \end{pmatrix}.$$

For any pair (i, j) we denote by A_{ij} the $(n-1) \times (n-1)$ submatrix of A obtained by erasing the i-th row and the j-th column of A, Firstly, the number $\det(A_{ij})$ is called the *minor* of the element a_{ij}. Then the *cofactor* α_{ij} of the element a_{ij} (or associated with a_{ij}) is defined as

$$\alpha_{ij} = (-1)^{i+j} \det(A_{ij}).$$

Exercise 5.1.8 With $A \in \mathbb{R}^{3,3}$ given by

$$A = \begin{pmatrix} 1 & 0 & -1 \\ 3 & -2 & -1 \\ 2 & 5 & 0 \end{pmatrix},$$

we easily compute for instance,

$$A_{11} = \begin{pmatrix} -2 & -1 \\ 5 & 0 \end{pmatrix}, \qquad A_{12} = \begin{pmatrix} 3 & -1 \\ 2 & 0 \end{pmatrix}$$

and

$$\alpha_{11} = (-1)^{1+1}|A_{11}| = 5, \qquad \alpha_{12} = (-1)^{1+2}|A_{12}| = -2.$$

Definition 5.1.9 Let $A = (a_{ij}) \in \mathbb{R}^{n,n}$. One defines its determinant by the formula

$$\det(A) = a_{11}\alpha_{11} + a_{12}\alpha_{12} + \cdots + a_{1n}\alpha_{1n}. \tag{5.4}$$

Such an expression is also referred to as the expansion of the determinant of the matrix A with respect to its first row.

The above definition is recursive: the determinant of a $n \times n$ matrix involves the determinants of a $(n-1) \times (n-1)$ matrices, starting from the definition of the determinant of a 2×2 matrix. The Definition 5.1.3 is indeed the expansion with respect to the first row as written in (5.4).

That the determinant $\det(A)$ of a matrix A can be equivalently defined in terms of a similar expansion with respect to any row or column of A is the content of the following important theorem, whose proof we omit.

Theorem 5.1.10 (Laplace) *For any* $i = 2, \ldots, n$ *it holds that*

$$\det(A) = a_{i1}\alpha_{i1} + a_{i2}\alpha_{i2} + \cdots + a_{in}\alpha_{in}. \tag{5.5}$$

This expression is called the expansion of the determinant of A *with respect to its* i*-th row.*

For any $j = 1, \ldots, n$, *it holds that*

$$\det(A) = a_{1j}\alpha_{1j} + a_{2j}\alpha_{2j} + \cdots + a_{nj}\alpha_{nj} \tag{5.6}$$

and this expression is the expansion of the determinant of A *with respect to its* j*-th column.*

The expansions (5.5) or (5.6) are called the *cofactor expansion* of the determinant with respect to the corresponding row or column.

Exercise 5.1.11 Let $I_n \in \mathbb{R}^{n,n}$ be the $n \times n$ unit matrix. It is immediate to compute

$$\det(I_n) = 1.$$

From the Laplace theorem the following statement is obvious.

Corollary 5.1.12 *Let* $A \in \mathbb{R}^{n,n}$. *Then* $\det({}^tA) = \det(A)$.

Also, from the Laplace theorem it is immediate to see that $\det(A) = 0$ if A has a null column or a null row. We can still think of the determinant of the matrix A as a function defined on its columns. If $A = (C_1, \cdots, C_n)$, one has $\det(A) = \det(C_1, \ldots, C_n)$, that is

$$\det : \mathbb{R}^n \times \cdots \times \mathbb{R}^n \to \mathbb{R}, \qquad (C_1, \ldots, C_n) \mapsto \det(A).$$

The following result, that can be proven by using the Definition 5.1.9, generalises properties already seen for the matrices of order two and three.

Proposition 5.1.13 *Let* $A = (C_1, \cdots, C_n) \in \mathbb{R}^{n,n}$. *One has the following properties:*

(i) For any $\lambda, \lambda' \in \mathbb{R}$ *and* $C_1' \in \mathbb{R}^n$, *it holds that*

$$\det\left(\lambda C_1 + \lambda' C_1', C_2, \ldots, C_n\right) = \lambda \det(C_1, C_2, \ldots, C_n) + \lambda' \det(C_1', C_2, \ldots, C_n).$$

Analogous properties hold for any other column of A.

(ii) If $A' = (C_{\sigma(1)}, \ldots, C_{\sigma(n)})$, *where* $\sigma = (\sigma(1), \ldots, \sigma(n))$ *is a permutation of the columns transforming* $A \mapsto A'$, *it holds that*

$$\det(A') = (-1)^\sigma \det(A),$$

where $(-1)^\sigma$ *is the parity of the permutation* σ, *that is* $(-1)^\sigma = 1$ *if* σ *is given by an even number of swaps, while* $(-1)^\sigma = -1$ *if* σ *is given by an odd number of swaps.*

Corollary 5.1.14 *Let* $A = (C_1, \cdots, C_n) \in \mathbb{R}^{n,n}$. *Then,*

(i) $\det(\lambda C_1, C_2, \ldots, C_n) = \lambda \det(A)$,

(ii) if $C_i = C_j$ *for any pair* i, j, *then* $\det(A) = 0$,

(iii) $\det(\alpha_2 C_2 + \cdots + \alpha_n C_n, C_2, \ldots, C_n) = 0$; *that is the determinant of a matrix* A *is zero if a column of* A *is a linear combination of its other columns,*

(iv) $\det(C_1 + \alpha_2 C_2 + \cdots + \alpha_n C_n, C_2, \ldots, C_n) = \det(A)$.

Proof (i) it follows from the Proposition 5.1.13, with $\lambda' = 0$,

(ii) if $C_i = C_j$, the odd permutation σ which swaps C_i with C_j does not change the matrix A; then from the Proposition 5.1.13, $\det(A) = -\det(A) \Rightarrow \det(A) = 0$,

(iii) from 5.1.13 we can write

$$\det(\alpha_2 C_2 + \cdots + \alpha_n C_n, C_2, \ldots, C_n) = \sum_{i=2}^{n} \alpha_i \det(C_i, C_2, \ldots, C_n) = 0$$

since, by point (ii), one has $\det(C_i, C_2, \ldots, C_n) = 0$ for any $i = 2, \ldots, n$,

(iv) from the previous point we have

$$\begin{aligned}\det(C_1 + \alpha_2 C_2 + \cdots + \alpha_n C_n, C_2, \ldots, C_n)\\ = \det(C_1, C_2, \ldots, C_n) + \sum_{i=2}^{n} \alpha_i \det(C_i, C_2, \ldots, C_n) = \det(A).\end{aligned}$$

This concludes the proof. □

Remark 5.1.15 From the Laplace theorem it follows that the determinant of A is an alternating and multilinear function even when it is defined via the expansion with respect to the rows of A.

We conclude this section with the next useful theorem, whose proof we omit.

Theorem 5.1.16 (Binet) *Given* $A, B \in \mathbb{R}^{n,n}$ *it holds that*

$$\det(AB) = \det(A)\det(B). \tag{5.7}$$

5.2 Computing Determinants via a Reduction Procedure

The Definition 5.1.9 and the Laplace theorem allow one to compute the determinant of any square matrix. In this section we illustate how the reduction procedure studied in the previous chapter can be used when computing a determinant. We start by considering upper triangular matrices.

Proposition 5.2.1 *Let* $A = (a_{ij}) \in \mathbb{R}^{n,n}$. *If* A *is diagonal then,*

$$\det(A) = a_{11}a_{22}\cdots a_{nn}.$$

More generally, if A *is an upper (respectively a lower) triangular matrix,* $\det(A) = a_{11}a_{22}\cdots a_{nn}$.

Proof The claim for a diagonal matrix is evident. With A an upper (respectively a lower) triangular matrix, by expanding $\det(A)$ with respect to the first column (respectively row) the submatrix A_{11} is upper (respectively lower) triangular. The result then follows by a recursive argument. □

Remark 5.2.2 In Sect. 4.4 we defined the type (s), (λ) and (D) elementary transformations on the rows of a matrix. If A is a square matrix, transformed under one of these transformations into the matrix A', we have the following results:

- (s) : $\det(A') = -\det(A)$ (Proposition 5.1.13),
- (λ) : $\det(A') = \lambda\det(A)$ (Corollary 5.1.14),
- (D) : $\det(A') = \det(A)$ (Corollary 5.1.14).

It is evident that the above relations are valid when A is mapped into A' with elementary transformations on its columns.

Exercise 5.2.3 Let us use row transformations on the matrix A:

$$A = \begin{pmatrix} 1 & 1 & -1 \\ 2 & 1 & 1 \\ 1 & 2 & 1 \end{pmatrix} \xrightarrow[R_3 \mapsto R_3 - R_1]{R_2 \mapsto R_2 - 2R_1} A'$$

$$A' = \begin{pmatrix} 1 & 1 & -1 \\ 0 & -1 & 3 \\ 0 & 1 & 2 \end{pmatrix} \xrightarrow[R'_3 \mapsto R'_3 + R'_2]{} A''$$

$$A'' = \begin{pmatrix} 1 & 1 & -1 \\ 0 & -1 & 3 \\ 0 & 0 & 5 \end{pmatrix}.$$

Since we have used only type (D) transformations, from the Remark 5.2.2 $\det(A) = \det(A'')$ and from Proposition 5.2.1 we have $\det(A'') = 1 \cdot (-1) \cdot 5 = -5$.

Exercise 5.2.4 Via a sequence of elementary transformations,

$$A = \begin{pmatrix} 0 & 1 & 1 \\ 1 & 2 & -1 \\ 1 & 1 & 1 \end{pmatrix} \xrightarrow{C_1 \leftrightarrow C_2} A'$$

$$A' = \begin{pmatrix} 1 & 0 & 1 \\ 2 & 1 & -1 \\ 1 & 1 & 1 \end{pmatrix} \xrightarrow[R'_3 \mapsto R'_3 - R'_1]{R'_2 \mapsto R'_2 - 2R'_1} A''$$

$$A'' = \begin{pmatrix} 1 & 0 & 1 \\ 0 & 1 & -3 \\ 0 & 1 & 0 \end{pmatrix} \xrightarrow{R''_3 \mapsto R''_3 - R''_2} A'''$$

$$A''' = \begin{pmatrix} 1 & 0 & 1 \\ 0 & 1 & -3 \\ 0 & 0 & 3 \end{pmatrix}.$$

Since we used once a type (s) transformation $\det(A) = -\det(A''') = -3$.

Remark 5.2.5 The sequence of transformations defined in the Exercise 5.2.3 does not alter the space of rows of the matrix A, that is $R(A) = R(A'')$. The sequence of transformations defined in the Exercise 5.2.4 does alter both the spaces of rows and of columns of the matrix A.

Proposition 5.2.6 *Let $A = (a_{ij}) \in \mathbb{R}^{n,n}$ be reduced by rows and without null rows. It holds that*

$$\det(A) = (-1)^{\sigma} a_{1,\sigma(1)} \cdots a_{n,\sigma(n)}$$

where $a_{i,\sigma(i)}$ is the pivot element of the i-th row and σ is the permutation of the columns mapping A into the corresponding (complete) upper triangular matrix.

Proof Let $B = (b_{ij}) \in \mathbb{R}^{n,n}$ be the complete upper triangular matrix obtained from A with the permutation σ. From the Proposition 5.1.13 we have $\det(A) = (-1)^{\sigma}\det(B)$, with $(-1)^{\sigma}$ the parity of σ. From the Proposition 5.2.1 we have $\det(B) = b_{11}b_{22}\cdots b_{nn}$, with $b_{11} = a_{1,\sigma(1)}, \ \ldots \ , b_{nn} = a_{n,\sigma(n)}$ by construction, thus obtaining the claim. □

The above proposition suggests that a sequence of type (D) transformations on the rows of a square matrix simplifies the computation of its determinant. We summarise this suggestion as a remark.

Remark 5.2.7 In order to compute the determinant of the matrix $A \in \mathbb{R}^{n,n}$:

- riduce A by row with only type (D) transformations to a matrix A'; this is always possible from the Proposition 4.4.3. Then $\det(A) = \det(A')$ from the Remark 5.2.2;
- compute the determinant of A'. Then,
  - if A' has a null row, from the Corollary 5.1.14 one has $\det(A') = 0$;
  - if A' has no null rows, from the Proposition 5.2.6 one has

$$\det(A') = (-1)^{\sigma} a'_{1,\sigma(1)} \cdots a'_{n,\sigma(n)}$$

  with $\sigma = (\sigma(1), \ldots, \sigma(n))$.

Again, the result continues to hold by exchanging rows with columns.

Exercise 5.2.8 With the above method we have the following equalities,

$$\begin{vmatrix} 1 & 2 & 1 & -1 \\ 0 & 1 & 1 & 1 \\ -1 & -1 & 1 & 1 \\ 1 & 2 & 0 & 1 \end{vmatrix} = \begin{vmatrix} 1 & 2 & 1 & -1 \\ 0 & 1 & 1 & 1 \\ 0 & 1 & 2 & 0 \\ 0 & 0 & -1 & 2 \end{vmatrix}$$

$$= \begin{vmatrix} 1 & 2 & 1 & -1 \\ 0 & 1 & 1 & 1 \\ 0 & 0 & 1 & -1 \\ 0 & 0 & -1 & 2 \end{vmatrix} = \begin{vmatrix} 1 & 2 & 1 & -1 \\ 0 & 1 & 1 & 1 \\ 0 & 0 & 1 & -1 \\ 0 & 0 & 0 & 1 \end{vmatrix} = 1.$$

5.3 Invertible Matrices

We now illustrate some use of the determinant in the study of invertible matrices.

Proposition 5.3.1 *Given $A \in \mathbb{R}^{n,n}$, it holds that*

$$\det(A) = 0 \quad \Leftrightarrow \quad \mathrm{rk}(A) < n.$$

Proof

($\Leftarrow$) : By hypothesis, the system of the n columns $C_1, \ldots, C_n$ of A is not free, so there is least a column of A which is a linear combination of the other columns. From the Corollary 5.1.14 it is then $\det(A) = 0$.

($\Rightarrow$) : Suppose $\mathrm{rk}(A) = n$. With this assumption A could be reduced by row to a matrix A' having no null rows since $\mathrm{rk}(A) = \mathrm{rk}(A') = n$. From the Proposition 5.2.6, $\det(A')$ is the product of the pivot elements in A' and since by hypothesis they would be non zero, we would have $\det(A') \neq 0$ and from the Remark 5.2.2 $\det(A) = \det(A') \neq 0$ thus contradicting the hypothesis. □

Remark 5.3.2 The equivalence in the above proposition can be stated as

$$\det(A) \neq 0 \quad \Leftrightarrow \quad \mathrm{rk}(A) = n.$$

Proposition 5.3.3 *A matrix $A = (a_{ij}) \in \mathbb{R}^{n,n}$ is invertible (or non-singular) if and only if*

$$\det(A) \neq 0.$$

Proof If A is invertible, the matrix inverse A^{-1} exists with $AA^{-1} = I_n$. From the Binet theorem, this yields $\det(A)\det(A^{-1}) = \det(I_n) = 1$ or $\det(A^{-1}) = (\det(A))^{-1} \neq 0$.

If $\det(A) \neq 0$, the inverse of A is the matrix $B = (b_{ij})$ with elements

$$b_{ij} = \frac{1}{\det(A)}\alpha_{ji}$$

and α_{ji} the cofactor of a_{ji} as in the Definition 5.1.7. Indeed, an explicit computation shows that

$$(AB)_{rs} = \sum_{k=1}^{n} a_{rk}b_{ks} = \frac{1}{\det(A)}\sum_{k=1}^{n} a_{rk}\alpha_{sk} = \begin{cases} \frac{\det(A)}{\det(A)} = 1 & \text{if } r = s \\ 0 & \text{if } r \neq s \end{cases}.$$

The result for $r = s$ is just the cofactor expansion of the determinant given by the Laplace theorem in Theorem 5.1.10, while the result for $r \neq s$ is known as the second Laplace theorem (whose discussion we omit). The above amounts to $AB = I_n$ so that A is invertible with $B = A^{-1}$. □

Notice that in the inverse matrix B there is an index transposition, that is up to the determinant factor, the element b_{ij} of B is given by the cofactor α_{ji} of A.

Exercise 5.3.4 Let us compute the inverse of the matrix

$$A = \begin{pmatrix} a & b \\ c & d \end{pmatrix}.$$

This is possible if and only if $|A| = ad - bc \neq 0$. In such a case,

$$A^{-1} = \frac{1}{|A|} \begin{pmatrix} \alpha_{11} & \alpha_{21} \\ \alpha_{12} & \alpha_{22} \end{pmatrix},$$

with $\alpha_{11} = d$, $\alpha_{21} = -b$, $\alpha_{12} = -c$, $\alpha_{22} = a$, so that we get the final result,

$$A^{-1} = \frac{1}{|A|} \begin{pmatrix} d & -b \\ -c & a \end{pmatrix}.$$

Exercise 5.3.5 Let us compute the inverse of the matrix A from the Exercise 5.1.4,

$$A = \begin{pmatrix} 1 & 0 & -1 \\ 1 & 1 & -1 \\ 2 & 1 & 0 \end{pmatrix}.$$

From the computation there $\det(A) = 2$, explicit computations show that

$$\begin{array}{lll} \alpha_{11} = (+)\,1 & \alpha_{12} = (-)\,2 & \alpha_{13} = (+)\,(-1) \\ \alpha_{21} = (-)\,1 & \alpha_{22} = (+)\,2 & \alpha_{23} = (-)\,1 \\ \alpha_{31} = (+)\,1 & \alpha_{32} = (-)\,0 & \alpha_{33} = (+)\,1 \end{array}.$$

It is then easy to find that

$$A^{-1} = \frac{1}{2} \begin{pmatrix} 1 & -1 & 1 \\ -2 & 2 & 0 \\ -1 & -1 & 1 \end{pmatrix}.$$

Chapter 6
Systems of Linear Equations

Linear equations and system of them are ubiquitous and an important tool in all of physics. In this chapter we shall present a systematic approach to them and to methods for their solutions.

6.1 Basic Notions

Definition 6.1.1 An equation in n *unknown variables* $x_1, \ldots, x_n$ with coefficients in $\mathbb{R}$ is called *linear* if it has the form

$$a_1 x_1 + \cdots + a_n x_n = b,$$

with $a_i \in \mathbb{R}$ and $b \in \mathbb{R}$. A *solution* for such a linear equation is an n-tuple of real numbers $(\alpha_1, \ldots, \alpha_n) \in \mathbb{R}^n$ which, when substituted for the unknowns, yield an 'identity', that is

$$a_1 \alpha_1 + \cdots + a_n \alpha_n = b.$$

Exercise 6.1.2 It is easy to see that the element $(2, 6, 1) \in \mathbb{R}^3$ is a solution for the equation with real coefficients given by

$$3x_1 - 2x_2 + 7x_3 = 1.$$

Clearly, this is not the only solution for the equation: the element $(\frac{1}{3}, 0, 0)$ is for instance a solution of the same equation.

Definition 6.1.3 A collection of m linear equations in the n unknown variables $x_1, \ldots, x_n$ and with real coefficients is called a *linear system* of m equations in n

G. Landi and A. Zampini, *Linear Algebra and Analytic Geometry for Physical Sciences*, Undergraduate Lecture Notes in Physics,
https://doi.org/10.1007/978-3-319-78361-1_6

unknowns. We shall adopt the following notation

$$\Sigma : \begin{cases} a_{11}x_1 + a_{12}x_2 + \ldots + a_{1n}x_n = b_1 \\ a_{21}x_1 + a_{22}x_2 + \ldots + a_{2n}x_n = b_2 \\ \vdots \qquad \vdots \qquad \vdots \qquad \vdots \\ a_{m1}x_1 + a_{m2}x_2 + \ldots + a_{mn}x_n = b_m \end{cases} .$$

A *solution* for a given linear system is an n-tuple $(\alpha_1, \ldots, \alpha_n)$ in $\mathbb{R}^n$ which simultaneously solves each equation of the system. The collection of the solutions of the system Σ is then a subset of $\mathbb{R}^n$, denoted by S_Σ and called the space of solutions of Σ.

A system Σ is called *compatible* or *solvable* if its space of solutions is non void, $S_\Sigma \neq \emptyset$; it will be said to be *incompatible* if $S_\Sigma = \emptyset$.

Exercise 6.1.4 The element $(1, -1) \in \mathbb{R}^2$ is a solution of the system

$$\begin{cases} x + y = 0 \\ x - y = 2 \end{cases} .$$

The following system

$$\begin{cases} x + y = 0 \\ x + y = 1 \end{cases} .$$

has no solutions.

In the present chapter we study conditions under which a linear system is compatible and in such a case find methods to determine its space of solutions. We shall make a systematic use of the matrix formalism described in the previous Chaps. 4 and 5.

Definition 6.1.5 There are two matrices naturally associated to the linear system Σ as given in the Definition 6.1.3:

1. the *matrix of the coefficients* of Σ, $A = (a_{ij}) \in \mathbb{R}^{m,n}$,
2. the *matrix of the inhomogeneous terms* of Σ, $B = {}^t(b_1, \ldots, b_m) \in \mathbb{R}^{m,1}$.

The *complete* or *augmented* matrix of the linear system Σ is given by

$$(A, B) = (a_{ij} \mid b_i) = \left(\begin{array}{cccc|c} a_{11} & a_{12} & \ldots & a_{1n} & b_1 \\ a_{21} & a_{22} & \ldots & a_{2n} & b_2 \\ \vdots & \vdots & & \vdots & \vdots \\ a_{m1} & a_{m2} & \ldots & a_{mn} & b_m \end{array} \right) .$$

By using these matrices the system Σ can be represented as follows

$$\Sigma : \begin{pmatrix} a_{11} & a_{12} & \dots & a_{1n} \\ a_{21} & a_{22} & \dots & a_{2n} \\ \vdots & \vdots & & \vdots \\ a_{m1} & a_{m2} & \dots & a_{mn} \end{pmatrix} \begin{pmatrix} x_1 \\ x_2 \\ \vdots \\ x_n \end{pmatrix} = \begin{pmatrix} b_1 \\ b_2 \\ \vdots \\ b_m \end{pmatrix}$$

or more succinctly as

$$\Sigma : AX = B$$

where the array of unknowns is written as $X = {}^t(x_1, \dots, x_n)$ and (abusing notations) thought to be an element in $\mathbb{R}^{n,1}$.

Definition 6.1.6 Two linear systems $\Sigma : AX = B$ and $\Sigma' : A'X = B'$ are called *equivalent* if their spaces of solutions coincide, that is $\Sigma \sim \Sigma'$ if $S_\Sigma = S_{\Sigma'}$. Notice that the vector of unknowns for the two systems is the same.

Remark 6.1.7 The linear systems $AX = B$ and $A'X = B'$ are trivially equivalent

- if (A', B') results from (A, B) by adding null rows,
- if (A', B') is given by a row permutation of (A, B).

The following linear systems are evidently equivalent:

$$\begin{cases} x + y = 0 \\ x - y = 2 \end{cases}, \qquad \begin{cases} x - y = 2 \\ x + y = 0 \end{cases}.$$

Remark 6.1.8 Notice that for a permutation of the columns of the matrix of its coefficients a linear system Σ changes to a system that is in general not equivalent to the starting one. As an example, consider the compatible linear system $AX = B$ given in Exercise 6.1.4. If the columns of A are swapped one has

$$(A, B) = \left(\begin{array}{cc|c} 1 & 1 & 0 \\ 1 & -1 & 2 \end{array}\right) \xrightarrow{\quad C_1 \leftrightarrow C_2 \quad} \left(\begin{array}{cc|c} 1 & 1 & 0 \\ -1 & 1 & 2 \end{array}\right) = (A', B).$$

One checks that the solution $(1, -1)$ of the starting system is not a solution for the system $A'X = B$.

6.2 The Space of Solutions for Reduced Systems

Definition 6.2.1 A linear system $AX = B$ is called *reduced* if the matrix A of its coefficients is reduced by rows in the sense of Sect. 4.4. Solving a reduced system is quite elementary, as the following exercises show.

Exercise 6.2.2 Let the linear system Σ be given by

$$\Sigma : \begin{cases} x + y + 2z = 4 \\ \quad y - 2z = -3 \\ \quad\quad z = 2 \end{cases} \quad \text{with} \quad (A, B) = \left(\begin{array}{ccc|c} 1 & 1 & 2 & 4 \\ 0 & 1 & -2 & -3 \\ 0 & 0 & 1 & 2 \end{array}\right).$$

It is reduced, and has the only solution $(x, y, z) = (-1, 1, 2)$. This is easily found by noticing that the third equation gives $z = 2$. By inserting this value into the second equation one has $y = 1$, and by inserting both these values into the first equation one eventually gets $x = -1$.

Exercise 6.2.3 To solve the linear system

$$\Sigma : \begin{cases} 2x + y + 2z + t = 1 \\ 2x + 3y - z = 3 \\ x + z = 0 \end{cases} \quad \text{with} \quad (A, B) = \left(\begin{array}{cccc|c} 2 & 1 & 2 & 1 & 1 \\ 2 & 3 & -1 & 0 & 3 \\ 1 & 0 & 1 & 0 & 0 \end{array}\right)$$

one proceeds as in the previous exercise. The last equation gives $z = -x$. By setting $x = \tau$, one gets the solutions $(x, y, z, t) = (\tau, -\tau + 1, -\tau, \tau)$ with $\tau \in \mathbb{R}$. Clearly Σ has an infinite number of solutions: the space of solutions for Σ is bijective to elements $\tau \in \mathbb{R}$.

Exercise 6.2.4 The linear system $\Sigma : AX = B$, with

$$(A, B) = \left(\begin{array}{ccc|c} 1 & 2 & 1 & 3 \\ 0 & -1 & 2 & 1 \\ 0 & 0 & 3 & 2 \\ 0 & 0 & 0 & 1 \end{array}\right),$$

is trivially not compatible since the last equation would give $0 = 1$.

Remark 6.2.5 If A is reduced by row, the Exercises 6.2.2 and 6.2.3 show that one first determines the value of the unknown corresponding to the pivot (special) element of the bottom row and then replaces such unknown by its value in the remaining equations. This amounts to *delete*, or *eliminate* one of the unknowns. Upon iterating this procedure one completely solves the system. This procedure is showed in the following displays where the pivot elements are bold typed:

$$(A, B) = \left(\begin{array}{ccc|c} \mathbf{1} & 1 & 2 & 4 \\ 0 & \mathbf{1} & -2 & -3 \\ 0 & 0 & \mathbf{1} & 2 \end{array}\right).$$

Here one determines z at first then y and finally x. As for the Exercise 6.2.3, one writes

$$(A, B) = \left(\begin{array}{cccc|c} 2 & 1 & 2 & \mathbf{1} & 1 \\ 2 & \mathbf{3} & -1 & 0 & 3 \\ 1 & 0 & \mathbf{1} & 0 & 0 \end{array}\right)$$

where one determines z, then y and after those one determines t.

The previous exercises suggest the following method that we describe as a proposition.

Proposition 6.2.6 (The method of eliminations) *Let $\Sigma : AX = B$ be a reduced system.*

(1) From the Remark 6.1.7 we may assume that (A, B) has no null rows.
(2) If A has null rows they correspond to equations like $0 = b_i$ with $b_i \neq 0$ since the augmented matrix (A, B) has no null rows. This means that the system is not compatible, $S_\Sigma = \emptyset$.
(3) If A has no null rows, then $m \leq n$. Since A is reduced, it has m pivot elements, so its rank is m. Starting from the bottom row one can then determine the unknown corresponding to the pivot element and then, by substituting such an unknown in the remaining equations, iterate the procedure thus determining the space of solutions.

We describe the general procedure when A is a complete upper triangular matrix.

$$(A, B) = \left(\begin{array}{cccccccccc|c} a_{11} & a_{12} & a_{13} & \dots & a_{1m} & * & \dots & * & a_{1n} & & b_1 \\ 0 & a_{22} & a_{23} & \dots & a_{2m} & * & \dots & * & a_{2n} & & b_2 \\ 0 & 0 & a_{33} & \dots & a_{3m} & * & \dots & * & a_{3n} & & b_3 \\ \vdots & \vdots & \vdots & & \vdots & \vdots & & \vdots & \vdots & & \vdots \\ 0 & 0 & 0 & \dots & a_{mm} & * & \dots & * & a_{mn} & & b_m \end{array}\right)$$

with all diagonal elements $a_{ii} \neq 0$. The equation corresponding to the bottom line of the matrix is

$$a_{mm}x_m + a_{mm+1}x_{m+1} + \cdots + a_{mn}x_n = b_m$$

with $a_{mm} \neq 0$. By dividing both sides of the equation by a_{mm}, one has

$$x_m = a_{mm}^{-1}(b_m - a_{mm+1}x_{m+1} + \cdots - a_{mn}x_n).$$

Then x_m is a function of $x_{m+1}, \dots, x_n$. From the $(m-1)$-th row one analogously obtains

$$x_{m-1} = a_{m-1m-1}^{-1}(b_{m-1} - a_{m-1m}x_m - a_{m-1m+1}x_{m+1} + \cdots - a_{m-1n}x_n).$$

By replacing x_m with its value (as a function of $x_{m+1}, \dots, x_n$) previously determined, one writes x_{m-1} as a function of the last unknowns $x_{m+1}, \dots, x_n$. The natural iterations of this process leads to write the unknowns $x_{m-2}, x_{m-3}, \dots, x_1$ as functions of the remaining ones $x_{m+1}, \dots, x_n$.

Remark 6.2.7 Since the m unknowns $x_1, \dots, x_m$ can be expressed as functions of the remaining ones, the $n - m$ unknowns $x_{m+1}, \dots, x_n$, the latter are said to be *free* unknowns. By choosing an arbitrary numerical value for them, $x_{m+1} = \lambda_1, \dots, x_n = \lambda_{n-m}$, with $\lambda_i \in \mathbb{R}$, one obtains a solution, since the matrix A is reduced, of the linear system. This allows one to define a bijection

$$\mathbb{R}^{n-m} \quad \Leftrightarrow \quad S_\Sigma$$

where n is the number of unknowns of Σ and $m = \mathrm{rk}(A)$. One usually labels this result by saying that the linear system has ∞^{n-m} solutions.

6.3 The Space of Solutions for a General Linear System

One of the possible methods to solve a general linear system $AX = B$ uses the notions of row reduction for a matrix as described in Sect. 4.4. From the definition at the beginning of that section one has the following proposition.

Theorem 6.3.1 *Let $\Sigma : AX = B$ be a linear system, and let (A', B') be a transformed by row matrix of (A, B). The linear systems Σ and the transformed one $\Sigma' : A'X = B'$ are equivalent.*

Proof We denote as usual $A = (a_{ij})$ and $B = {}^t(b_1, \dots, b_m)$. If (A', B') is obtained from (A, B) under a type (e) elementary transformation, the claim is obvious as seen in Remark 6.1.7. If (A', B') is obtained from (A, B) under a type (λ) transformation by the row R_i the claim follows by noticing that, for any $\lambda \neq 0$, the linear equation

$$a_{i1}x_1 + \cdots + a_{in}x_n = b_i$$

is equivalent to the equation

$$\lambda a_{i1}x_1 + \cdots + \lambda a_{in}x_n = \lambda b_i.$$

Let now (A', B') be obtained from (A, B) via a type (D) elementary transformation,

$$R_i \mapsto R_i + \lambda R_j$$

with $j \neq i$. To be definite we take $i = 2$ and $j = 1$. We then have

$$(A', B') = \begin{pmatrix} R_1 \\ R_2 + \lambda R_1 \\ \vdots \\ R_m \end{pmatrix}.$$

Let us assume that $\alpha = (\alpha_1, \dots, \alpha_n)$ is a solution for Σ, that is

$$a_{i1}\alpha_1 + \cdots + a_{in}\alpha_n = b_i$$

for any $i = 1, \dots, m$. That all but the second equation of Σ' are solved by α is obvious; it remains to verify whether α solves also the second equation in it that is, to show that

$$(a_{21} + \lambda a_{11})x_1 + \cdots + (a_{2n} + \lambda a_{1n})x_n = b_2 + \lambda b_1.$$

If we add the equation for $i = 2$ to λ times the equation for $i = 1$, we obtain

$$(a_{21} + \lambda a_{11})\alpha_1 + \cdots + (a_{2n} + \lambda a_{1n})\alpha_n = b_2 + \lambda b_1$$

thus $(\alpha_1, \ldots, \alpha_n)$ is a solution for Σ' and $S_\Sigma \subseteq S_{\Sigma'}$. The inclusion $S_{\Sigma'} \subseteq S_\Sigma$ is proven in an analogous way. □

By using the above theorem one proves a general method to solve linear systems known as *Gauss' elimination method* or *Gauss' algorithm.*

Theorem 6.3.2 *The space S_Σ of the solutions of the linear system $\Sigma : AX = B$ is determined via the following steps.*

(1) Reduce by rows the matrix (A, B) to (A', B') with A' reduced by row.
(2) Using the method given in the Proposition 6.2.6 determine the space $S_{\Sigma'}$ of the solutions for the system $\Sigma' : A'X = B'$.
(3) From the Theorem 6.3.1 it is $\Sigma \sim \Sigma'$ that is $S_\Sigma = S_{\Sigma'}$.

Exercise 6.3.3 Let us solve the following linear system

$$\Sigma = \begin{cases} 2x + y + z = 1 \\ x - y - z = 0 \\ x + 2y + 2z = 1 \end{cases}$$

whose complete matrix is

$$(A, B) = \left(\begin{array}{ccc|c} 2 & 1 & 1 & 1 \\ 1 & -1 & -1 & 0 \\ 1 & 2 & 2 & 1 \end{array}\right).$$

By reducing such a matrix by rows, we have

$$(A, B) \xrightarrow[R_3 \mapsto R_3 - 2R_1]{R_2 \mapsto R_2 + R_1} \left(\begin{array}{ccc|c} 2 & 1 & 1 & 1 \\ 3 & 0 & 0 & 1 \\ -3 & 0 & 0 & -1 \end{array}\right)$$

$$\xrightarrow[R_3 \mapsto R_3 + R_2]{} \left(\begin{array}{ccc|c} 2 & 1 & 1 & 1 \\ 3 & 0 & 0 & 1 \\ 0 & 0 & 0 & 0 \end{array}\right) = (A', B').$$

Since A' is reduced the linear system $\Sigma' : A'X = B'$ is reduced and then solvable by the Gauss' method. We have

$$\Sigma' : \begin{cases} 2x + y + z = 1 \\ 3x = 1 \end{cases} \implies \begin{cases} y + z = \frac{1}{3} \\ x = \frac{1}{3} \end{cases}.$$

It is now clear that one unknown is free so the linear system has ∞^1 solutions. By choosing $z = \lambda$ the space S_Σ of solutions for Σ is

$$S_\Sigma = \{(x, y, z) \in \mathbb{R}^3 \mid (x, y, z) = (\tfrac{1}{3}, \tfrac{1}{3} - \lambda, \lambda),\ \lambda \in \mathbb{R}\}.$$

On the other end, by choosing $y = \alpha$ the space S_Σ can be written as

$$S_\Sigma = \{(x, y, z) \in \mathbb{R}^3 \mid (x, y, z) = (\tfrac{1}{3}, \alpha, \tfrac{1}{3} - \alpha),\ \alpha \in \mathbb{R}\}.$$

It is obvious that we are representing the same subset $S_\Sigma \subset \mathbb{R}^3$ in two different ways.

Notice that the number of free unknowns is the difference between the total number of unknowns and the rank of the matrix A.

Exercise 6.3.4 Let us solve the following linear system,

$$\Sigma : \begin{cases} x + y - z = 0 \\ 2x - y = 1 \\ y + 2z = 2 \end{cases}$$

whose complete matrix is

$$(A, B) = \left(\begin{array}{ccc|c} 1 & 1 & -1 & 0 \\ 2 & -1 & 0 & 1 \\ 0 & 1 & 2 & 2 \end{array}\right).$$

The reduction procedure gives

$$(A, B) \xrightarrow{R_2 \mapsto R_2 - 2R_1} \left(\begin{array}{ccc|c} 1 & 1 & -1 & 0 \\ 0 & -3 & 2 & 1 \\ 0 & 1 & 2 & 2 \end{array}\right)$$
$$\xrightarrow[R_3 \mapsto R_3 - R_2]{} \left(\begin{array}{ccc|c} 1 & 1 & -1 & 0 \\ 0 & -3 & 2 & 1 \\ 0 & 4 & 0 & 1 \end{array}\right) = (A', B').$$

Since A' is reduced the linear system $\Sigma' : A'X = B'$ is reduced with no free unknowns. This means that $S_{\Sigma'}$ (and then S_Σ) has $\infty^0 = 1$ solution. The Gauss' method provides us a way to find such a solution, namely

$$\Sigma' : \begin{cases} x - y + z = 0 \\ -3y + 2z = 1 \\ 4y = 1 \end{cases} \implies \begin{cases} x - z = -\frac{1}{4} \\ 2z = \frac{7}{4} \\ y = \frac{1}{4} \end{cases}.$$

This gives $S_\Sigma = \{(x, y, z) = (\frac{5}{8}, \frac{1}{4}, \frac{7}{8})\}$. Once more the number of free unknowns is the difference between the total number of unknowns and the rank of the matrix A.

The following exercise shows how to solve a linear system with one coefficient given by a real parameter instead of a fixed real number. By solving such a system we mean to analyse the conditions on the parameter under which the system is solvable and to provide its space of solutions as depending on the possible values of the parameter.

Exercise 6.3.5 Let us study the following linear system,

$$\Sigma_\lambda : \begin{cases} x + 2y + z + t = -1 \\ x + y - z + 2t = 1 \\ 2x + \lambda y + \lambda t = 0 \\ -\lambda y - 2z + \lambda t = 2 \end{cases}$$

with $\lambda \in \mathbb{R}$. When the complete matrix for such a system is reduced, particular care must be taken for some critical values of λ. We have

$$(A, B) = \left(\begin{array}{cccc|c} 1 & 2 & 1 & 1 & -1 \\ 1 & 1 & -1 & 2 & 1 \\ 2 & \lambda & 0 & \lambda & 0 \\ 0 & -\lambda & -2 & \lambda & 2 \end{array}\right)$$

$$\xrightarrow[R_3 \mapsto R_3 - 2R_1]{R_2 \mapsto R_2 - R_1} \left(\begin{array}{cccc|c} 1 & 2 & 1 & 1 & -1 \\ 0 & -1 & -2 & 1 & 2 \\ 0 & \lambda - 4 & -2 & \lambda - 2 & 2 \\ 0 & -\lambda & -2 & \lambda & 2 \end{array}\right)$$

$$\xrightarrow[R_4 \to R_4 - R_2]{R_3 \mapsto R_3 - R_2} \left(\begin{array}{cccc|c} 1 & 2 & 1 & 1 & -1 \\ 0 & -1 & -2 & 1 & 2 \\ 0 & \lambda - 3 & 0 & \lambda - 3 & 0 \\ 0 & -\lambda + 1 & 0 & \lambda - 1 & 0 \end{array}\right) = (A', B').$$

The transformations $R_3 \mapsto R_3 + R_4$, then $R_3 \mapsto \frac{1}{2} R_3$ and finally $R_4 \mapsto R_4 + (1 - \lambda) R_3$ give a further reduction of (A', B') as

$$\left(\begin{array}{cccc|c} 1 & 2 & 1 & 1 & -1 \\ 0 & -1 & -2 & 1 & 2 \\ 0 & -1 & 0 & \lambda - 2 & 0 \\ 0 & -\lambda + 1 & 0 & \lambda - 1 & 0 \end{array}\right) \mapsto \left(\begin{array}{cccc|c} 1 & 2 & 1 & 1 & -1 \\ 0 & -1 & -2 & 1 & 2 \\ 0 & -1 & 0 & \lambda - 2 & 0 \\ 0 & 0 & 0 & a_{44} & 0 \end{array}\right) = (A'', B'')$$

with $a_{44} = (1 - \lambda)(\lambda - 3)$. Notice that the last transformation is meaningful for any $\lambda \in \mathbb{R}$. In the reduced form (A'', B'') we have that R_4 is null if and only if either $\lambda = 3$ or $\lambda = 1$. For such values of the parameter λ either R_3 or R_4 in A' is indeed null. We can now conclude that Σ_λ is solvable for any value of $\lambda \in \mathbb{R}$ and we have

- If $\lambda \in \{1, 3\}$ then $a_{44} = 0$, so $\mathrm{rk}(A) = 3$ and Σ_λ has ∞^1 solutions,
- If $\lambda \notin \{1, 3\}$ then $a_{44} \neq 0$, so $\mathrm{rk}(A) = 4$ and Σ_λ has a unique solution.

We can now study the following three cases:

(a) $\lambda \notin \{1, 3\}$, that is

$$\Sigma_\lambda : \begin{cases} x + 2y + z + t = -1 \\ -y - 2z + t = 2 \\ -y + (\lambda - 2)t = 0 \\ (\lambda - 3)(\lambda - 1)t = 0 \end{cases} .$$

From our assumption, we have that $a_{44} = (\lambda - 3)(\lambda - 1) \neq 0$ so we get $t = 0$. By using the Gauss' method we then write

$$\begin{cases} x = 0 \\ z = -1 \\ y = 0 \\ t = 0 \end{cases} .$$

This shows that for $\lambda \neq 1, 3$ the space S_{Σ_λ} does not depend on λ.

(b) If $\lambda = 1$ we can delete the fourth equation since it is a trivial identity. We have then

$$\Sigma_{\lambda=1} : \begin{cases} x + 2y + z + t = -1 \\ -y - 2z + t = 2 \\ y + t = 0 \end{cases} .$$

The Gauss' method gives us

$$\begin{cases} x = 0 \\ z = t - 1 \\ y = -t \end{cases}$$

and this set of solutions can be written as

$$\{(x, y, z, t) \in \mathbb{R}^4 \mid (x, y, z, t) = (0, -\alpha, \alpha - 1, \alpha),\ \alpha \in \mathbb{R}\}.$$

(c) If $\lambda = 3$ the non trivial part of the system turns out to be

$$\Sigma_{\lambda=3} : \begin{cases} x + 2y + z + t = -1 \\ -y - 2z + t = 2 \\ -y + t = 0 \end{cases}$$

and we write the solutions as

$$\begin{cases} x = -3t \\ z = -1 \\ y = t \end{cases}$$

or equivalently $S_{\Sigma_{\lambda=3}} = \{(x, y, z, t) \in \mathbb{R}^4 \mid (x, y, z, t) = (-3\alpha, \alpha, -1, \alpha),\ \alpha \in \mathbb{R}\}$.

What we have discussed can be given in the form of the following theorem which provides general conditions under which a linear system is solvable.

Theorem 6.3.6 (Rouché–Capelli). *The linear system* $\Sigma : AX = B$ *is solvable if and only if* $\mathrm{rk}(A) = \mathrm{rk}(A, B)$. *In such a case, denoting* $\mathrm{rk}(A) = \mathrm{rk}(A, B) = \rho$ *and with* n *the number of unknowns in* Σ, *the following holds true:*

(a) the number of free unknowns is $n - \rho$,
(b) the $n - \rho$ *free unknowns have to be selected in such a way that the remaining* ρ *unknowns correspond to linearly independent columns of* A.

Proof By noticing that the linear system Σ can be written as

$$x_1 C_1 + \cdots + x_n C_n = B$$

with $C_1, \ldots, C_n$ the columns of A, we see that Σ is solvable if and only if B is a linear combination of these columns that is if and only if the linear span of the columns of A coincides with the linear span of the columns of (A, B). This condition is fulfilled if and only if $\mathrm{rk}(A) = \mathrm{rk}(A, B)$.

Suppose then that the system is solvable.

(a) Let $\Sigma' : A'X = B'$ be the system obtained from (A, B) by reduction by rows. From the Remark 6.2.7 the system Σ' has $n - \mathrm{rk}(A')$ free unknowns. Since $\Sigma \sim \Sigma'$ and $\mathrm{rk}(A) = \mathrm{rk}(A')$ the claim follows.
(b) Possibly with a swap of the columns in $A = (C_1, \ldots, C_n)$ (which amounts to renaming the unknown), the result that we aim to prove is the following:

$$x_{\rho+1}, \ldots, x_n \text{ are free } \Leftrightarrow C_1, \ldots, C_\rho \text{ are linearly independent.}$$

Let us at first suppose that $C_1, \ldots, C_\rho$ are linearly independent, and set $\overline{A} = (C_1, \ldots, C_\rho)$. By a possible reduction and a swapping of some equations, with $\mathrm{rk}(\overline{A}) = \mathrm{rk}(A, B) = \rho$, the matrix for the system can be written as

$$(A', B') = \begin{pmatrix} a_{11} & a_{12} & a_{13} & \ldots & a_{1\rho} & * & \ldots & * & b_1 \\ 0 & a_{22} & a_{23} & \ldots & a_{2\rho} & * & \ldots & * & b_2 \\ 0 & 0 & a_{33} & \ldots & a_{3\rho} & * & \ldots & * & b_3 \\ \vdots & \vdots & \vdots & & \vdots & \vdots & & \vdots & \vdots \\ 0 & 0 & 0 & \ldots & a_{\rho\rho} & * & \ldots & * & b_\rho \\ 0 & 0 & 0 & \ldots & 0 & 0 & \ldots & 0 & 0 \\ \vdots & \vdots & \vdots & & \vdots & \vdots & & \vdots & \vdots \\ 0 & 0 & 0 & \ldots & 0 & 0 & \ldots & 0 & 0 \end{pmatrix}.$$

The claim—that $x_{\rho+1}, \ldots, x_n$ can be taken to be free—follows easily from the Gauss' method.
On the other hand, let us assume that $x_{\rho+1}, \ldots, x_n$ are free unknowns for the linear system and let us also suppose that $C_1, \ldots, C_\rho$ are linearly dependent. This

would result in the rank of $\overline{A}$ be less that ρ and there would exist a reduction of (A, B) for which the matrix of the linear system turns out to be

$$(A', B') = \begin{pmatrix} a_{11} & \dots & a_{1\rho} & * \dots * \\ \vdots & & \vdots & \vdots \quad \vdots \\ a_{\rho-1 1} & \dots & a_{\rho-1\,\rho} & * \dots * \\ 0 & \dots & 0 & * \dots * \\ \vdots & & \vdots & \vdots \quad \vdots \\ 0 & \dots & 0 & * \dots * \end{pmatrix}.$$

Since $\mathrm{rk}(A', B') = \mathrm{rk}(A, B) = \rho$ there would then be a non zero row R_i in (A', B') with $i \geq \rho$. The equation corresponding to such an R_i, not depending on the first ρ unknowns, would provide a relation among the $x_{\rho+1}, \dots, x_n$, which would then be not free. □

Remark 6.3.7 If the linear system $\Sigma : AX = B$, with n unknowns and m equations is solvable with $\mathrm{rk}(A) = \rho$, then

(i) Σ is equivalent to a linear system Σ' with ρ equations arbitrarily chosen among the m equations in Σ, provided they are linearly independent.
(ii) there is a bijection between the space S_Σ and $\mathbb{R}^{n-\rho}$.

Exercise 6.3.8 Let us solve the following linear system depending on a parameter $\lambda \in \mathbb{R}$,

$$\Sigma : \begin{cases} \lambda x + z = -1 \\ x + (\lambda - 1)y + 2z = 1 \\ x + (\lambda - 1)y + 3z = 0 \end{cases}.$$

We reduce by rows the complete matrix corresponding to Σ as

$$(A, B) = \left(\begin{array}{ccc|c} \lambda & 0 & 1 & -1 \\ 1 & \lambda - 1 & 2 & 1 \\ 1 & \lambda - 1 & 3 & 0 \end{array}\right)$$

$$\xrightarrow[R_3 \mapsto R_3 - 3R_1]{R_2 \mapsto R_2 - 2R_1} \left(\begin{array}{ccc|c} \lambda & 0 & 1 & -1 \\ 1 - 2\lambda & \lambda - 1 & 0 & 3 \\ 1 - 3\lambda & \lambda - 1 & 0 & 3 \end{array}\right)$$

$$\xrightarrow[R_3 \mapsto R_3 - R_2]{} \left(\begin{array}{ccc|c} \lambda & 0 & 1 & -1 \\ 1 - 2\lambda & \lambda - 1 & 0 & 3 \\ -\lambda & 0 & 0 & 0 \end{array}\right) = (A', B').$$

Depending on the values of the parameter λ we have the following cases.

(a) If $\lambda = 1$, the matrix A' is not reduced. We then write

$$(A', B') = \left(\begin{array}{ccc|c} 1 & 0 & 1 & -1 \\ -1 & 0 & 0 & 3 \\ -1 & 0 & 0 & 0 \end{array}\right) \xrightarrow[R_3 \mapsto R_3 - R_2]{} \left(\begin{array}{ccc|c} 1 & 0 & 1 & -1 \\ -1 & 0 & 0 & 3 \\ 0 & 0 & 0 & -3 \end{array}\right).$$

The last row gives the equation $0 = -3$ and in this case the system has no solution.

(b) If $\lambda \neq 1$ the matrix A' is reduced, so we have:

- If $\lambda \neq 0$, then $\mathrm{rk}(A) = 3 = \mathrm{rk}(A, B)$, so the linear system $\Sigma_{\lambda=0}$ has a unique solution. With $\lambda \notin \{0, 1\}$ the reduced system is

$$\Sigma' : \begin{cases} \lambda x + z = -1 \\ (1 - 2\lambda)x + (\lambda - 1)y = 3 \\ -\lambda x = 0 \end{cases}$$

and the Gauss' method gives $S_{\Sigma_\lambda} = (x, y, z) = (0, 3/(\lambda - 1), -1)$.

- If $\lambda = 0$ the system we have to solve is

$$\Sigma' : \begin{cases} z = -1 \\ x - y = 3 \end{cases}$$

whose solutions are given as

$$S_{\Sigma_{\lambda=0}} = \{(x, y, z) \in \mathbb{R}^3 \mid (x, y, z) = (\alpha + 3, \alpha, -1)\, \alpha \in \mathbb{R}\}.$$

Exercise 6.3.9 Let us show that the following system of vectors,

$$v_1 = (1, 1, 0), \quad v_2 = (0, 1, 1), \quad v_3 = (1, 0, 1),$$

is free and then write $v = (1, 1, 1)$ as a linear combination of v_1, v_2, v_3.

We start by recalling that v_1, v_2, v_3 are linearly independent if and only if the rank of the matrix whose columns are the vectors themselves is 3. We have the following reduction,

$$(v_1\, v_2\, v_3) = \begin{pmatrix} 1 & 0 & 1 \\ 1 & 1 & 0 \\ 0 & 1 & 1 \end{pmatrix} \mapsto \begin{pmatrix} 1 & 0 & 1 \\ 0 & 1 & -1 \\ 0 & 1 & 1 \end{pmatrix} \mapsto \begin{pmatrix} 1 & 0 & 1 \\ 0 & 1 & -1 \\ 0 & 0 & 2 \end{pmatrix}.$$

The number of non zero rows of the reduced matrix is 3 so the vectors v_1, v_2, v_3 are linearly independent. Then they are a basis for $\mathbb{R}^3$, so the following relation,

$$xv_1 + yv_2 + zv_3 = v$$

is fullfilled by a unique triple (x, y, z) of coefficients for any $v \in \mathbb{R}^3$. Such a triple is the unique solution of the linear system whose complete matrix is $(A, B) = (v_1\ v_2\ v_3\ v)$. For the case we are considering in this exercise we have

$$(A, B) = \left(\begin{array}{ccc|c} 1 & 0 & 1 & 1 \\ 1 & 1 & 0 & 1 \\ 0 & 1 & 1 & 1 \end{array}\right).$$

Using for (A, B) the same reduction we used above for A we have

$$(A, B) \quad \mapsto \quad \left(\begin{array}{ccc|c} 1 & 0 & 1 & 1 \\ 0 & 1 & -1 & 0 \\ 0 & 1 & 1 & 1 \end{array}\right) \quad \mapsto \quad \left(\begin{array}{ccc|c} 1 & 0 & 1 & 1 \\ 0 & 1 & -1 & 0 \\ 0 & 0 & 2 & 1 \end{array}\right).$$

The linear system we have then to solve is

$$\begin{cases} x + z = 1 \\ y - z = 0 \\ 2z = 1 \end{cases}$$

giving $(x, y, x) = \frac{1}{2}(1, 1, 1)$. One can indeed directly compute that

$$\tfrac{1}{2}(1, 1, 0) + \tfrac{1}{2}(0, 1, 1) + \tfrac{1}{2}(1, 0, 1) = (1, 1, 1).$$

Exercise 6.3.10 Let us consider the matrix

$$M_\lambda = \begin{pmatrix} \lambda & 1 \\ 1 & \lambda \end{pmatrix}$$

with $\lambda \in \mathbb{R}$. We compute its inverse using the theory of linear systems.

We can indeed write the problem in terms of the linear system

$$\begin{pmatrix} \lambda & 1 \\ 1 & \lambda \end{pmatrix} \begin{pmatrix} x & y \\ z & t \end{pmatrix} = \begin{pmatrix} 1 & 0 \\ 0 & 1 \end{pmatrix},$$

that is

$$\Sigma : \begin{cases} \lambda x + z = 1 \\ x + \lambda z = 0 \\ \lambda y + t = 0 \\ y + \lambda t = 1 \end{cases}.$$

We reduce the complete matrix of the linear system as follows:

$$(A, B) = \left(\begin{array}{cccc|c} \lambda & 0 & 1 & 0 & 1 \\ 1 & 0 & \lambda & 0 & 0 \\ 0 & \lambda & 0 & 1 & 0 \\ 0 & 1 & 0 & \lambda & 1 \end{array}\right)$$

$$\xrightarrow{R_2 \mapsto R_2 - \lambda R_1} \left(\begin{array}{cccc|c} \lambda & 0 & 1 & 0 & 1 \\ 1-\lambda^2 & 0 & 0 & 0 & -\lambda \\ 0 & \lambda & 0 & 1 & 0 \\ 0 & 1 & 0 & \lambda & 1 \end{array}\right)$$

$$\xrightarrow[R_4 \mapsto R_4 - \lambda R_3]{} \left(\begin{array}{cccc|c} \lambda & 0 & 1 & 0 & 1 \\ 1-\lambda^2 & 0 & 0 & 0 & -\lambda \\ 0 & \lambda & 0 & 1 & 0 \\ 0 & 1-\lambda^2 & 0 & 0 & 1 \end{array}\right) = (A', B').$$

The elementary transformations we used are well defined for any real value of λ. We start by noticing that if $1 - \lambda^2 = 0$ that is $\lambda = \pm 1$, we have

$$(A', B') = \left(\begin{array}{cccc|c} \pm 1 & 0 & 1 & 0 & 1 \\ 0 & 0 & 0 & 0 & \mp 1 \\ 0 & \pm 1 & 0 & 1 & 0 \\ 0 & 0 & 0 & 0 & 1 \end{array}\right).$$

The second and the fourth rows of this matrix show that the corresponding linear system is incompatible. This means that when $\lambda = \pm 1$ the matrix M_λ is not invertible (as we would immediately see by computing its determinant).

We assume next that $1 - \lambda^2 \neq 0$. In such a case we have $\mathrm{rk}(A) = \mathrm{rk}(A, B) = 4$, so there exists a unique solution for the linear system. We write it in the reduced form as

$$\Sigma' : \begin{cases} \lambda x + z = 1 \\ (1 - \lambda^2)x = -\lambda \\ \lambda y + t = 0 \\ (1 - \lambda^2)y = 1 \end{cases}.$$

Its solution is then

$$\begin{cases} z = 1/(1 - \lambda^2) \\ x = -\lambda/(1 - \lambda^2) \\ t = -\lambda/(1 - \lambda^2) \\ y = 1/(1 - \lambda^2) \end{cases},$$

that we write in matrix form as

$$M_\lambda^{-1} = \frac{1}{(1-\lambda^2)} \begin{pmatrix} -\lambda & 1 \\ 1 & -\lambda \end{pmatrix}.$$

6.4 Homogeneous Linear Systems

We analyse now an interesting class of linear systems (for easy of notation we write $0 = 0_{\mathbb{R}^m}$).

Definition 6.4.1 A linear system $\Sigma : AX = B$ is called *homogeneous* if $B = 0$.

Remark 6.4.2 A linear system $\Sigma : AX = 0$ with $A \in \mathbb{R}^{m,n}$ is always solvable since the null n-tuple (the null vector in $\mathbb{R}^n$) gives a solution for Σ, albeit a trivial one. This also follows form the Rouché-Capelli theorem since one obviously has $\mathrm{rk}(A) = \mathrm{rk}(A, 0)$. The same theorem allows one to conclude that such a trivial solution is indeed the only solution for Σ if and only if $n = \rho = \mathrm{rk}(A)$.

Theorem 6.4.3 *Let* $\Sigma : AX = 0$ *be a homogeneous linear system with* $A \in \mathbb{R}^{m,n}$. *Then* S_Σ *is a vector subspace of* $\mathbb{R}^n$ *with* $\dim S_\Sigma = n - \mathrm{rk}(A)$.

Proof From the Proposition 2.2.2 we have to show that if $X_1, X_2 \in S_\Sigma$ with λ_1, $\lambda_2 \in \mathbb{R}$, then $\lambda_1 X_1 + \lambda_2 X_2$ is in S_Σ. Since by hypothesis we have $AX_1 = 0$ and $AX_2 = 0$ we have also $\lambda_1(AX_1) + \lambda_2(AX_2) = 0$. From the properties of the matrix calculus we have in turn $\lambda_1(AX_1) + \lambda_2(AX_2) = A(\lambda_1 X_1 + \lambda_2 X_2)$, thus giving $\lambda_1 X_1 + \lambda_2 X_2$ in S_Σ. We conclude that S_Σ is a vector subspace of $\mathbb{R}^n$.

With $\rho = \mathrm{rk}(A)$, from the Rouché-Capelli theorem we know that Σ has $n - \rho$ free unknowns. This number coincides with the dimension of S_Σ. To show this fact we determine a basis made up of $n - \rho$ elements. Let us assume for simplicity that the free unknowns are the last ones $x_{\rho+1}, \ldots, x_n$. Any solution of Σ can then be written as

$$(*, \ldots, *, \ x_{\rho+1}, \ldots, x_n)$$

where the ρ symbols $*$ stand for the values of $x_1, \ldots, x_\rho$ corresponding to each possible value of $x_{\rho+1}, \ldots, x_n$. We let now the $(n - \rho)$-dimensional 'vector' $x_{\rho+1}, \ldots, x_n$ range over all elements of the canonical basis of $\mathbb{R}^{n-\rho}$ and write the corresponding elements in S_Σ as

$$\begin{aligned} v_1 &= (*, \ldots, *, 1, 0, \ldots, 0) \\ v_2 &= (*, \ldots, *, 0, 1, \ldots, 0) \\ &\vdots \\ v_{n-\rho} &= (*, \ldots, *, 0, 0, \ldots, 1). \end{aligned}$$

The rank of the matrix $(v_1, \ldots, v_{n-\rho})$ (that is the matrix whose rows are these vectors) is clearly equal to $n - \rho$, since its last $n - \rho$ columns are linearly independent. This means that its rows, the vectors $v_1, \ldots, v_{n-\rho}$, are linearly independent. It is easy to see that such rows generate S_Σ so they are a basis for it and $\dim(S_\Sigma) = n - \rho$. □

It is clear that the general reduction procedure allows one to solve any homogeneous linear system Σ. Since the space S_Σ is in this case a linear space, one can determine a basis for it. The proof of the previous theorem provides indeed an easy method to get such a basis for S_Σ. Once the elements in S_Σ are written in terms of the $n-\rho$ free unknowns a basis for S_Σ is given by fixing for these unknowns the values corresponding to the elements of the canonical basis in $\mathbb{R}^{n-\rho}$.

Exercise 6.4.4 Let us solve the following homogeneous linear system,

$$\Sigma : \begin{cases} x_1 - 2x_3 + x_5 + x_6 = 0 \\ x_1 - x_2 - x_3 + x_4 - x_5 + x_6 = 0 \\ x_1 - x_2 + 2x_4 - 2x_5 + 2x_6 = 0 \end{cases}$$

and let us determine a basis for its space of solutions. The corresponding A matrix is

$$A = \begin{pmatrix} 1 & 0 & -2 & 0 & 1 & 1 \\ 1 & -1 & -1 & 1 & -1 & 1 \\ 1 & -1 & 0 & 2 & -2 & 2 \end{pmatrix}.$$

We reduce it as follows

$$A \quad \mapsto \quad \begin{pmatrix} 1 & 0 & -2 & 0 & 1 & 1 \\ 0 & -1 & 1 & 1 & -2 & 0 \\ 0 & -1 & 2 & 2 & -3 & 1 \end{pmatrix} \quad \mapsto \quad \begin{pmatrix} 1 & 0 & -2 & 0 & 1 & 1 \\ 0 & -1 & 1 & 1 & -2 & 0 \\ 0 & 0 & 1 & 1 & -1 & 1 \end{pmatrix} = A'.$$

Thus $\mathrm{rk}(A) = \mathrm{rk}(A') = 3$. Since the first three rows in A' (and then in A) are linearly independent we choose x_4, x_5, x_6 to be the free unknowns. One clearly has $\Sigma \sim \Sigma' : A'X = 0$ so we can solve

$$\Sigma' : \begin{cases} x_1 - 2x_3 + x_5 + x_6 = 0 \\ x_2 - x_3 - x_4 + 2x_5 = 0 \\ x_3 + x_4 - x_5 + x_6 = 0 \end{cases}.$$

By setting $x_4 = a$, $x_5 = b$ and $x_6 = c$ we have

$$S_\Sigma = \{(x_1, ..., x_6) = (-2a + b - 3c, -b - c, -a + b - c, a, b, c) \mid a, b, c \in \mathbb{R}\}.$$

To determine a basis for S_Σ we let (a, b, c) be the vectors $(1, 0, 0)$, $(0, 1, 0)$, $(0, 0, 1)$ of the canonical basis in $\mathbb{R}^3$ since $n - \rho = 6 - 3 = 3$. With this choice we get the following basis

$$\begin{aligned} v_1 &= (-2, 0, -1, 1, 0, 0) \\ v_2 &= (1, -1, 1, 0, 1, 0) \\ v_3 &= (-3, -1, -1, 0, 0, 1). \end{aligned}$$

Chapter 7
Linear Transformations

Together with the theory of linear equations and matrices, the notion of linear transformations is crucial in both classical and quantum physics. In this chapter we introduce them and study their main properties.

7.1 Linear Transformations and Matrices

We have already seen that differently looking sets may have the same vector space structure. In this chapter we study mappings between vector spaces which are, in a proper sense, compatible with the vector space structure. The action of such maps will be represented by matrices.

Example 7.1.1 Let $A = \begin{pmatrix} a & b \\ c & d \end{pmatrix} \in \mathbb{R}^{2,2}$. Let us define the map $f : \mathbb{R}^2 \to \mathbb{R}^2$ by

$$f(X) = AX$$

where $X = (x, y)$ is a (column) vector representing a generic element in $\mathbb{R}^2$ and AX denotes the usual row by column product, that is

$$f\begin{pmatrix} x \\ y \end{pmatrix} = \begin{pmatrix} a & b \\ c & d \end{pmatrix}\begin{pmatrix} x \\ y \end{pmatrix}.$$

With $X = (x_1, x_2)$ and $Y = (y_1, y_2)$ two elements in $\mathbb{R}^2$, using the properties of the matrix calculus it is easy to show that

$$f(X + Y) = A(X + Y) = AX + AY = f(X) + f(Y)$$

G. Landi and A. Zampini, *Linear Algebra and Analytic Geometry for Physical Sciences*, Undergraduate Lecture Notes in Physics,
https://doi.org/10.1007/978-3-319-78361-1_7

as well as, with $\lambda \in \mathbb{R}$, that

$$f(\lambda X) = A(\lambda X) = \lambda A(X) = \lambda f(X).$$

This example is easily generalised to matrices of arbitrary dimensions.

Exercise 7.1.2 Given $A = (a_{ij}) \in \mathbb{R}^{m,n}$ one considers the map $f : \mathbb{R}^n \to \mathbb{R}^m$

$$f(X) = AX,$$

with $X = {}^t(x_1, \ldots, x_n)$ and AX the usual row by column product. The above properties are easily generalised so this map satisfies the identities $f(X+Y) = f(X) + f(Y)$ for any $X, Y \in \mathbb{R}^n$ and $f(\lambda X) = \lambda f(X)$ for any $X \in \mathbb{R}^n$, $\lambda \in \mathbb{R}$.

Example 7.1.3 Let $A = \begin{pmatrix} 1 & 2 & 1 \\ 1 & -1 & 0 \end{pmatrix} \in \mathbb{R}^{2,3}$. The associated map $f : \mathbb{R}^3 \to \mathbb{R}^2$ is given by

$$f((x, y, z)) = \begin{pmatrix} 1 & 2 & 1 \\ 1 & -1 & 0 \end{pmatrix} \begin{pmatrix} x \\ y \\ z \end{pmatrix} = \begin{pmatrix} x + 2y + z \\ x - y \end{pmatrix}.$$

The above lines motivate the following.

Definition 7.1.4 Let V and W be two vector spaces over $\mathbb{R}$. A map $f : V \to W$ is called *linear* if the following properties hold:

(L1) $f(X+Y) = f(X) + f(Y)$ for all $X, Y \in V$,
(L2) $f(\lambda X) = \lambda f(X)$ for all $X \in V$, $\lambda \in \mathbb{R}$.

The proof of the following identities is immediate.

Proposition 7.1.5 *If $f : V \to W$ is a linear map then,*

(a) $f(0_V) = 0_W$,
(b) $f(-v) = -f(v)$ *for any* $v \in V$,
(c) $f(a_1 v_1 + \cdots + a_p v_p) = a_1 f(v_1) + \cdots + a_p f(v_p)$, *for any* $v_1, \ldots, v_p \in V$ *and* $a_1, \ldots, a_p \in \mathbb{R}$.

Proof (a) Since $0_V = 0_{\mathbb{R}} 0_V$ the (L2) defining property gives

$$f(0_V) = f(0_{\mathbb{R}} 0_V) = 0_{\mathbb{R}} f(0_V) = 0_W.$$

(b) Since $-v = (-1)v$, again from (L2) we have

$$f(-v) = f((-1)v) = (-1)f(v) = -f(v).$$

(c) This is proved by induction on p. If $p = 2$ the claim follows directly from (L1) and (L2) with

$$f(a_1v_1 + a_2v_2) = f(a_1v_1) + f(a_2v_2) = a_1 f(v_1) + a_2 f(v_2).$$

Let us assume it to be true for $p-1$. By setting $w = a_1v_1 + \cdots + a_{p-1}v_{p-1}$, we have

$$f(a_1v_1 + \cdots + a_pv_p) = f(w + a_pv_p) = f(w) + f(a_pv_p) = f(w) + a_p f(v_p)$$

(the first equality follows from (L1), the second from (L2)). From the induction hypothesis, we have $f(w) = a_1 f(v_1) + \cdots + a_{p-1} f(v_{p-1})$, so

$$f(a_1v_1 + \cdots + a_pv_p) = f(w) + a_p f(v_p) = a_1 f(v_1) + \cdots + a_{p-1} f(v_{p-1}) + a_p f(v_p),$$

which is the statement for p.

□

Example 7.1.6 The Example 7.1.1 and the Exercise 7.1.2 show how one associates a linear map between $\mathbb{R}^n$ and $\mathbb{R}^m$ to a matrix $A \in \mathbb{R}^{m,n}$. This construction can be generalised by using bases for vector spaces V and W.

Let us consider a basis $\mathcal{B} = (v_1, \ldots, v_n)$ for V and a basis $\mathcal{C} = (w_1, \ldots, w_m)$ for W. Given the matrix $A = (a_{ij}) \in \mathbb{R}^{m,n}$ we define $f : V \to W$ as follows. For any $v \in V$ we have uniquely $v = x_1v_1 + \cdots + x_nv_n$, that is $v = (x_1, \ldots, x_n)_{\mathcal{B}}$. With $X = {}^t(x_1, \ldots, x_n)$, we consider the vector $AX \in \mathbb{R}^m$ with $AX = {}^t(y_1, \ldots, y_m)_{\mathcal{C}}$. We write then

$$f(v) = y_1w_1 + \cdots + y_mw_m$$

which can be written as

$$f((x_1, \ldots, x_n)_{\mathcal{B}}) = \left(A \begin{pmatrix} x_1 \\ \vdots \\ x_n \end{pmatrix} \right)_{\mathcal{C}}.$$

Exercise 7.1.7 Let us consider the matrix $A = \begin{pmatrix} 1 & 2 & 1 \\ 1 & -1 & 0 \end{pmatrix} \in \mathbb{R}^{2,3}$, with $V = \mathbb{R}[X]_2$ and $W = \mathbb{R}[X]_1$. With respect to the bases $\mathcal{B} = (1, X, X^2)$ for V and $\mathcal{C} = (1, X)$ for W the map corresponding to A as in the previous example is

$$f(a + bX + cX^2) = (A\,{}^t(a, b, c))_{\mathcal{C}}$$

that is

$$f(a + bX + cX^2) = (a + 2b + c, a - b)_{\mathcal{C}} = a + 2b + c + (a - b)X.$$

Proposition 7.1.8 *The map $f : V \to W$ defined in the Example 7.1.6 is linear.*

Proof Let $v, v' \in V$ with $v = (x_1, \ldots, x_n)_{\mathcal{B}}$ and $v' = (x'_1, \ldots, x'_n)_{\mathcal{B}}$. From the Remark 2.4.16 we have

$$v + v' = (x_1 + x'_1, \ldots, x_n + x'_n)_{\mathcal{B}}$$

so we get

$$\begin{aligned} f(v + v') &= (A^t(x_1 + x'_1, \ldots, x_n + x'_n))_{\mathcal{C}} \\ &= (A^t(x_1, \ldots, x_n))_{\mathcal{C}} + (A^t(x'_1, \ldots, x'_n))_{\mathcal{C}} \\ &= f(v) + f(v') \end{aligned}$$

(notice that the second equality follows from the Proposition 4.1.10). Along the same line one shows easily that for any $\lambda \in \mathbb{R}$ one has $f(\lambda v) = \lambda f(v)$. □

The following definition (a rephrasing of Example 7.1.6) plays a central role in the theory of linear transformations.

Definition 7.1.9 With V and W two vector spaces over $\mathbb{R}$ and bases $\mathcal{B} = (v_1, \ldots, v_n)$ for V and $\mathcal{C} = (w_1, \ldots, w_m)$ for W, consider a matrix $A = (a_{ij}) \in \mathbb{R}^{m,n}$. The linear map

$$f_A^{\mathcal{C},\mathcal{B}} \; : \; V \; \to \; W$$

defined by

$$V \ni v = x_1 v_1 + \cdots + x_n v_n \;\mapsto\; f_A^{\mathcal{B},\mathcal{C}}(v) = y_1 w_1 + \cdots + y_m w_m \;\in\; W$$

with

$${}^t(y_1, \ldots, y_m) = A\,{}^t(x_1, \ldots, x_n),$$

is the *linear map corresponding to the matrix A with respect to the basis $\mathcal{B}$ e $\mathcal{C}$.*

Remark 7.1.10 Denoting $f_A^{\mathcal{C},\mathcal{B}} = f$, one immediately sees that the n columns in A provide the components with respect to $\mathcal{C}$ in W of the vectors $f(v_1), \ldots, f(v_n)$, with $(v_1, \ldots, v_n)$ the basis $\mathcal{B}$ for V. One has

$$v_1 = 1v_1 + 0v_2 + \cdots + 0v_n = (1, 0, \ldots, 0)_{\mathcal{B}},$$

thus giving

$$\begin{aligned} f(v_1) = (A\,{}^t(1, 0, \ldots, 0))_{\mathcal{C}} &= {}^t(a_{11}, \ldots, a_{m1})_{\mathcal{C}} \\ &= f(v_1) = a_{11} w_1 + \cdots + a_{m1} w_m. \end{aligned}$$

It is straightforward now to show that $f(v_j) = (a_{1j}, \ldots, a_{mj})_{\mathcal{C}}$ for any index j.

If $A = (a_{ij}) \in \mathbb{R}^{m,n}$ and $f \;:\; \mathbb{R}^n \;\to\; \mathbb{R}^m$ is the linear map defined by $f(X) = AX$, then the columns of A give the images under f of the vectors $(e_1, \ldots e_n)$ of the canonical basis $\mathcal{E}_n$ in $\mathbb{R}^n$. This can be written as

$$A \;=\; \big(f(e_1)\; f(e_2)\; \cdots\; f(e_n)\big)\,.$$

Exercise 7.1.11 Let us consider the matrix

$$A = \begin{pmatrix} 1 & 1 & -1 \\ 0 & 1 & 2 \\ 1 & 1 & 0 \end{pmatrix},$$

with $\mathcal{B} = \mathcal{C} = \mathcal{E}_3$ the canonical basis in $\mathbb{R}^3$, and the corresponding linear map $f = f_A^{\mathcal{E}_3,\mathcal{E}_3} : \mathbb{R}^3 \to \mathbb{R}^3$. If $(x, y, z) \in \mathbb{R}^3$ then $f((x, y, z)) = A^t(x, y, z)$. The action of f is then given by

$$f((x, y, z)) = (x + y - z, y + 2z, x + y).$$

Being $\mathcal{B}$ the canonical basis, it is also

$$f(e_1) = (1, 0, 1), \quad f(e_2) = (1, 1, 1), \quad f(e_3) = (-1, 2, 0).$$

We see that $f(e_1), f(e_2), f(e_3)$ are the columns of A. This is not an accident: as mentioned the columns of A are, in the general situation, the components of $f(e_1), f(e_2), f(e_3)$ with respect to a basis $\mathcal{C}$—in this case the canonical one.

The Proposition 7.1.8 shows that, given a matrix A, the map $f_A^{\mathcal{C},\mathcal{B}}$ is linear. Our aim is now to prove that for *any* linear map $f : V \to W$ there exists a matrix A such that $f = f_A^{\mathcal{B},\mathcal{C}}$, with respect to two given bases $\mathcal{B}$ and $\mathcal{C}$ for V and W respectively.

In order to determine such a matrix we use the Remark 7.1.10: given a matrix A the images under $f_A^{\mathcal{C},\mathcal{B}}$ of the elements in the basis $\mathcal{B}$ of V are given by the column elements in A. This suggests the following definition.

Definition 7.1.12 Let $\mathcal{B} = (v_1, \ldots, v_n)$ be a basis for the real vector space V and $\mathcal{C} = (w_1, \ldots, w_m)$ a basis for the real vector space W. Let $f : V \to W$ be a linear map. The matrix *associated* to f with respect to the basis $\mathcal{B}$ and $\mathcal{C}$, that we denote by $M_f^{\mathcal{C},\mathcal{B}}$, is the element in $\mathbb{R}^{m,n}$ whose columns are given by the components with respect to $\mathcal{C}$ of the images under f of the basis elements in $\mathcal{B}$. That is, the matrix $M_f^{\mathcal{C},\mathcal{B}} = A = (a_{ij})$ is given by

$$\begin{aligned} f(v_1) &= a_{11}w_1 + \cdots + a_{m1}w_m \\ &\vdots \\ f(v_n) &= a_{1n}w_1 + \cdots + a_{mn}w_m, \end{aligned}$$

which can be equivalently written as

$$M_f^{\mathcal{C},\mathcal{B}} = (f(v_1), \quad \ldots \quad , f(v_n)).$$

Such a definition inverts the one given in the Definition 7.1.9. This is the content of the following proposition, whose proof we omit.

Proposition 7.1.13 *Let V be a real vector space with basis $\mathcal{B} = (v_1, \ldots, v_n)$ and W a real vector space with basis $\mathcal{C} = (w_1, \ldots, w_m)$. The following results hold.*

(i) *If $f : V \to W$ is a linear map, by setting $A = M_f^{\mathcal{C},\mathcal{B}}$ it holds that*

$$f_A^{\mathcal{C},\mathcal{B}} = f.$$

(ii) *If $A \in \mathbb{R}^{m,n}$, by setting $f = f_A^{\mathcal{C},\mathcal{B}}$ it holds that*

$$M_f^{\mathcal{C},\mathcal{B}} = A.$$

Proposition 7.1.14 *Let V and W be two real vector spaces with $(v_1, \ldots, v_n)$ a basis for V. For any choice of $\{u_1, \ldots, u_n\}$ of n elements in W there exists a unique linear map $f : V \to W$ such that $f(v_j) = u_j$ for any $j = 1, \ldots, n$.*

Proof To define such a map one uses that any vector $v \in V$ can be written uniquely as

$$v = a_1 v_1 + \cdots + a_n v_n$$

with respect to the basis $(v_1, \ldots, v_n)$. By setting

$$f(v) = a_1 f(v_1) + \cdots + a_n f(v_n) = a_1 u_1 + \cdots + a_n u_n$$

we have a linear (by construction) map f that satisfies the required condition $f(v_j) = u_j$ for any $j \in 1, \ldots, n$.

Let us now suppose this map is not unique and that there exists a second linear map $g : V \to W$ with $g(v_j) = u_j$. From the Proposition 7.1.5 we could then write

$$g(v) = a_1 g(v_1) + \cdots + a_n g(v_n) = a_1 u_1 + \cdots + a_n u_n = f(v),$$

thus getting $g = f$. □

What we have discussed so far gives two equivalent ways to define a linear map between two vector spaces V and W.

I. Once a basis $\mathcal{B}$ for V, a basis $\mathcal{C}$ for W and a matrix $A = (a_{ij}) \in \mathbb{R}^{m,n}$ are fixed, from the Proposition 7.1.13 we know that the linear map $f_A^{\mathcal{C},\mathcal{B}}$ is uniquely determined.

II. Once a basis $\mathcal{B} = (v_1, \ldots, v_n)$ for V and n vectors $\{u_1, \ldots, u_n\}$ in W are fixed, we know from the Proposition 7.1.14 that there exists a unique linear map $f : V \to W$ with $f(v_j) = u_j$ for any $j = 1, \ldots, n$.

From now on, if $V = \mathbb{R}^n$ and $\mathcal{B} = \mathcal{E}$ is its canonical basis we shall denote by $f((x_1, \ldots, x_n))$ what we have previously denoted as $f((x_1, \ldots, x_n)_{\mathcal{B}})$. Analogously, with $\mathcal{C} = \mathcal{E}$ the canonical basis for $W = \mathbb{R}^m$ we shall write $(y_1, \ldots, y_m)$ instead of $(y_1, \ldots, y_m)_{\mathcal{C}}$.

With such a notation, if $f : \mathbb{R}^n \to \mathbb{R}^m$ is the linear map which, with respect to the canonical basis for both vector spaces corresponds to the matrix A, its action is written as

$$f((x_1, \ldots, x_n)) = A^t(x_1, \ldots, x_n)$$

or equivalently

$$f((x_1, \ldots, x_n)) = (a_{11}x_1 + \cdots + a_{1n}x_n, \ldots, a_{m1}x_1 + \cdots + a_{mn}x_n).$$

Exercise 7.1.15 Let $f_0 : V \to W$ be the *null (zero) map*, that is $f_0(v) = 0_W$ for any $v \in V$. With $\mathcal{B}$ and $\mathcal{C}$ arbitrary bases for V and W respectively, it is clearly

$$M_{f_0}^{\mathcal{C},\mathcal{B}} = 0_{\mathbb{R}^{m,n}},$$

that is the null matrix.

Exercise 7.1.16 If $\mathrm{id}_V(v) = v$ is the identity map on V then, using any basis $\mathcal{B} = (v_1, \ldots, v_n)$ for V, one has the following expression

$$\mathrm{id}_V(v_j) = v_j = (0, \ldots, 0, \underbrace{1}_{j}, 0, \ldots, 0)_{\mathcal{B}}$$

for any $j = 1, \ldots, n$. That is $M_{\mathrm{id}_V}^{\mathcal{B},\mathcal{B}}$ is the identity matrix I_n. Notice that $M_{\mathrm{id}_V}^{\mathcal{C},\mathcal{B}} \neq I_n$ if $\mathcal{B} \neq \mathcal{C}$.

Exercise 7.1.17 Let us consider for $\mathbb{R}^3$ both the canonical basis $\mathcal{E}_3 = (e_1, e_2, e_3)$ and the basis $\mathcal{B} = (v_1, v_2, v_3)$ with

$$v_1 = (0, 1, 1), \ v_2 = (1, 0, 1), \ v_3 = (1, 1, 0).$$

A direct computation gives

$$M_{\mathrm{id}}^{\mathcal{E}_3,\mathcal{B}} = \begin{pmatrix} 0 & 1 & 1 \\ 1 & 0 & 1 \\ 1 & 1 & 0 \end{pmatrix}, \qquad M_{\mathrm{id}}^{\mathcal{E}_3,\mathcal{B}} = \frac{1}{2}\begin{pmatrix} -1 & 1 & 1 \\ 1 & -1 & 1 \\ 1 & 1 & -1 \end{pmatrix}$$

and each of these matrices turns out to be the inverse of the other, that is $M_{\mathrm{id}}^{\mathcal{E}_3,\mathcal{B}} M_{\mathrm{id}}^{\mathcal{B},\mathcal{E}_3} = I_n$.

7.2 Basic Notions on Maps

Before we proceed we recall in a compact and direct way some of the basic notions concerning injectivity, surjectivity and bijectivity of mappings between sets.

Definition 7.2.1 Let X and Y be two non empty sets and $f : X \to Y$ a map between them. The element $f(x)$ in Y is called the *image* under f of the element $x \in X$. The set

$$\mathrm{Im}(f) = \{y \in Y \mid \exists\, x \in X \; : \; y = f(x)\}$$

is called the *image* (or *range*) of f in Y. The set (that might be empty)

$$f^{-1}(y) = \{x \in X \; : \; f(x) = y\}.$$

defines the *pre-image* of the element $y \in Y$.

Definition 7.2.2 Let X and Y be two non empty sets, with a map $f : X \to Y$. One says that:

(i) f is *injective* if, for any pair $x_1, x_2 \in X$ with $x_1 \neq x_2$, it is $f(x_1) \neq f(x_2)$,
(ii) f is *surjective* if $\mathrm{Im}(f) = Y$,
(iii) f is *bijective* if f is both injective and surjective.

Definition 7.2.3 Let $f : X \to Y$ and $g : Y \to Z$ be two maps. The composition of g with f is the map

$$g \circ f : X \to Z$$

defined as $(g \circ f)(x) = g(f(x))$ for any $x \in X$.

Definition 7.2.4 A map $f : X \to Y$ is *invertible* if there exists a map $g : Y \to X$ such that $g \circ f = \mathrm{id}_X$ and $f \circ g = \mathrm{id}_Y$. In such a case the map g is called the *inverse* of f and denoted by f^{-1}. It is possible to prove that, if f is invertible, then f^{-1} is unique.

Proposition 7.2.5 *A map $f : X \to Y$ is invertible if and only if it is bijective. In such a case the map f^{-1} is invertible as well, with $(f^{-1})^{-1} = f$.*

7.3 Kernel and Image of a Linear Map

Injectivity and surjectivity of a linear map are measured by two vector subspaces that we now introduce and study.

Definition 7.3.1 Consider a linear map $f : V \to W$. The set

$$V \supseteq \ker(f) = \{v \in V \; : \; f(v) = 0_W\}$$

is called the *kernel* of f, while the set

$$W \supseteq \mathrm{Im}(f) = \{w \in W : \exists\, v \in V : w = f(v)\}$$

is called the *image* of f.

Theorem 7.3.2 *Given a linear map $f : V \to W$, the set $\ker(f)$ is a vector subspace in V and $\mathrm{Im}(f)$ is a vector subspace in W.*

Proof We recall the Proposition 2.2.2. Given $v, v' \in \ker(f)$ and $\lambda, \lambda' \in \mathbb{R}$ we need to compute $f(\lambda v + \lambda' v')$. Since $f(v) = 0_W = f(v')$ by hypothesis, from the Proposition 7.1.5 we have $f(\lambda v + \lambda' v') = \lambda f(v) + \lambda' f(v') = 0_W$. This shows that $\ker(f)$ is a vector subspace in V.

Analogously, let $w, w' \in \mathrm{Im}(f)$ and $\lambda, \lambda' \in \mathbb{R}$. From the hypothesis there exist $v, v' \in V$ such that $w = f(v)$ and $w' = f(v')$; thus we can write $\lambda w + \lambda' w' = \lambda f(v) + \lambda' f(v') = f(\lambda v + \lambda' v') \in \mathrm{Im}(f)$ again from he Proposition 7.1.5. This shows that $\mathrm{Im}(f)$ is a vector subspace in W. □

Having proved that $\mathrm{Im}(f)$ and $\ker(f)$ are vector subspaces we look for a system of generators for them. Such a task is easier for the image of f as the following lemma shows.

Lemma 7.3.3 *With $f : V \to W$ a linear map, one has that $\mathrm{Im}(f) = \mathcal{L}(f(v_1), \dots, f(v_n))$, where $\mathcal{B} = (v_1, \dots, v_n)$ is an arbitrary basis for V. The map f is indeed surjective if and only if $f(v_1), \dots, f(v_n)$ generate W.*

Proof Let $w \in \mathrm{Im}(f)$, that is $w = f(v)$ for some $v \in V$. Being $\mathcal{B}$ a basis for V, one has $v = a_1 v_1 + \cdots + a_n v_n$ and since f is linear, one has $w = a_1 f(v_1) + \cdots + a_n f(v_n)$, thus giving $w \in \mathcal{L}(f(v_1), \dots, f(v_n))$. We have then $\mathrm{Im}(f) \subseteq \mathcal{L}(f(v_1), \dots, f(v_n))$. The opposite inclusion is obvious since $\mathrm{Im}(f)$ is a vector subspace in W and contains the vectors $f(v_1), \dots, f(v_n)$.

The last statement is the fact that f is surjective (Definition 7.2.2) if and only if $\mathrm{Im}(f) = W$. □

Exercise 7.3.4 Let us consider the linear map $f : \mathbb{R}^3 \to \mathbb{R}^2$ given by

$$f((x, y, z)) = (x + y - z, x - y + z).$$

From the lemma above, the vector subspace $\mathrm{Im}(f)$ is generated by the images under f of an arbitrary basis in $\mathbb{R}^3$. With the canonical basis $\mathcal{E} = (e_1, e_2, e_3)$ we have $\mathrm{Im}(f) = \mathcal{L}(f(e_1), f(e_2), f(e_3))$, with

$$f(e_1) = (1, 1), \qquad f(e_2) = (1, -1), \qquad f(e_3) = (-1, 1).$$

It is immediate to see that $\mathrm{Im}(f) = \mathbb{R}^2$, that is f is surjective.

Lemma 7.3.5 *Let $f : V \to W$ be a linear map between two real vector spaces. Then,*

(i) f is injective if and only if $\ker(f) = \{0_V\}$,
(ii) if f is injective and $(v_1, \ldots, v_n)$ is a basis for V, the vectors $f(v_1), \ldots, f(v_n)$ are linearly independent.

Proof (i) Let us assume that f is injective and $v \in \ker(f)$, that is $f(v) = 0_W$. From the Proposition 7.1.5 we know that $f(0_V) = 0_W$. Since f is injective it must be $v = 0_V$, that is $\ker(f) = \{0_V\}$.
Viceversa, let us assume that $\ker(f) = 0_V$ and let us consider two vectors v_1, v_2 such that $f(v_1) = f(v_2)$. Since f is linear this reads $0_W = f(v_1) - f(v_2) = f(v_1 - v_2)$, that is $v_1 - v_2 \in \ker(f)$ which, being the latter the null vector subspace, thus gives $v_1 = v_2$.

(ii) In order to study the linear independence of the system of vectors $\{f(v_1), \ldots, f(v_n)\}$ let us take scalars $\lambda_1, \ldots, \lambda_n \in \mathbb{R}$ such that $\lambda_1 f(v_1) + \cdots + \lambda_n f(v_n) = 0_W$. Being f linear, this gives $f(\lambda_1 v_1 + \cdots + \lambda_n v_n) = 0_W$ and then $\lambda_1 v_1 + \cdots + \lambda_n v_n \in \ker(f)$. Since f is injective, from (i) we have $\ker(f) = \{0_V\}$ so it is $\lambda_1 v_1 + \cdots + \lambda_n v_n = 0_V$. Being $(v_1, \ldots, v_n)$ a basis for V, we have that $\lambda_1 = \cdots = \lambda_n = 0_{\mathbb{R}}$ thus proving that also $f(v_1), \ldots, f(v_n)$ are linearly independent.

□

Exercise 7.3.6 Let us consider the linear map $f : \mathbb{R}^2 \to \mathbb{R}^3$ given by

$$f((x, y)) = (x + y, x - y, 2x + 3y).$$

The kernel of f is given by

$$\ker(f) = \{(x, y) \in \mathbb{R}^2 \mid f((x, y)) = (x + y, x - y, 2x + 3y) = (0, 0, 0)\}$$

so we have to solve the linear system

$$\begin{cases} x + y = 0 \\ x - y = 0 \\ 2x + 3y = 0 \end{cases}.$$

Its unique solution is $(0, 0)$ so $\ker(f) = \{0_{\mathbb{R}^2}\}$ and we can conclude, from the lemma above, that f is injective. From the same lemma we also know that the images under f of a basis for $\mathbb{R}^2$ make a linearly independent set of vectors. If we take the canonical basis for $\mathbb{R}^2$ with $e_1 = (1, 0)$ and $e_2 = (0, 1)$, we have

$$f(e_1) = (1, 1, 2), \quad f(e_2) = (1, -1, 3).$$

7.4 Isomorphisms

Definition 7.4.1 Let V and W be two real vector spaces. A bijective linear map $f : V \to W$ is called an *isomorphism*. Two vector spaces are said to be *isomorphic* if there exists an isomorphism between them. If $f : V \to W$ is an isomorphism we write $V \cong W$.

Proposition 7.4.2 *If the map $f : V \to W$ is an isomorphism, such is its inverse $f^{-1} : W \to V$.*

Proof From the Proposition 7.2.5 we have that f is invertible, with an invertible inverse map f^{-1}. We need to prove that f^{-1} is linear. Let us consider two arbitrary vectors $w_1, w_2 \in W$ with $v_1 = f^{-1}(w_1)$ and $v_2 = f^{-1}(w_2)$ in V; this is equivalent to $w_1 = f(v_1)$ and $w_2 = f(v_2)$. Let us consider also $\lambda_1, \lambda_2 \in \mathbb{R}$. Since f is linear we can write

$$\lambda_1 w_1 + \lambda_2 w_2 = f(\lambda_1 v_1 + \lambda_2 v_2).$$

For the action of f^{-1} is then

$$f^{-1}(\lambda_1 w_1 + \lambda_2 w_2) = \lambda_1 v_1 + \lambda_2 v_2 = \lambda_1 f^{-1}(w_1) + \lambda_2 f^{-1}(w_2),$$

which amounts to say that f^{-1} is a linear map. □

In order to characterise isomorphisms we first prove a preliminary result.

Lemma 7.4.3 *Let $f : V \to W$ be a linear map with $(v_1, \dots, v_n)$ a basis for V. The map f is an isomorphism if and only if $(f(v_1), \dots, f(v_n))$ is a basis for W.*

Proof If f is an isomorphism, it is both injective and surjective. From the Lemma 7.3.3 the system $f(v_1), \dots, f(v_n)$ generates W, while from the Lemma 7.3.5 such a system is linearly independent. This means that $(f(v_1), \dots, f(v_n))$ is a basis for W.

Let us now assume that the vectors $(f(v_1), \dots, f(v_n))$ are a basis for W. From the Proposition 7.1.14 there exists a linear map $g : W \to V$ such that $g(f(v_j)) = v_j$ for any $j = 1, \dots, n$. This means that the linear maps $g \circ f$ and id_V coincide on the basis $(v_1, \dots, v_n)$ in V and then (again from Proposition 7.1.14) they coincide, that is $g \circ f = \mathrm{id}_V$. Along the same lines it is easy to show that $f \circ g = \mathrm{id}_W$, so we have $g = f^{-1}$; the map f is then invertible so it is an isomorphism. □

Theorem 7.4.4 *Let V and W be two real vector spaces. They are isomorphic if and only if* $\dim(V) = \dim(W)$.

Proof Let us assume V and W to be isomorphic, that is there exists an isomorphism $f : V \to W$. From the previous lemma, if $(v_1, \dots, v_n)$ is a basis for V, then $(f(v_1), \dots, f(v_n))$ is a basis for W and this gives $\dim(V) = n = \dim(W)$.

Let us now assume $n = \dim(V) = \dim(W)$ and try to define an isomorphism $f : V \to W$. By fixing a basis $\mathcal{B} = (v_1, \dots, v_n)$ for V and a basis $\mathcal{C} = (w_1, \dots, w_n)$

for W, we define the linear map $f(v_j) = w_j$ for any j. Such a linear map exists and it is unique from the Proposition 7.1.14. From the lemma above, f is an isomorphism since it maps the basis $\mathcal{B}$ to the basis $\mathcal{C}$ for W. □

Corollary 7.4.5 *If V is a real vector space with* $\dim(V) = n$*, then* $V \cong \mathbb{R}^n$*. Any choice of a basis $\mathcal{B}$ for V induces the natural isomorphism*

$$\alpha : V \xrightarrow{\cong} \mathbb{R}^n \quad \textit{given by} \quad (x_1, \dots, x_n)_{\mathcal{B}} \mapsto (x_1, \dots, x_n).$$

Proof The first claim follows directly from the Theorem 7.4.4 above. Once the basis $\mathcal{B} = (v_1, \dots, v_n)$ is chosen the map α is defined as the linear map such that $\alpha(v_j) = e_j$ for any $j = 1, \dots, n$. From the Lemma 7.4.3 such a map α is an isomorphism. It is indeed immediate to check that the action of α on any vector in V is given by $\alpha : (x_1, \dots, x_n)_{\mathcal{B}} \mapsto (x_1, \dots, x_n)$. □

Exercise 7.4.6 Let $V = \mathbb{R}[X]_2$ be the space of the polynomials whose degree is not higher than 2. As we know, V has dimension 3 and a basis for it is given by $\mathcal{B} = (1, X, X^2)$. The isomorphism $\alpha : \mathbb{R}[X]_2 \xrightarrow{\cong} \mathbb{R}^3$ corresponding to such a basis reads

$$a + bX + cX^2 \quad \mapsto \quad (a, b, c).$$

It is simple to check whether a given system of polynomials is a basis for $\mathbb{R}[X]_2$. As an example we consider

$$p_1(X) = 3X - X^2, \quad p_2(X) = 1 + X, \quad p_3(X) = 2 + 3X^2.$$

By setting $v_1 = \alpha(p_1) = (0, 3, -1)$, $v_2 = \alpha(p_2) = (1, 1, 0)$ and $v_3 = \alpha(p_3) = (2, 0, 3)$, it is clear that the rank of the matrix whose columns are the vectors v_1, v_2, v_3 is 3, thus proving that (v_1, v_2, v_3) is a basis for $\mathbb{R}^3$. Since α is an isomorphism, the inverse $\alpha^{-1} : \mathbb{R}^3 \to \mathbb{R}[X]_2$ is an isomorphism as well: the vectors $\alpha^{-1}(v_1)$, $\alpha^{-1}(v_2)$, $\alpha^{-1}(v_2)$ provide a basis for $\mathbb{R}[X]_2$ and coincide with the given polynomials $p_1(X)$, $p_2(X)$, $p_3(X)$.

Theorem 7.4.4 shows that a linear isomorphism exists only if its domain has the same dimension of its image. A condition that characterises isomorphism can then be introduced only for vector spaces with the same dimensions. This is done in the following sections.

7.5 Computing the Kernel of a Linear Map

We have seen that isomorphisms can be defined only between spaces with the same dimension. Being not an isomorphism indeed means for a linear map to fail to be injective or surjective. In this section and the following one we characterise injectivity and surjectivity of a linear map via the study of its kernel and its image. In particular, we shall describe procedures to exhibit bases for such spaces.

Proposition 7.5.1 *Let $f : V \to W$ be a linear map between real vector spaces, and $\dim(V) = n$. Fix a basis $\mathcal{B}$ for V and a basis $\mathcal{C}$ for W, with associated matrix $A = M_f^{\mathcal{C},\mathcal{B}}$. By denoting $\Sigma : AX = 0$ the linear system associated to A, the following hold:*

(i) $S_\Sigma \cong \ker(f)$ *via the isomorphism* $(x_1, \dots, x_n) \mapsto (x_1, \dots, x_n)_\mathcal{B}$,
(ii) $\dim(\ker(f)) = n - \mathrm{rk}(A)$,
(iii) *if* $(v_1, \dots, v_p)$ *is a basis for* S_Σ, *the vectors* $\big((v_1)_\mathcal{B}, \dots, (v_p)_\mathcal{B}\big)$ *are a basis for* $\ker(f)$.

Proof (i) With the given hypothesis, from the definition of the kernel of a linear map we can write

$$\begin{aligned}\ker(f) &= \{v \in V \ : \ f(v) = 0_W\} \\ &= \left\{ v = (x_1, \dots, x_n)_\mathcal{B} \in V \ : \ \left(A \begin{pmatrix} x_1 \\ \vdots \\ x_n \end{pmatrix}\right)_\mathcal{C} = \begin{pmatrix} 0 \\ \vdots \\ 0 \end{pmatrix}_\mathcal{C} \right\} \\ &= \{(x_1, \dots, x_n)_\mathcal{B} \in V \ : \ (x_1, \dots, x_n) \in S_\Sigma\}\end{aligned}$$

with S_Σ denoting the space of solutions for Σ. As in Corollary 7.4.5 we can then write down the isomorphism $S_\Sigma \to \ker(f)$ given by

$$(x_1, \dots, x_n) \quad \mapsto \quad (x_1, \dots, x_n)_\mathcal{B}.$$

(ii) From the isomorphism of the previous point we then have

$$\dim(\ker(f)) = \dim(S_\Sigma) = n - \mathrm{rk}(A)$$

where the last equality follows from the Theorem 6.4.3.

(iii) From the Lemma 7.4.3 we know that, under the isomorphism $S_\Sigma \to \ker(f)$, a basis for S_Σ is mapped into a basis for $\ker(f)$.

□

Exercise 7.5.2 Consider the linear map $f : \mathbb{R}^3 \to \mathbb{R}^3$ defined by

$$f((x, y, z)_\mathcal{B}) = (x + y - z, x - y + z, 2x)_\mathcal{E}$$

where $\mathcal{B} = \big((1, 1, 0), (0, 1, 1), (1, 0, 1)\big)$ and $\mathcal{E}$ is the canonical basis for $\mathbb{R}^3$. We determine $\ker(f)$ and compute a basis for it with respect to both $\mathcal{B}$ and $\mathcal{E}$. Start by considering the matrix associated to the linear map f with the given basis,

$$A = M_f^{\mathcal{E},\mathcal{B}} = \begin{pmatrix} 1 & 1 & -1 \\ 1 & -1 & 1 \\ 2 & 0 & 0 \end{pmatrix}.$$

To solve the linear system $\Sigma : AX = 0$ we reduce the matrix A by rows:

$$A \mapsto \begin{pmatrix} 1 & 1 & -1 \\ 2 & 0 & 0 \\ 2 & 0 & 0 \end{pmatrix} \mapsto \begin{pmatrix} 1 & 1 & -1 \\ 2 & 0 & 0 \\ 0 & 0 & 0 \end{pmatrix}$$

and the space S_Σ of the solutions of Σ is then given by

$$S_\Sigma = \{(0, a, a) : a \in \mathbb{R}\} = \mathcal{L}((0, 1, 1)).$$

This reads

$$\ker(f) = \{(0, a, a)_\mathcal{B} : a \in \mathbb{R}\},$$

with a basis given by the vector $(0, 1, 1)_\mathcal{B}$. With the explicit expression of the elements of $\mathcal{B}$,

$$(0, 1, 1)_\mathcal{B} = (0, 1, 1) + (1, 0, 1) = (1, 1, 2).$$

This shows that the basis vector for $\ker(f)$ given by $(0, 1, 1)$ on the basis $\mathcal{B}$ is the same as the basis vector $(1, 1, 2)$ with respect to the canonical basis $\mathcal{E}$ for $\mathbb{R}^3$.

Exercise 7.5.3 With canonical bases $\mathcal{E}$, consider the linear map $f : \mathbb{R}^3 \to \mathbb{R}^3$ given by

$$f((x, y, z)) = (x + y - z, x - y + z, 2x).$$

To determine the space $\ker(f)$, we observe that the matrix associated to f is the same matrix of the previous exercise, so the linear system $\Sigma : AX = 0$ has solutions $S_\Sigma = \mathcal{L}((0, 1, 1)) = \ker(f)$, since $\mathcal{E}$ is the canonical basis.

Since the kernel of a linear map is the preimage of the null vector in the image space, we can generalise the above procedure to compute the preimage of any element $w \in W$. We denote it as $f^{-1}(w) = \{v \in V : f(v) = w\}$, with $\ker(f) = f^{-1}(0_W)$. Notice that we denote the preimage of a set under f by writing f^{-1} also when f is not invertible.

Proposition 7.5.4 *Consider a real vector space V with basis $\mathcal{B}$ and a real vector space W with basis $\mathcal{C}$. Let $f : V \to W$ be a linear map with $A = M_f^{\mathcal{C},\mathcal{B}}$ its corresponding matrix. Given any $w = (y_1, \ldots, y_m)_\mathcal{C} \in W$, it is*

$$f^{-1}(w) = \{(x_1, \ldots, x_n)_\mathcal{B} \in V : A\,{}^t(x_1, \ldots, x_n) = {}^t(y_1, \ldots, y_m)\}.$$

Proof It is indeed true that, with $v = (x_1, \ldots, x_n)_\mathcal{B}$, one has

$$f(v) = f((x_1, \ldots, x_n)_{\mathcal{B}}) = (A\,{}^t(x_1, \ldots, x_n))_{\mathcal{C}}.$$

The equality $f(v) = w$ is the equality of components, given by $A\,{}^t(x_1, \ldots, x_n) = {}^t(y_1, \ldots, y_m)$, on the basis $\mathcal{C}$. □

Remark 7.5.5 This fact can be expressed via linear systems. Given $w \in W$, its preimage $f^{-1}(w)$ is made of vectors in V whose components with respect to $\mathcal{B}$ solve the linear system $AX = B$, where B is the column of the components of w with respect to $\mathcal{C}$.

Exercise 7.5.6 Consider the linear map $f : \mathbb{R}^3 \to \mathbb{R}^3$ given in the Exercise 7.5.2. We compute $f^{-1}(w)$ for $w = (1, 1, 1)$. We have then to solve the system $\Sigma : AX = B$, with $B = {}^t(1, 1, 1)$. We reduce the matrix (A, B) as follows

$$(A, B) = \begin{pmatrix} 1 & 1 & -1 & 1 \\ 1 & -1 & 1 & 1 \\ 2 & 0 & 0 & 1 \end{pmatrix} \mapsto \begin{pmatrix} 1 & 1 & -1 & 1 \\ 2 & 0 & 0 & 2 \\ 2 & 0 & 0 & 1 \end{pmatrix} \mapsto \begin{pmatrix} 1 & 1 & -1 & 1 \\ 1 & 0 & 0 & 1 \\ 0 & 0 & 0 & 1 \end{pmatrix}.$$

This shows that the system Σ has no solution, that is $w \notin \mathrm{Im}(f)$.

Next, let us compute $f^{-1}(u)$ for $u = (2, 0, 2)$, so we have the linear system $\Sigma : AX = B$ with $B = {}^t(2, 0, 2)$. Reducing by row, we have

$$(A, B) = \begin{pmatrix} 1 & 1 & -1 & 2 \\ 1 & -1 & 1 & 0 \\ 2 & 0 & 0 & 2 \end{pmatrix} \mapsto \begin{pmatrix} 1 & 1 & -1 & 2 \\ 2 & 0 & 0 & 2 \\ 2 & 0 & 0 & 2 \end{pmatrix} \mapsto \begin{pmatrix} 1 & 1 & -1 & 2 \\ 2 & 0 & 0 & 2 \\ 0 & 0 & 0 & 0 \end{pmatrix}.$$

The system Σ is then equivalent to

$$\begin{cases} x = 1 \\ y = z + 1 \end{cases}$$

whose space of solutions is $S_\Sigma = \{(1, a + 1, a) \; : \; a \in \mathbb{R}\}$. We can then write

$$f^{-1}(2, 0, 2) = \{(1, a + 1, a)_{\mathcal{B}} \; : \; a \in \mathbb{R}\} = \{(a + 1, a + 2, 2a + 1) \; : \; a \in \mathbb{R}\}.$$

7.6 Computing the Image of a Linear Map

We next turn to the study of the image of a linear map.

Proposition 7.6.1 *Let $f : V \to W$ be a linear map between real vector spaces, with $\dim(V) = n$ and $\dim(W) = m$. Fix a basis $\mathcal{B}$ for V and a basis $\mathcal{C}$ for W, with associated matrix $A = M_f^{\mathcal{C},\mathcal{B}}$ and with $C(A)$ its space of columns. The following results hold:*

(i) $\mathrm{Im}(f) \cong C(A)$ via the isomorphism $(y_1, \ldots, y_m)_{\mathcal{C}} \mapsto (y_1, \ldots, y_m)$,

(ii) $\dim(\mathrm{Im}(f)) = \mathrm{rk}(A)$,

(iii) *if* $(w_1, \dots, w_r)$ *is a basis for* $C(A)$*, then* $((w_1)_{\mathcal{C}}, \dots, (w_r)_{\mathcal{C}})$ *is a basis for* $\mathrm{Im}(f)$.

Proof (i) With the given hypothesis, from the definition of the image of a linear map we can write

$$\begin{aligned}
\mathrm{Im}(f) &= \{w \in W \,:\, \exists v \in V \,:\, w = f(v)\} \\
&= \left\{ w = (y_1, \dots, y_m)_{\mathcal{C}} \in W \,:\, \exists (x_1, \dots, x_n)_{\mathcal{B}} \in V \,:\, \begin{pmatrix} y_1 \\ \vdots \\ y_m \end{pmatrix}_{\mathcal{C}} = \left(A \begin{pmatrix} x_1 \\ \vdots \\ x_n \end{pmatrix} \right)_{\mathcal{C}} \right\} \\
&= \left\{ (y_1, \dots, y_m)_{\mathcal{C}} \in W \,:\, \exists (x_1, \dots, x_n) \in \mathbb{R}^n \,:\, \begin{pmatrix} y_1 \\ \vdots \\ y_m \end{pmatrix} = A \begin{pmatrix} x_1 \\ \vdots \\ x_n \end{pmatrix} \right\}.
\end{aligned}$$

Representing the matrix A by its columns, that is $A = (C_1 \ \cdots \ C_n)$, we have

$$A \begin{pmatrix} x_1 \\ \vdots \\ x_n \end{pmatrix} = x_1 C_1 + \cdots + x_n C_n.$$

We can therefore write

$$\begin{aligned}
\mathrm{Im}(f) &= \left\{ (y_1, \dots, y_m)_{\mathcal{C}} \in W \,:\, \exists (x_1, \dots, x_n) \in \mathbb{R}^n \,:\, \begin{pmatrix} y_1 \\ \vdots \\ y_m \end{pmatrix} = x_1 C_1 + \cdots x_n C_n \right\} \\
&= \left\{ (y_1, \dots, y_m)_{\mathcal{C}} \in W \,:\, \begin{pmatrix} y_1 \\ \vdots \\ y_m \end{pmatrix} \in C(A) \right\}.
\end{aligned}$$

We have then the isomorphism $C(A) \to \mathrm{Im}(f)$ defined by

$$(y_1, \dots, y_m) \quad \mapsto \quad (y_1, \dots, y_m)_{\mathcal{C}}$$

(compare this with the one in the Corollary 7.4.5).

(ii) Being $\mathrm{Im}(f) \cong C(A)$, it is $\dim(\mathrm{Im}(f)) = \dim(C(A)) = \mathrm{rk}(A)$.

(iii) The claim follows from (i) and the Lemma 7.4.3.

□

Remark 7.6.2 To determine a basis for $C(A)$ as in (iii) above, one can proceed as follows.

(a) If the rank of A is known, one has to select n linearly independent columns: they will give a basis for $C(A)$.

(b) If the rank of A is not known, by denoting A' the matrix obtained from A by reduction by columns, a basis for $C(A)$ is given by the r non zero columns of A'.

Exercise 7.6.3 Let $f : \mathbb{R}^3 \to \mathbb{R}^3$ be the linear map with associated matrix

$$A = M_f^{\mathcal{C},\mathcal{E}} = \begin{pmatrix} 1 & -1 & 2 \\ 0 & 1 & -3 \\ 2 & -1 & 1 \end{pmatrix}$$

for the canonical basis $\mathcal{E}$ and basis $\mathcal{C} = (w_1, w_2, w_3)$, with $w_1 = (1, 1, 0)$, $w_2 = (0, 1, 1)$, $w_3 = (1, 0, 1)$. We reduce A by columns

$$A \xrightarrow[C_3 \mapsto C_3 - 2C_1]{C_2 \mapsto C_2 + C_1} \begin{pmatrix} 1 & 0 & 0 \\ 0 & 1 & -3 \\ 2 & 1 & -3 \end{pmatrix} \xrightarrow{C_3 \mapsto C_3 + 3C_1} \begin{pmatrix} 1 & 0 & 0 \\ 0 & 1 & 0 \\ 2 & 1 & 0 \end{pmatrix} = A'.$$

Being A' reduced by columns, its non zero columns yield a basis for the space $C(A)$. Thus, $C(A) = C(A') = \mathcal{L}((1, 0, 2), (0, 1, 1))$. From the Proposition 7.6.1 a basis for $\mathrm{Im}(f)$ is given by the pair (u_1, u_2),

$$u_1 = (1, 0, 2)_{\mathcal{C}} = w_1 + 2w_3 = (3, 1, 2)$$
$$u_2 = (0, 1, 1)_{\mathcal{C}} = w_2 + w_3 = (1, 1, 2).$$

Clearly, $\dim(\mathrm{Im}(f)) = 2 = \mathrm{rk}(A)$.

From the previous results we have the following theorem.

Theorem 7.6.4 *Let $f : V \to W$ be a linear map. It holds that*

$$\dim(\ker(f)) + \dim(\mathrm{Im}(f)) = \dim(V).$$

Proof Let A be any matrix associated to f (that is irrespective of the bases chosen in V and W). From the Proposition 7.5.1 one has $\dim(\ker(f)) = \dim(V) - \mathrm{rk}(A)$, while from the Proposition 7.6.1 one has $\dim(\mathrm{Im}(f)) = \mathrm{rk}(A)$. The claim follows. □

From this theorem, the next corollary follows easily.

Corollary 7.6.5 *Let $f : V \to W$ be a linear map, with* $\dim(V) = \dim(W)$. *The following statements are equivalent.*

(i) f is injective,
(ii) f is surjective,
(iii) f is an isomorphism.

Proof Clearly it is sufficient to prove the equivalence (i) $\Leftrightarrow$ (ii). From the Lemma 7.3.5 we know that f is injective if and only if $\dim(\ker(f)) = 0$. We also known that f is surjective if and only if $\dim(\mathrm{Im}(f)) = \dim(W)$. Since $\dim(V) = \dim(W)$ by hypothesis, the statement thus follows from the Theorem 7.6.4. □

7.7 Injectivity and Surjectivity Criteria

In this section we study conditions for injectivity and surjectivity of a linear map through properties of its associated matrix.

Proposition 7.7.1 (Injectivity criterion) *Let $f : V \to W$ be a linear map. Then f is injective if and only if* $\mathrm{rk}(A) = \dim(V)$ *for any matrix A associated to f (that is, irrespective of the bases with respect to which the matrix A is given).*

Proof From (i) in the Lemma 7.3.5 we know that f is injective if and only if $\ker(f) = \{0_V\}$, which means $\dim(\ker(f)) = 0$. From the Proposition 7.5.1 we have that $\dim(\ker(f)) = \dim(V) - \mathrm{rk}(A)$ for any matrix A associated to f. We then have that f is injective if and only if $\dim(V) - \mathrm{rk}(A) = 0$. □

Exercise 7.7.2 Let $f : \mathbb{R}[X]_2 \to \mathbb{R}^{2,2}$ be the linear map associated to the matrix

$$A = \begin{pmatrix} 2 & 1 & 0 \\ -1 & 0 & 1 \\ 2 & 1 & 1 \\ 1 & 0 & 0 \end{pmatrix}$$

with respect to two given basis. Since$A$ is already reduced by column, $\mathrm{rk}(A) = 3$, the number of its non zero columns. Being $\dim(\mathbb{R}[X]_2) = 3$ we have, from the Proposition 7.7.1, that f is injective.

Proposition 7.7.3 (Surjectivity criterion) *Let $f : V \to W$ be a linear map. The map f is surjective if and only if* $\mathrm{rk}(A) = \dim(W)$ *for any matrix associated to f (again irrespective of the bases with respect to which the matrix A is given).*

Proof This follows directly from the Proposition 7.6.1. □

Exercise 7.7.4 Let $f : \mathbb{R}^3 \to \mathbb{R}^2$ be the linear map given by

$$f(x, y, z) = (x + y - z, 2x - y + 2z).$$

With $\mathcal{E}$ the canonical basis in $\mathbb{R}^3$ and $\mathcal{C}$ the canonical basis in $\mathbb{R}^2$, we have

$$A = M_f^{\mathcal{C},\mathcal{E}} = \begin{pmatrix} 1 & 1 & -1 \\ 2 & -1 & 2 \end{pmatrix} :$$

by reducing by rows,

$$A \mapsto \begin{pmatrix} 1 & 1 & -1 \\ 3 & 0 & 1 \end{pmatrix} = A'.$$

We know that $\mathrm{rk}(A) = \mathrm{rk}(A') = 2$, the number of non zero rows in A'. Being $\dim(\mathbb{R}^2) = 2$, the map f is surjective from the Proposition 7.7.3.

We have seen in the Proposition 7.4.2 that if a linear map f is an isomorphism, then its domain and image have the same dimension. Injectivity and surjectivity of a linear map provide necessary conditions on the relative dimensions of the domain and the image of the map.

Remark 7.7.5 Let $f : V \to W$ be a linear map. One has:

(a) If f is injective, then $\dim(V) \leq \dim(W)$. This claim easily follows from the Lemma 7.3.5, since the images under f of a basis for V gives linearly independent vectors in W.
(b) If f is surjective, then $\dim(V) \geq \dim(W)$. This claim follows from the Lemma 7.3.3, since the images under f of a basis for V generate (that is they linearly span) W.

Remark 7.7.6 Let $f : V \to W$ be a linear map, with A its corresponding matrix with respect to any basis. One has:

(a) With $\dim(V) < \dim(W)$, f is injective if and only if $\mathrm{rk}(A)$ is maximal;
(b) With $\dim(V) > \dim(W)$, f is surjective if and only if $\mathrm{rk}(A)$ is maximal;
(c) With $\dim(V) = \dim(W)$, f is an isomorphism if and only if $\mathrm{rk}(A)$ is maximal.

Exercise 7.7.7 The following linear maps are represented with respect to canonical bases.

(1) Let the map $f : \mathbb{R}^3 \to \mathbb{R}^4$ be defined by

$$(x, y, z) \quad \mapsto \quad (x - y + 2z, y + z, -x + z, 2x + y).$$

To compute the rank of the corresponding matrix A with respect to the canonical basis, as usual we reduce it by rows. We have

$$A = \begin{pmatrix} 1 & -1 & 2 \\ 0 & 1 & 1 \\ -1 & 0 & 1 \\ 2 & 1 & 0 \end{pmatrix} \mapsto \begin{pmatrix} 1 & -1 & 2 \\ 0 & 1 & 1 \\ 0 & 0 & 1 \\ 0 & 0 & 0 \end{pmatrix},$$

and the rank of A is maximal, $\mathrm{rk}(A) = 3$. Since $\dim(V) < \dim(W)$ we have that f is injective.

(2) Let the map $f : \mathbb{R}^4 \to \mathbb{R}^3$ be defined by

$$(x, y, z, t) \quad \mapsto \quad (x - y + 2z + t, y + z + 3t, x - y + 2z + 2t).$$

We proceed as above and compute, via the following reduction,

$$A = \begin{pmatrix} 1 & -1 & 2 & 1 \\ 0 & 1 & 1 & 3 \\ 1 & -1 & 2 & 2 \end{pmatrix} \mapsto \begin{pmatrix} 1 & -1 & 2 & 1 \\ 0 & 1 & 1 & 3 \\ 0 & 0 & 0 & 1 \end{pmatrix},$$

that $\mathrm{rk}(A) = 3$. Since $\mathrm{rk}(A)$ is maximal, with $\dim(V) > \dim(W)$, f turns out to be surjective.

(3) Let $f : \mathbb{R}^3 \to \mathbb{R}^3$ be represented as before by the matrix

$$A = \begin{pmatrix} 1 & 2 & 1 \\ 2 & 1 & 1 \\ 1 & -1 & 2 \end{pmatrix},$$

which, by reduction, becomes

$$A \mapsto \begin{pmatrix} 1 & 2 & 1 \\ 0 & -3 & -1 \\ 0 & 0 & 2 \end{pmatrix},$$

whose rank is clearly maximal. Thus f is an isomorphism since $\dim(V) = \dim(W)$.

7.8 Composition of Linear Maps

We rephrase the general Definition 7.2.3 of composing maps.

Definition 7.8.1 Let $f : V \to W$ and $g : W \to Z$ be two linear maps between real vector spaces. The composition between g and f is the map

$$g \circ f : X \to Z$$

defined as $(g \circ f)(v) = g(f(v))$, for any $v \in X$.

Proposition 7.8.2 *If $f : V \to W$ and $g : W \to Z$ are two linear maps, the composition map $g \circ f : V \to Z$ is linear as well.*

Proof For any $v, v' \in V$ and $\lambda, \lambda' \in \mathbb{R}$, the linearity of both f and g allows one to write:

$$\begin{aligned} (g \circ f)(\lambda v + \lambda' v') &= g(f(\lambda v + \lambda' v')) \\ &= g(\lambda f(v) + \lambda' f(v')) \\ &= \lambda g(f(v)) + \lambda' g(f(v')) \\ &= \lambda (g \circ f)(v) + \lambda' (g \circ f)(v'), \end{aligned}$$

showing the linearity of the composition map. □

The following proposition, whose proof we omit, characterises the matrix corresponding to the composition of two linear maps.

Proposition 7.8.3 *Let V, W, Z be real vector spaces with basis $\mathcal{B}$, $\mathcal{C}$, $\mathcal{D}$ respectively. Given linear maps $f : V \to W$ and $g : W \to Z$, the corresponding matrices with respect to the given bases are related by*

$$M_{g\circ f}^{\mathcal{D},\mathcal{B}} = M_g^{\mathcal{D},\mathcal{C}} \cdot M_f^{\mathcal{C},\mathcal{B}}.$$

The following theorem characterises an isomorphism in terms of its corresponding matrix.

Theorem 7.8.4 *Let $f : V \to W$ be a linear map. The map f is an isomorphism if and only if, for any choice of the bases $\mathcal{B}$ for V and $\mathcal{C}$ for W, the corresponding matrix $M_f^{\mathcal{C},\mathcal{B}}$ with respect to the given bases is invertible, with*

$$M_{f^{-1}}^{\mathcal{B},\mathcal{C}} = \left(M_f^{\mathcal{C},\mathcal{B}}\right)^{-1}.$$

Proof Let us assume that f is an isomorphism: we can then write $\dim(V) = \dim(W)$, so $M_f^{\mathcal{C},\mathcal{B}}$ is a square matrix whose size is $n \times n$ (say). From the Proposition 7.4.2 we know that f^{-1} exists as a linear map whose corresponding matrix, with the given bases, will be $M_{f^{-1}}^{\mathcal{B},\mathcal{C}}$. From the Proposition 7.8.3 we can write

$$M_{f^{-1}}^{\mathcal{B},\mathcal{C}} \cdot M_f^{\mathcal{C},\mathcal{B}} = M_{f^{-1}\circ f}^{\mathcal{B},\mathcal{B}} = M_{\mathrm{id}_V}^{\mathcal{B},\mathcal{B}} = I_n \quad \Rightarrow \quad M_{f^{-1}}^{\mathcal{B},\mathcal{C}} = \left(M_f^{\mathcal{C},\mathcal{B}}\right)^{-1}.$$

We set now $A = M_f^{\mathcal{C},\mathcal{B}}$. By hypothesis A is a square invertible matrix, with inverse A^{-1}, so we can consider the linear map

$$g = f_{A^{-1}}^{\mathcal{B},\mathcal{C}} : W \to V.$$

In order to show that g is the inverse of f, consider the matrix corresponding to $g \circ f$ with respect to the basis $\mathcal{B}$. From the Proposition 7.8.3,

$$M_{g\circ f}^{\mathcal{B},\mathcal{B}} = M_g^{\mathcal{B},\mathcal{C}} \cdot M_f^{\mathcal{C},\mathcal{B}} = A^{-1} \cdot A = I_n.$$

Since linear maps are in bijection with matrices, we have that $g \circ f = \mathrm{id}_V$. Along the same lines we can show that $f \circ g = \mathrm{id}_W$, thus proving $g = f^{-1}$. □

Exercise 7.8.5 Consider the linear map $f : \mathbb{R}^3 \to \mathbb{R}^3$ defined by

$$f((x, y, z)) = (x - y + z, 2y + z, z).$$

With the canonical basis $\mathcal{E}$ for $\mathbb{R}^3$ the corresponding matrix is

$$A = M_f^{\mathcal{E},\mathcal{E}} = \begin{pmatrix} 1 & -1 & 1 \\ 0 & 2 & 1 \\ 0 & 0 & 1 \end{pmatrix}.$$

Since $\mathrm{rk}(A) = 3$, f is an isomorphism, with f^{-1} the linear map corresponding to A^{-1}. From the Proposition 5.3.3, we have

$$A^{-1} = \frac{1}{\det(A)} \left(\alpha_{ji}\right) = \begin{pmatrix} 1 & 1/2 & -3/2 \\ 0 & 1/2 & -1/2 \\ 0 & 0 & 1 \end{pmatrix} = M^{\mathcal{E},\mathcal{E}}_{f^{-1}}.$$

7.9 Change of Basis in a Vector Space

In this section we study how to relate the components of the *same* vector in a vector space with respect to *different* bases. This problem has a natural counterpart in physics, where different bases for the same vector space represent different reference systems. Thus different observers measuring observables of the same physical system in a compatible way.

Example 7.9.1 We start by considering the vector space $\mathbb{R}^2$ with two bases given by

$$\mathcal{E} = \left(e_1 = (1, 0),\ e_2 = (0, 1)\right), \qquad \mathcal{B} = \left(b_1 = (1, 2),\ b_2 = (3, 4)\right).$$

Any vector $v \in \mathbb{R}^2$ will then be written as

$$v = (x_1, x_2)_{\mathcal{B}} = (y_1, y_2)_{\mathcal{E}},$$

or, more explicitly,

$$v = x_1 b_1 + x_2 b_2 = y_1 e_1 + y_2 e_2.$$

By writing the components of the elements in $\mathcal{B}$ in the basis $\mathcal{E}$, that is

$$b_1 = e_1 + 2e_2, \qquad b_2 = 3e_1 + 4e_2,$$

we have

$$\begin{aligned} y_1 e_1 + y_2 e_2 &= x_1(e_1 + 2e_2) + x_2(3e_1 + 4e_2) \\ &= (x_1 + 3x_2)e_1 + (2x_1 + 4x_2)e_2. \end{aligned}$$

We have then obtained

$$y_1 = x_1 + 3x_2, \qquad y_2 = 2x_1 + 4x_2.$$

These expression can be written in matrix form

$$\begin{pmatrix} y_1 \\ y_2 \end{pmatrix} = \begin{pmatrix} x_1 + 3x_2 \\ 2x_1 + 4x_2 \end{pmatrix} \quad \Leftrightarrow \quad \begin{pmatrix} y_1 \\ y_2 \end{pmatrix} = \begin{pmatrix} 1 & 3 \\ 2 & 4 \end{pmatrix} \begin{pmatrix} x_1 \\ x_2 \end{pmatrix}.$$

Such a relation can be written as

$$\begin{pmatrix} y_1 \\ y_2 \end{pmatrix} = A \begin{pmatrix} x_1 \\ x_2 \end{pmatrix},$$

where

$$A = \begin{pmatrix} 1 & 3 \\ 2 & 4 \end{pmatrix}.$$

Notice that the columns of A above are given by the components of the vectors in $\mathcal{B}$ with respect to the basis $\mathcal{E}$. We have the following general result.

Proposition 7.9.2 *Let V be a real vector space with $\dim(V) = n$. Let $\mathcal{B}$ and $\mathcal{C}$ be two bases for V and denote by $(x_1, \ldots, x_n)_\mathcal{B}$ and $(y_1, \ldots, y_n)_\mathcal{C}$ the component of the same vector v with respect to them. It is*

$$\begin{pmatrix} y_1 \\ \vdots \\ y_n \end{pmatrix} = M^{\mathcal{C},\mathcal{B}}_{\mathrm{id}_V} \begin{pmatrix} x_1 \\ \vdots \\ x_n \end{pmatrix}.$$

Such an expression will also be written as

$$^t(y_1, \ldots, y_n) = M^{\mathcal{C},\mathcal{B}}_{\mathrm{id}_V} \cdot {}^t(x_1, \ldots, x_n).$$

Proof This is clear, by recalling the Definition 7.1.12 and the Proposition 7.1.13. □

Definition 7.9.3 The matrix $M^{\mathcal{C},\mathcal{B}} = M^{\mathcal{C},\mathcal{B}}_{\mathrm{id}_V}$ is called the *matrix of the change of basis* from $\mathcal{B}$ to $\mathcal{C}$. The columns of this matrix are given by the components with respect to $\mathcal{C}$ of the vectors in $\mathcal{B}$.

Exercise 7.9.4 Let $\mathcal{B} = (v_1, v_2, v_3)$ and $\mathcal{C} = (w_1, w_2, w_3)$ two different bases for $\mathbb{R}^3$, with

$$\begin{aligned} v_1 &= (0, 1, -1), \quad v_2 = (1, 0, -1), \quad v_3 = (2, -2, 2), \\ w_1 &= (0, 1, 1), \quad w_2 = (1, 0, 1), \quad w_3 = (1, 1, 0). \end{aligned}$$

We consider the vector $v = (1, -1, 1)_\mathcal{B}$ and we wish to determine its components with respect to $\mathcal{C}$. The solution to the linear system

$$\begin{aligned} v_1 &= a_{11}w_1 + a_{21}w_2 + a_{31}w_3 \\ v_2 &= a_{12}w_1 + a_{22}w_2 + a_{32}w_3 \\ v_3 &= a_{13}w_1 + a_{23}w_2 + a_{33}w_3 \end{aligned}$$

give the entries for the matrix of the change of basis, which is found to be

$$M^{\mathcal{C},\mathcal{B}} = \begin{pmatrix} 0 & -1 & -1 \\ -1 & 0 & 3 \\ 1 & 1 & -1 \end{pmatrix}.$$

We can then write

$$v = \begin{pmatrix} y_1 \\ y_2 \\ y_3 \end{pmatrix}_{\mathcal{C}} = \left(M^{\mathcal{C},\mathcal{B}} \begin{pmatrix} x_1 \\ x_2 \\ x_3 \end{pmatrix} \right)_{\mathcal{C}} = \left(\begin{pmatrix} 0 & -1 & -1 \\ -1 & 0 & 3 \\ 1 & 1 & -1 \end{pmatrix} \begin{pmatrix} 1 \\ -1 \\ 1 \end{pmatrix} \right)_{\mathcal{C}} = \begin{pmatrix} 0 \\ 2 \\ -1 \end{pmatrix}_{\mathcal{C}}.$$

Theorem 7.9.5 *Let $\mathcal{B}$ and $\mathcal{C}$ be two bases for the vector space V over $\mathbb{R}$. The matrix $M^{\mathcal{C},\mathcal{B}}$ is invertible, with*

$$\left(M^{\mathcal{C},\mathcal{B}}\right)^{-1} = M^{\mathcal{B},\mathcal{C}}.$$

Proof This easily follows by applying the Theorem 7.8.4 to $M^{\mathcal{C},\mathcal{B}} = M^{\mathcal{C},\mathcal{B}}_{\mathrm{id}_V}$, since $\mathrm{id}_V = \mathrm{id}_V^{-1}$. □

Theorem 7.9.6 *Let $A \in \mathbb{R}^{n,n}$ be an invertible matrix. Denoting by $v_1, \ldots, v_n$ the column vectors in A and setting $\mathcal{B} = (v_1, \ldots, v_n)$, it holds that:*

(i) *$\mathcal{B}$ is a basis for $\mathbb{R}^n$,*
(ii) *$A = M^{\mathcal{B},\mathcal{E}}$ with $\mathcal{E}$ the canonical basis in $\mathbb{R}^n$.*

Proof (i) From the Remark 7.7.6, we know that A has maximal rank, that is $\mathrm{rk}(A) = n$. Being the column vectors in A, the system $v_1, \ldots, v_n$ is then free. A system of n linearly independent vectors in $\mathbb{R}^n$ is indeed a basis for $\mathbb{R}^n$ (see the Corollary 2.5.5 in Chap. 2).
(ii) It directly follows from the Definition 7.9.3.

□

Remark 7.9.7 From the Theorems 7.9.5 and 7.9.6 we have that the group $\mathrm{GL}(n, \mathbb{R})$ of invertible matrices of order n, is the same as (the group of) matrices providing change of basis in $\mathbb{R}^n$.

Exercise 7.9.8 The matrix

$$A = \begin{pmatrix} 1 & 1 & -1 \\ 0 & 1 & 2 \\ 0 & 0 & 1 \end{pmatrix}$$

is invertible since $\mathrm{rk}(A) = 3$ (the matrix A is reduced by rows, so its rank is the number of non zero columns). The column vectors in A, that is

$$v_1 = (1, 0, 0), \quad v_2 = (1, 1, 0), \quad v_3 = (-1, 2, 1),$$

form a basis for $\mathbb{R}^3$. It is also clear that $A = M^{\mathcal{E},\mathcal{B}} = M^{\mathcal{E},\mathcal{B}}_{\mathrm{id}_{\mathbb{R}^3}}$, with $\mathcal{B} = (v_1, v_2, v_3)$.

We next turn to study how $M_f^{\mathcal{C},\mathcal{B}}$, the matrix associated to a linear map $f : V \to W$ with respect to the bases $\mathcal{B}$ for V and $\mathcal{C}$ in W, is transformed under a change of basis in V and W. In the following pages we shall denote by $V_{\mathcal{B}}$ the vector space V referred to its basis $\mathcal{B}$.

Theorem 7.9.9 *Let $\mathcal{B}$ and $\mathcal{B}'$ be two bases for the real vector space V and $\mathcal{C}$ and $\mathcal{C}'$ two bases for the real vector space W. With $f : V \to W$ a linear map, one has that*

$$M_f^{\mathcal{C}',\mathcal{B}'} = M^{\mathcal{C}',\mathcal{C}} \cdot M_f^{\mathcal{C},\mathcal{B}} \cdot M^{\mathcal{B},\mathcal{B}'}.$$

Proof The *commutative diagram*

$$\begin{array}{ccc} V_{\mathcal{B}} & \xrightarrow{\quad f \quad} & W_{\mathcal{C}} \\ \mathrm{id}_V \uparrow & & \downarrow \mathrm{id}_W \\ V'_{\mathcal{B}} & \xrightarrow[\quad f \quad]{} & W'_{\mathcal{C}} \end{array}$$

shows the claim: going from $V'_{\mathcal{B}}$ to $W'_{\mathcal{C}}$ along the bottom line is equivalent to going around the diagram, that is

$$f = \mathrm{id}_W \circ f \circ \mathrm{id}_V.$$

Such a relation can be translated in a matrix form,

$$M_f^{\mathcal{C}',\mathcal{B}'} = M_{\mathrm{id}_W \circ f \circ \mathrm{id}_V}^{\mathcal{C}',\mathcal{B}'}$$

and, by recalling the Proposition 7.8.3, we have

$$M_{\mathrm{id}_W \circ f \circ \mathrm{id}_V}^{\mathcal{C}',\mathcal{B}'} = M_{\mathrm{id}_W}^{\mathcal{C}',\mathcal{C}} \cdot M_f^{\mathcal{C},\mathcal{B}} \cdot M_{\mathrm{id}_V}^{\mathcal{B},\mathcal{B}'},$$

which proves the claim. □

Exercise 7.9.10 Consider the linear map $f : \mathbb{R}^2_{\mathcal{B}} \to \mathbb{R}^3_{\mathcal{C}}$ whose corresponding matrix is

$$A = M_f^{\mathcal{C},\mathcal{B}} = \begin{pmatrix} 1 & 2 \\ -1 & 0 \\ 1 & 1 \end{pmatrix}$$

with respect to $\mathcal{B} = \big((1, 1), (0, 1)\big)$ and $\mathcal{C} = \big((1, 1, 0), (1, 0, 1), (0, 1, 1)\big)$. We determine the matrix $B = M_f^{\mathcal{E}_3,\mathcal{E}_2}$, with $\mathcal{E}_2$ the canonical basis for $\mathbb{R}^2$ and $\mathcal{E}_3$ the canonical basis for $\mathbb{R}^3$. The commutative diagram above turns out to be

$$
\begin{array}{ccc}
\mathbb{R}^2{}_{\mathcal{B}} & \xrightarrow{f_A^{\mathcal{C},\mathcal{B}}} & \mathbb{R}^3{}_{\mathcal{C}} \\
\mathrm{id}_{\mathbb{R}^2} \uparrow & & \downarrow \mathrm{id}_{\mathbb{R}^3} \\
\mathbb{R}^2{}_{\mathcal{E}_2} & \xrightarrow[f_B^{\mathcal{E}_3,\mathcal{E}_2}]{} & \mathbb{R}^3{}_{\mathcal{E}_3}
\end{array},
$$

which reads

$$B = M_f^{\mathcal{E}_3,\mathcal{E}_2} = M^{\mathcal{E}_3,\mathcal{C}} \, A \, M^{\mathcal{B},\mathcal{E}_2}.$$

We have to compute the matrices $M^{\mathcal{E}_3,\mathcal{C}}$ and $M^{\mathcal{B},\mathcal{E}_2}$. Clearly,

$$M^{\mathcal{E}_3,\mathcal{C}} = \begin{pmatrix} 1 & 1 & 0 \\ 1 & 0 & 1 \\ 0 & 1 & 1 \end{pmatrix},$$

and, from the Theorem 7.9.5, it is

$$M^{\mathcal{B},\mathcal{E}_2} = \left(M^{\mathcal{E}_2,\mathcal{B}}\right)^{-1} = \left(\begin{pmatrix} 1 & 0 \\ 1 & 1 \end{pmatrix}\right)^{-1} = \begin{pmatrix} 1 & 0 \\ -1 & 1 \end{pmatrix}$$

(the last equality follows from the Proposition 5.3.3). We have then

$$B = \begin{pmatrix} 1 & 1 & 0 \\ 1 & 0 & 1 \\ 0 & 1 & 1 \end{pmatrix} \begin{pmatrix} 1 & 2 \\ -1 & 0 \\ 1 & 1 \end{pmatrix} \begin{pmatrix} 1 & 0 \\ -1 & 1 \end{pmatrix} = \begin{pmatrix} -2 & 2 \\ -1 & 3 \\ -1 & 1 \end{pmatrix}.$$

We close this section by studying how to construct linear maps with specific properties.

Exercise 7.9.11 We ask whether there is a linear map $f : \mathbb{R}^3 \to \mathbb{R}^3$ which fulfils the conditions $f(1, 0, 2) = 0$ and $\mathrm{Im}(f) = \mathcal{L}((1, 1, 0), (2, -1, 0))$. Also, if such a map exists, is it unique?

We start by setting $v_1 = (1, 0, 2)$, $v_2 = (1, 1, 0)$, $v_3 = (2, -1, 0)$. Since a linear map is characterised by its action on the elements of a basis and v_1 is required to be in the kernel of f, we complete v_1 to a basis for $\mathbb{R}^3$. By using the elements of the canonical basis $\mathcal{E}_3$, we may take the set (v_1, e_1, e_2), which is indeed a basis: the matrix

$$\begin{pmatrix} 1 & 0 & 2 \\ 1 & 0 & 0 \\ 0 & 1 & 0 \end{pmatrix},$$

whose rows are given by (v_1, e_1, e_2) has rank 3 (the matrix is already reduced by rows). So we can take the basis $\mathcal{B} = (v_1, e_1, e_2)$ and define

$$\begin{aligned} f(v_1) &= 0_{\mathbb{R}^3}, \\ f(e_1) &= v_2, \\ f(e_2) &= v_3. \end{aligned}$$

Such a linear map satisfies the required conditions, since $\ker(f) = \{v_1\}$ and

$$\mathrm{Im}(f) = \mathcal{L}(f(v_1), f(e_1), f(e_2)) = \mathcal{L}(v_2, v_3).$$

With respect to the bases $\mathcal{E}$ and $\mathcal{B}$ we have

$$M_f^{\mathcal{E},\mathcal{B}} = \begin{pmatrix} 0 & 1 & 2 \\ 0 & 1 & -1 \\ 0 & 0 & 0 \end{pmatrix}.$$

In order to understand whether the required conditions can be satisfied by a different linear map f, we start by analysing whether the set (v_1, v_2, v_3) itself provides a basis for $\mathbb{R}^3$. As usual, we reduce by rows the matrix associated to the vectors,

$$A = \begin{pmatrix} 1 & 0 & 2 \\ 1 & 1 & 0 \\ 2 & -1 & 0 \end{pmatrix} \mapsto \begin{pmatrix} 1 & 0 & 2 \\ 1 & 1 & 0 \\ 3 & 0 & 0 \end{pmatrix}.$$

Such a reduction gives $\mathrm{rk}(A) = 3$, that is $\mathcal{C} = (v_1, v_2, v_3)$ is a basis for $\mathbb{R}^3$. Then, let $g : \mathbb{R}^3 \to \mathbb{R}^3$ be defined by

$$\begin{aligned} g(v_1) &= 0_{\mathbb{R}^3}, \\ g(v_2) &= v_2, \\ g(v_3) &= v_3. \end{aligned}$$

Also the linear map g satisfies the conditions we set at the beginning and the matrix

$$M_g^{\mathcal{C},\mathcal{C}} = \begin{pmatrix} 0 & 0 & 0 \\ 0 & 1 & 0 \\ 0 & 0 & 1 \end{pmatrix}$$

represents its action by the basis $\mathcal{C}$. It seems clear that f and g are different. In order to prove this claim, we shall see that their corresponding matrices with respect to the same pair of bases differ. We need then to find $M_g^{\mathcal{E},\mathcal{B}}$. We know that $M_g^{\mathcal{E},\mathcal{B}} = M_g^{\mathcal{E},\mathcal{C}} M^{\mathcal{C},\mathcal{B}}$, with

$$M_g^{\mathcal{E},\mathcal{C}} = \begin{pmatrix} 0 & 1 & 2 \\ 0 & 1 & -1 \\ 0 & 0 & 0 \end{pmatrix},$$

since the column vectors in $M_g^{\mathcal{E},\mathcal{C}}$ are given by $g(v_1), g(v_2), g(v_3)$. The columns of the matrix $M_{\mathcal{C},\mathcal{B}}$ are indeed the components with respect to $\mathcal{C}$ of the vectors in $\mathcal{B}$, that is

$$\begin{aligned} v_1 &= v_1, \\ e_1 &= \frac{1}{3} v_2 + \frac{1}{3} v_3, \\ e_2 &= \frac{2}{3} v_2 - \frac{1}{3} v_3. \end{aligned}$$

Thus we have

$$M^{\mathcal{C},\mathcal{B}} = \tfrac{1}{3} \begin{pmatrix} 3 & 0 & 0 \\ 0 & 1 & 2 \\ 0 & 1 & -1 \end{pmatrix},$$

and in turn,

$$M_g^{\mathcal{E},\mathcal{B}} = M_g^{\mathcal{E},\mathcal{C}} \, M^{\mathcal{C},\mathcal{B}} = \begin{pmatrix} 0 & 1 & 2 \\ 0 & 1 & -1 \\ 0 & 0 & 0 \end{pmatrix} \tfrac{1}{3} \begin{pmatrix} 3 & 0 & 0 \\ 0 & 1 & 2 \\ 0 & 1 & -1 \end{pmatrix} = \begin{pmatrix} 0 & 1 & 0 \\ 0 & 0 & 1 \\ 0 & 0 & 0 \end{pmatrix}.$$

This shows that $M_g^{\mathcal{E},\mathcal{B}} \neq M_f^{\mathcal{E},\mathcal{B}}$, so that $g \neq f$.

Chapter 8
Dual Spaces

8.1 The Dual of a Vector Space

Let us consider two finite dimensional real vector spaces V and W, and denote by $\mathrm{Lin}(V \to W)$ the collection of all linear maps $f : V \to W$. It is easy to show that $\mathrm{Lin}(V \to W)$ is itself a vector space over $\mathbb{R}$. This is with respect to a sum $(f_1 + f_2)$ and a product by a scalar (λf), for any $f_1, f_2, f \in \mathrm{Lin}(V \to W)$ and $\lambda \in \mathbb{R}$, defined *pointwise*, that is by

$$\begin{aligned}(f_1 + f_2)(v) &= f_1(v) + f_2(v)\\ (\lambda f)(v) &= \lambda f(v)\end{aligned}$$

for any $v \in V$. If $\mathcal{B}$ is a basis for V (of dimension n) and $\mathcal{C}$ a basis for W (of dimension m), the map $\mathrm{Lin}(V \to W) \to \mathbb{R}^{m,n}$ given by

$$f \quad \mapsto \quad M_f^{\mathcal{C},\mathcal{B}}$$

is an isomorphism of real vector spaces and the following relations

$$\begin{aligned}M_{f_1+f_2}^{\mathcal{C},\mathcal{B}} &= M_{f_1}^{\mathcal{C},\mathcal{B}} + M_{f_2}^{\mathcal{C},\mathcal{B}}\\ M_{\lambda f}^{\mathcal{C},\mathcal{B}} &= \lambda M_f^{\mathcal{C},\mathcal{B}}\end{aligned} \tag{8.1}$$

hold (see the Proposition 4.1.4). It is then clear that $\dim(\mathrm{Lin}(V \to W)) = mn$.

In particular, the vector space of linear maps from a vector space V to $\mathbb{R}$, that is the set of *linear forms* on V, deserves a name of its own.

Definition 8.1.1 Given a finite dimensional vector space V, the space of linear maps $\mathrm{Lin}(V \to \mathbb{R})$ is called the *dual space* to V and is denoted by $V^* = \mathrm{Lin}(V \to \mathbb{R})$.

The next result follows from the general discussion above.

G. Landi and A. Zampini, *Linear Algebra and Analytic Geometry for Physical Sciences*, Undergraduate Lecture Notes in Physics,
https://doi.org/10.1007/978-3-319-78361-1_8

Proposition 8.1.2 *Given a finite dimensional real vector space V, its dual space V^* is a real vector space with* $\dim(V^*) = \dim(V)$.

Let $\mathcal{B} = (b_1, \ldots, b_n)$ be a basis for V. We define elements $\{\varphi_i\}_{i=1,\ldots,n}$ in V^* by

$$\varphi_i(b_j) = \delta_{ij} \quad \text{with} \quad \begin{cases} 1 & \text{if } i = j \\ 0 & \text{if } i \neq j \,. \end{cases} \tag{8.2}$$

With $V \ni v = x_1 b_1 + \ldots + x_n b_n$, we have for the components that $x_i = \varphi_i(v)$. If $f \in V^*$ we write

$$\begin{aligned} f(v) &= f(b_1)\, x_1 + \ldots + f(b_n)\, x_n \\ &= f(b_1)\, \varphi_1(v) + \ldots + f(b_n)\, \varphi_n(v) \\ &= \big(f(b_1)\varphi_1 + \ldots + f(b_n)\varphi_n\big)(v). \end{aligned}$$

This shows that the action of f upon the vector v is the same as the action on v of the linear map $f = f(b_1)\varphi_1 + \ldots + f(b_n)\varphi_n$, that is we have that $V^* = \mathcal{L}(\varphi_1, \ldots, \varphi_n)$. It is indeed immediate to prove that, with respect to the linear structure in V^*, the linear maps φ_i are linearly independent, so they provide a basis for V^*. We have then sketched the proof of the following proposition.

Proposition 8.1.3 *Given a basis $\mathcal{B}$ for a n-dimensional real vector space V, the elements φ_i defined in* (8.2) *provide a basis for V^*. Such a basis, denoted $\mathcal{B}^*$, is called the* dual basis *to $\mathcal{B}$.*

We can also write

$$f(v) = (f(b_1) \ \ldots \ f(b_n)) \begin{pmatrix} x_1 \\ \vdots \\ x_n \end{pmatrix}. \tag{8.3}$$

Referring to the Definition 7.1.12 (and implicitly fixing a basis for $W \cong \mathbb{R}$), the relation (8.3) provides us the single row matrix $M_f^{\mathcal{B}} = (f(b_1) \ \ldots \ f(b_n))$ associated to f with respect to the basis $\mathcal{B}$ for V. Its entries are the image under f of the basis elements in $\mathcal{B}$. The proof of the proposition above shows that such entries are the components of $f \in V^*$ with respect to the dual basis $\mathcal{B}^*$.

Let $\mathcal{B}'$ be another basis for V, with elements $\{b'_i\}_{i=1,\ldots,n}$. With

$$v = x_1 b_1 + \ldots + x_n b_n = x'_1 b'_1 + \ldots + x'_n b'_n$$

we can write, following the Definition 7.9.3,

$$x'_k = \sum_{s=1}^{n} (M^{\mathcal{B}',\mathcal{B}})_{ks} x_s, \qquad b_i = \sum_{j=1}^{n} (M^{\mathcal{B}',\mathcal{B}})_{ji} b'_j$$

or, in a matrix notation,

$$\begin{pmatrix} x'_1 \\ \vdots \\ x'_n \end{pmatrix} = M^{\mathcal{B}',\mathcal{B}} \begin{pmatrix} x_1 \\ \vdots \\ x_n \end{pmatrix}, \qquad (b_1 \dots b_n)\, M^{\mathcal{B},\mathcal{B}'} = (b'_1 \dots b'_n). \tag{8.4}$$

From the Theorem 7.9.9 we have the matrix associated to f with respect to $\mathcal{B}'$

$$M_f^{\mathcal{B}'} = M_f^{\mathcal{B}} M^{\mathcal{B},\mathcal{B}'},$$

which we write as

$$(f(b_1) \dots f(b_n))\, M^{\mathcal{B},\mathcal{B}'} = (f(b'_1) \dots f(b'_n)). \tag{8.5}$$

Since the entries of $M_f^{\mathcal{B}'}$ provide the components of the element $f \in V^*$ with respect to the basis $\mathcal{B}'^*$, a comparison between (8.4) and (8.5) shows that, under a change of basis $\mathcal{B} \mapsto \mathcal{B}'$ for V and the corresponding change of the dual basis in V^*, the components of a vector in V^* are transformed under a map which is the *inverse* of the map that transforms the components of a vector in V.

The above is usually referred to by saying that the transformation law for vectors in V^* is *contravariant* with respect to the *covariant* one for vectors in V. In Sect. 13.3 we shall describe these facts with an important physical example, the study of the electromagnetic field.

If we express $f \in V^*$ with respect to the dual bases $\mathcal{B}^*$ and $\mathcal{B}'^*$ as

$$f(v) = \sum_{i=1}^{n} f(b_i)\varphi_n = \sum_{k=1}^{n} f(v'_k)\varphi'_k$$

and consider the rule for the change of basis, we have

$$\sum_{k,i=1}^{n} (M^{\mathcal{B}',\mathcal{B}})_{ki} f(b'_k)\varphi_i = \sum_{k=1}^{n} f(v'_k)\varphi'_k.$$

Since this must be valid for any $f \in V^*$, we can write the transformation law $\mathcal{B}^* \mapsto \mathcal{B}'^*$:

$$\varphi'_k = \sum_{i=1}^{n} (M^{\mathcal{B}',\mathcal{B}})_{ki}\varphi_i \qquad \text{that is} \qquad \begin{pmatrix} \varphi'_1 \\ \vdots \\ \varphi'_n \end{pmatrix} = M^{\mathcal{B}',\mathcal{B}} \begin{pmatrix} \varphi_1 \\ \vdots \\ \varphi_n \end{pmatrix}.$$

It is straightforward to extend to the complex case, *mutatis mutandis*, all the results of the present chapter given above. In particular, one has the following natural definition.

Definition 8.1.4 Let V be a finite dimensional complex vector space. The set $V^* = \mathrm{Lin}(V \to \mathbb{C})$ is called the *dual* space to V.

Indeed, the space V^* is a complex vector space, with $\dim(V^*) = \dim(V)$, and a natural extension of (8.2) to the complex case allows one to introduce a dual basis $\mathcal{B}^*$ for any basis $\mathcal{B}$ of V.

Also, we could consider linear maps between finite dimensional *complex* vector spaces. In the next section we shall explicitly consider linear transformations of the complex vector space $\mathbb{C}^n$.

8.2 The Dirac's Bra-Ket Formalism

Referring to Sect. 3.4 let us denote by $H^n = (\mathbb{C}^n, \cdot)$ the canonical hermitian vector space. Following Dirac (and by now a standard practice in textbooks on quantum mechanics), the hermitian product is denoted as

$$\langle\ |\ \rangle : \mathbb{C}^n \times \mathbb{C}^n \to \mathbb{C}', \qquad \langle z|w\rangle = \bar{z}_1 w_1 + \cdots + \bar{z}_n w_n,$$

for any $z = (z_1, \ldots, z_n), w = (w_1, \ldots, w_n) \in \mathbb{C}^n$. Thus its properties (see the Proposition 3.4.2) are written as follows. For any $z, w, v \in \mathbb{C}^n$ and $a, b \in \mathbb{C}$,

(i) $\langle w|z\rangle = \overline{\langle z|w\rangle}$,
(ii) $\langle az + bw|v\rangle = \bar{a}\langle z|v\rangle + \bar{b}\langle w|v\rangle$ while $\langle v|az + bw\rangle = a\langle v|z\rangle + b\langle v|w\rangle$,
(iii) $\langle z|z\rangle \geq 0$,
(iv) $\langle z|z\rangle = 0 \quad \Leftrightarrow \quad z = (0, \ldots, 0) \in \mathbb{C}^n$.

Since the hermitian product is bilinear (for the sum), for any fixed $w \in H^n$, the mapping

$$f_w : v \mapsto \langle w|z\rangle$$

provides indeed a linear map from $\mathbb{C}^n$ to $\mathbb{C}$, that is f_w is an element of the dual space $(\mathbb{C}^n)^*$. Given a hermitian basis $\mathcal{B} = \{e_1, \ldots, e_n\}$ for H^n, with $w = (w_1, \ldots, w_n)_{\mathcal{B}}$ and $z = (z_1, \ldots, z_n)_{\mathcal{B}}$, one has

$$f_w(z) = \bar{w}_1 z_1 + \ldots + \bar{w}_n z_n.$$

The corresponding dual basis $\mathcal{B}^* = \{\varepsilon_1, \ldots, \varepsilon_n\}$ for $(\mathbb{C}^n)^*$ is defined in analogy to (8.2) for the real case by taking $\varepsilon_i(e_j) = \delta_{ij}$. In terms of the hermitian product, these linear maps can be defined as $\varepsilon_i(z) = \langle e_i|z\rangle$. Then, to any $w = (w_1, \ldots, w_n)_{\mathcal{B}}$ we can associate an element $f_w = \bar{w}_1\varepsilon_1 + \ldots + \bar{w}_n\varepsilon_n$ in $(\mathbb{C}^n)^*$, whose action on $\mathbb{C}^n$ can be written as

$$f_w(v) = \langle w|v\rangle.$$

Thus, *via the hermitian product*, to any vector $w \in \mathbb{C}^n$ one associates a unique dual element $f_w \in (\mathbb{C}^n)^*$; viceversa, to any element $f \in (\mathbb{C}^n)^*$ one associates a unique element $w \in \mathbb{C}^n$ in such a way that $f = f_w$:

$$w = w_1 e_1 + \ldots + w_n e_n \quad \leftrightarrow \quad f_w = \bar{w}_1 \varepsilon_1 + \ldots + \bar{w}_n \varepsilon_n.$$

Remark 8.2.1 Notice that this bijection between $\mathbb{C}^n$ and $(\mathbb{C}^n)^*$ is anti-linear (for the product by complex numbers), since we have to complex conjugate the components of the vectors in order to satisfy the defining requirement of the hermitian product in H^n, that is

$$f_{\lambda w} = \bar{\lambda} f_w, \qquad \text{for} \quad \lambda \in \mathbb{C}, w \in \mathbb{C}^n.$$

For the canonical euclidean space E^n one could proceed in a similar manner and in such a case the bijection between E^n and its dual $(E^n)^*$ given by the euclidean product is linear.

Given the bijection above, Dirac's idea was to *split* the hermitian product bracket. Any element $w \in H^n$ provides a *ket* element $|w\rangle$ and a *bra* element $\langle w| \in (\mathbb{C}^n)^*$. A basis for H^n is then written as made of elements $|e_j\rangle$ while the bra elements $\langle e_j|$ form the dual basis for $(\mathbb{C}^n)^*$, with

$$\begin{aligned} w = w_1 e_1 + \ldots + w_n e_n \quad &\leftrightarrow \quad |w\rangle = w_1 |e_1\rangle + \ldots + w_n |e_n\rangle, \\ f_w = \bar{w}_1 \varepsilon_1 + \ldots + \bar{w}_n \varepsilon_n \quad &\leftrightarrow \quad \langle w| = \bar{w}_1 \langle e_1| + \ldots + \bar{w}_n \langle e_n|. \end{aligned}$$

The action of a bra element on a ket element is just given as a *bra-ket juxtapposition*, with

$$f_w(z) = \langle w|z\rangle \in \mathbb{C}.$$

We are now indeed allowed to define a *ket-bra juxtaposition*, that is we have elements $T = |z\rangle\langle w|$. The action of such a T from the left upon a $|u\rangle$, is then defined as

$$T \,:\, |u\rangle \mapsto |z\rangle\langle w|u\rangle.$$

Since $\langle w|u\rangle$ is a complex number, we see that for this action the element T maps a ket vector *linearly* into a ket vector, so T is a linear map from H^n to H^n.

Definition 8.2.2 With $z, w \in H^n$, the ket-bra element $T = |z\rangle\langle w|$ is the linear operator whose action is defined as $v \mapsto T(v) = \langle w|v\rangle z = (w \cdot v)z$.

It is then natural to consider linear combination of the form $T = \sum_{k,s=1}^{n} T_{ks} |e_k\rangle\langle e_s|$ with $T_{ks} \in \mathbb{C}$ the entries of a matrix $T \in \mathbb{C}^{n,n}$ so to compute

$$\begin{aligned} T|e_j\rangle &= \sum_{k,s=1}^{n} T_{ks} |e_k\rangle\langle e_s|e_j\rangle = \sum_{k=1}^{n} T_{kj} |e_k\rangle \\ T_{kj} &= \langle e_k|T(e_j)\rangle. \end{aligned} \tag{8.6}$$

In order to relate this formalism to the one we have already developed in this chapter, consider a linear map $\phi : H^n \to H^n$ and its associated matrix $M_{\phi}^{\mathcal{B},\mathcal{B}} = (a_{ks})$ with respect to a given hermitian basis $\mathcal{B} = (e_1, \ldots, e_n)$. From the Propositions 7.1.13 and 7.1.14 it is easy to show that one has

$$a_{kj} = e_k \cdot (A(e_j)) = \langle e_k | A(e_j) \rangle. \tag{8.7}$$

The analogy between (8.6) and (8.7) shows that, for a fixed basis of H^n, the action of a linear map ϕ with associated matrix $A = M_{\phi}^{\mathcal{B},\mathcal{B}} = (a_{ks})$ is equivalently written as the action of the operator

$$T_A (= T_{\phi}) = \sum_{k,s=1}^{n} a_{ks} |e_k\rangle\langle e_s|$$

in the Dirac's notation. The association $A \to T_A$ is indeed an isomorphism of (complex) vector space of dimension n^2.

Next, let ϕ, ψ be two linear maps on H^n with associated matrices A, B with respect to the hermitian basis $\mathcal{B}$. They correspond to the operators that we write as $T_A = \sum_{r,s=1}^{n} a_{rs} |e_r\rangle\langle e_s|$ and $T_B = \sum_{j,k=1}^{n} b_{jk} |e_j\rangle\langle e_k|$. With a natural juxtaposition we write the composition of the linear maps as

$$\begin{aligned} \phi \circ \psi &= \sum_{r,s=1}^{n} \sum_{j,k=1}^{n} a_{rs} b_{jk} |e_r\rangle\langle e_s|e_j\rangle\langle e_k| \\ &= \sum_{r,k=1}^{n} (\sum_{j=1}^{n} a_{rj} b_{jk}) |e_r\rangle\langle e_k| . \end{aligned}$$

We see that the matrix associated, via the isomorphism $A \to T_A$ above, to the composition $\phi \circ \psi$ has entries (r, k) given by $\sum_{j=1}^{n} a_{rj} b_{jk}$, thus coinciding with the row by column product between the matrices A and B associated to ϕ and ψ, that is

$$T_{AB} = T_A T_B .$$

Thus, the Proposition 7.8.3 for composition of matrices associated to linear maps is valid when we represent linear maps on H^n using the Dirac's notation.

All of this section has clearly a real version and could be repeated for the (real) euclidean space E^n with its linear maps and associated real matrices $T \in \mathbb{R}^{n,n}$.

Chapter 9
Endomorphisms and Diagonalization

Both in classical and quantum physics, and in several branches of mathematics, it is hard to overestimate the role that the notion of diagonal action of a linear map has. The aim of this chapter is to introduce this topic which will be crucial in all the following chapters.

9.1 Endomorphisms

Definition 9.1.1 Let V be a real vector space. A linear map $\phi : V \to V$ is called an *endomorphism* of V. The set of all endomorphisms of V is denoted $\mathrm{End}(V)$. Non invertible endomorphisms are also called *singular* or *degenerate*.

As seen in Sect. 8.1, the set $\mathrm{End}(V)$ is a real vector space with $\dim(\mathrm{End}(V)) = n^2$ if $\dim(V) = n$.

The question we address now is whether there exists a class of bases of the vector space V, with respect to which a matrix $M_{\phi}^{\mathcal{B},\mathcal{B}}$ has a particular (diagonal, say) form. We start with a definition.

Definition 9.1.2 The matrices $A, B \in \mathbb{R}^{n,n}$ are called *similar* if there exists a real vector space V and an endomorphism $\phi \in \mathrm{End}(V)$ such that $A = M_{\phi}^{\mathcal{B},\mathcal{B}}$ and $B = M_{\phi}^{\mathcal{C},\mathcal{C}}$, where $\mathcal{B}$ and $\mathcal{C}$ are bases for V. We denote similar matrices by $A \sim B$.

Similarity between matrices can be described in a purely algebraic way.

Proposition 9.1.3 *The matrices $A, B \in \mathbb{R}^{n,n}$ are similar if and only if there exists an invertible matrix $P \in \mathrm{GL}(n)$, such that $P^{-1}AP = B$.*

G. Landi and A. Zampini, *Linear Algebra and Analytic Geometry for Physical Sciences*, Undergraduate Lecture Notes in Physics,
https://doi.org/10.1007/978-3-319-78361-1_9

Proof Let us assume $A \sim B$: we then have a real vector space V, bases $\mathcal{B}$ and $\mathcal{C}$ for it and an endomorphism $\phi \in \text{End}(V)$ such that $A = M_{\phi}^{\mathcal{B},\mathcal{B}}$ e $B = M_{\phi}^{\mathcal{C},\mathcal{C}}$. From the Theorem 7.9.9 we have

$$B = M^{\mathcal{C},\mathcal{B}} \, A \, M^{\mathcal{B},\mathcal{C}}.$$

Since the matrix $M^{\mathcal{C},\mathcal{B}}$ is invertible, with $(M^{\mathcal{C},\mathcal{B}})^{-1} = M^{\mathcal{B},\mathcal{C}}$, the claim follows with $P = M^{\mathcal{B},\mathcal{C}}$.

Next, let us assume there exists a matrix $P \in \text{GL}(n)$ such that $P^{-1}AP = B$. From the Theorem 7.9.6 and the Remark 7.9.7 we know that the invertible matrix P gives a change of basis in $\mathbb{R}^n$: there exists a basis $\mathcal{C}$ for $\mathbb{R}^n$ (the columns of P), with $P = M^{\mathcal{E},\mathcal{C}}$ and $P^{-1} = M^{\mathcal{C},\mathcal{E}}$. Let $\phi = f_A^{\mathcal{E},\mathcal{E}}$ be the endomorphism in $\mathbb{R}^n$ corresponding to the matrix A with respect to the canonical bases, $A = M_{\phi}^{\mathcal{E},\mathcal{E}}$. We then have

$$\begin{aligned} B &= P^{-1} \, A \, P \\ &= M^{\mathcal{C},\mathcal{E}} \, M_{\phi}^{\mathcal{E},\mathcal{E}} \, M^{\mathcal{E},\mathcal{C}} \\ &= M_{\phi}^{\mathcal{C},\mathcal{C}}. \end{aligned}$$

This shows that B corresponds to the endomorphism ϕ with respect to the different basis $\mathcal{C}$, that is A and B are similar. □

Remark 9.1.4 The similarity we have introduced is an equivalence relation in $\mathbb{R}^{n,n}$, since it is

(a) reflexive, that is $A \sim A$ since $A = I_n A I_n$,
(b) symmetric, that is $A \sim B \Rightarrow B \sim A$ since

$$P^{-1} \, A \, P = B \quad \Rightarrow \quad P \, B \, P^{-1} = A,$$

(c) transitive, that is $A \sim B$ and $B \sim C$ imply $A \sim C$, since $P^{-1} \, A \, P = B$ and $Q^{-1} \, B \, Q = C$ clearly imply $Q^{-1} \, P^{-1} \, A \, P \, Q = (PQ)^{-1} \, A(PQ) = C$.

If $A \in \mathbb{R}^{n,n}$, we denote its equivalence class by similarity as $[A] = \{B \in \mathbb{R}^{n,n} : B \sim A\}$.

Proposition 9.1.5 *Let matrices $A, B \in \mathbb{R}^{n,n}$ be similar. Then*

$$\det(B) = \det(A) \qquad \textit{and} \qquad \text{tr}(B) = \text{tr}(A).$$

Proof From Proposition 9.1.3, we know there exists an invertible matrix $P \in \text{GL}(n)$, such that $P^{-1}AP = B$. From the Binet Theorem 5.1.16 and the Proposition 4.5.2 we can write

$$\begin{aligned} \det(B) &= \det(P^{-1}AP) \\ &= \det(P^{-1}) \det(A) \det(P) = \det(P^{-1}) \det(P) \det(A) \\ &= \det(A) \end{aligned}$$

and $\text{tr}(B) = \text{tr}(P^{-1}AP) = \text{tr}(PP^{-1}A) = \text{tr}(A)$. □

A natural question is whether, for a given A, the equivalence class $[A]$ contains a diagonal element (equivalently, whether A is similar to a diagonal matrix).

Definition 9.1.6 A matrix $A \in \mathbb{R}^{n,n}$ is called *diagonalisable* if it is similar to a diagonal (Δ say) matrix, that is if there is a diagonal matrix Δ in the equivalence class $[A]$.

Such a definition has a counterpart in terms of endomorphisms.

Definition 9.1.7 An endomorphism $\phi \in \mathrm{End}(V)$ is called *simple* if there exists a basis $\mathcal{B}$ for V such that the matrix $M_\phi^{\mathcal{B},\mathcal{B}}$ is diagonalisable.

We expect that for an endomorphism to be simple is an intrinsic property which does not depend on the basis with respect to which its corresponding matrix is given. The following proposition confirms this point.

Proposition 9.1.8 *Let V be a real vector space, with $\phi \in \mathrm{End}(V)$. The following are equivalent:*

(i) ϕ is simple, there is a basis $\mathcal{B}$ for V such that $M_\phi^{\mathcal{B},\mathcal{B}}$ is diagonalisable,
(ii) there exists a basis $\mathcal{C}$ for V such that $M_\phi^{\mathcal{C},\mathcal{C}}$ is diagonal,
(iii) given any basis $\mathcal{D}$ for V, the matrix $M_\phi^{\mathcal{D},\mathcal{D}}$ is diagonalisable.

Proof (i) $\Rightarrow$ (ii): Since $M_\phi^{\mathcal{B},\mathcal{B}}$ is similar to a diagonal matrix Δ, from the proof of the Proposition 9.1.3 we know that there is a basis $\mathcal{C}$ with respect to which $\Delta = M_\phi^{\mathcal{C},\mathcal{C}}$ is diagonal.

(ii) $\Rightarrow$ (iii): Let $\mathcal{C}$ be a basis of V such that $M_\phi^{\mathcal{C},\mathcal{C}} = \Delta$ is diagonal. For any basis $\mathcal{D}$ we have then $M_\phi^{\mathcal{D},\mathcal{D}} \sim \Delta$, thus $M_\phi^{\mathcal{D},\mathcal{D}}$ is diagonalisable.

(iii) $\Rightarrow$ (i): obvious.

□

9.2 Eigenvalues and Eigenvectors

Remark 9.2.1 Let $\phi : V \to V$ be a simple endomorphism, with $\Delta = M_\phi^{\mathcal{C},\mathcal{C}}$ a diagonal matrix associated to ϕ. It is then

$$\Delta = \begin{pmatrix} \lambda_1 & 0 & 0 & \cdots & 0 \\ 0 & \lambda_2 & 0 & \cdots & 0 \\ \vdots & \vdots & \vdots & & \vdots \\ 0 & 0 & 0 & \cdots & \lambda_n \end{pmatrix},$$

for scalars $\lambda_j \in \mathbb{R}$, with $j = 1, \ldots, n$. By setting $\mathcal{C} = (v_1, \ldots, v_n)$, we write then $\phi(v_j) = \lambda_j v_j$.

The vectors of the basis $\mathcal{C}$ and the scalars λ_j plays a prominent role in the analysis of endomorphisms. This motivates the following definition.

Definition 9.2.2 Let $\phi \in \text{End}(V)$ with V a real vector space. If there exists a non zero vector $v \in V$ and a scalar $\lambda \in \mathbb{R}$, such that

$$\phi(v) = \lambda v,$$

then λ is called an *eigenvalue* of ϕ and v is called an *eigenvector* of ϕ associated to λ. The *spectrum* of an endomorphism is the collection of its eigenvalues.

Remark 9.2.3 Let $\phi \in \text{End}(V)$ and $\mathcal{C}$ be a basis of V. With the definition above, the content of the Remark 9.2.1 can be rephrased as follow:

(a) $M_{\phi}^{\mathcal{C},\mathcal{C}}$ is diagonal if and only if $\mathcal{C}$ is a basis of eigenvectors for ϕ,
(b) ϕ is simple if and only if V has a basis of eigenvectors for ϕ (from the Definition 9.1.7).

Notice that each eigenvector v for an endomorphism ϕ is uniquely associated to an eigenvalue λ of ϕ. On the other hand, more than one eigenvector can be associated to a given eigenvalue λ. It is indeed easy to see that, if v is associated to λ, also αv, with $\alpha \in \mathbb{R}$, is associated to the same λ since $\phi(\alpha v) = \alpha\phi(v) = \alpha(\lambda v) = \lambda(\alpha v)$.

Proposition 9.2.4 *If V is a real vector space , and $\phi \in \text{End}(V)$, the set*

$$V_{\lambda} = \{v \in V : \phi(v) = \lambda v\}$$

is a vector subspace in V.

Proof We explicitly check that V_{λ} is closed under linear combinations. With $v_1,\ v_2 \in V_{\lambda}$ and $a_1,\ a_2 \in \mathbb{R}$, we can write

$$\phi(a_1 v_1 + a_2 v_2) = a_1\phi(v_1) + a_2\phi(v_2) = a_1\lambda v_1 + a_2\lambda v_2 = \lambda(a_1 v_1 + a_2 v_2),$$

showing that V_{λ} is a vector subspace of V □

Definition 9.2.5 If $\lambda \in \mathbb{R}$ is an eigenvalue of $\phi \in \text{End}(V)$, the space V_{λ} is called the *eigenspace* corresponding to λ.

Remark 9.2.6 It is easy to see that if $\lambda \in \mathbb{R}$ is not an eigenvalue for the endomorphism ϕ, then the set $V_{\lambda} = \{v \in V \mid \phi(v) = \lambda v\}$ contains only the zero vector. It is indeed clear that, if V_{λ} contains the zero vector only, then λ is not an eigenvalue for ϕ. We have that $\lambda \in \mathbb{R}$ is an eigenvalue for ϕ if and only if $\dim(V_{\lambda}) \geq 1$.

Exercise 9.2.7 Let $\phi \in \text{End}(\mathbb{R}^2)$ be defined by $\phi((x, y)) = (y, x)$. Is $\lambda = 2$ an eigenvalue for ϕ? The corresponding set V_2 would then be

$$V_2 = \{v \in \mathbb{R}^2 : \phi(v) = 2v\} = \{(x, y) \in \mathbb{R}^2 : (y, x) = 2(x, y)\},$$

that is, V_2 would be given by the solutions of the system

$$\begin{cases} y = 2x \\ x = 2y \end{cases} \Rightarrow \begin{cases} y = 2x \\ x = 4x \end{cases} \Rightarrow \begin{cases} x = 0 \\ y = 0 \end{cases}.$$

Since $V_2 = \{(0, 0)\}$, we conclude that $\lambda = 2$ is *not* an eigenvalue for ϕ.

Exercise 9.2.8 The endomorphism $\phi \in \mathrm{End}(\mathbb{R}^2)$ given by $\phi((x, y)) = (2x, 3y)$ is simple since the corresponding matrix with respect to the canonical basis $\mathcal{E} = (e_1, e_2)$ is diagonal,

$$M_\phi^{\mathcal{E},\mathcal{E}} = \begin{pmatrix} 2 & 0 \\ 0 & 3 \end{pmatrix}.$$

Its eigenvalues are $\lambda_1 = 2$ (with eigenvector e_1) and $\lambda_2 = 3$ (with eigenvector e_2). The corresponding eigenspaces are then $V_2 = \mathcal{L}(e_1)$ and $V_3 = \mathcal{L}(e_2)$.

Exercise 9.2.9 We consider again the endomorphism $\phi((x, y)) = (y, x)$ in $\mathbb{R}^2$ given in the Exercise 9.2.7. We wonder whether it is simple. We start by noticing that its corresponding matrix with respect to the canonical basis is the following,

$$M_\phi^{\mathcal{E},\mathcal{E}} = \begin{pmatrix} 0 & 1 \\ 1 & 0 \end{pmatrix},$$

which is not diagonal. We look then for a basis (if it exists) with respect to which the matrix corresponding to ϕ is diagonal. By recalling the Remark 9.2.3 we look for a basis of $\mathbb{R}^2$ made up of eigenvectors for ϕ. In order for $v = (a, b)$ to be an eigenvector for ϕ, there must exist a real scalar λ such that $\phi((a, b)) = \lambda(a, b)$,

$$\begin{cases} b = \lambda a \\ a = \lambda b \end{cases}.$$

It follows that the eigenvalues, if they exist, must fulfill the condition $\lambda^2 = 1$. For $\lambda = 1$ the corresponding eigenspace is

$$V_1 = \{(x, y) \in \mathbb{R}^2 : \phi((x, y)) = (x, y)\} = \{(x, x) \in \mathbb{R}^2\} = \mathcal{L}((1, 1)).$$

And for $\lambda = -1$ the corresponding eigenspace is

$$V_{-1} = \{(x, y) \in \mathbb{R}^2 : \phi((x, y)) = -(x, y)\} = \{(x, -x) \in \mathbb{R}^2\} = \mathcal{L}((1, -1)).$$

Since the vectors $(1, 1), (1, -1)$ form a basis $\mathcal{B}$ for $\mathbb{R}^2$ with respect to which the matrix of ϕ is

$$M_\phi^{\mathcal{B},\mathcal{B}} = \begin{pmatrix} 1 & 0 \\ 0 & -1 \end{pmatrix},$$

we conclude that ϕ is simple. We expect $M_{\phi}^{\mathcal{B},\mathcal{B}} \sim M_{\phi}^{\mathcal{E},\mathcal{E}}$, since they are associated to the same endomorphism; the algebraic proof of this claim is easy. By defining

$$P = M^{\mathcal{E},\mathcal{B}} = \begin{pmatrix} 1 & 1 \\ 1 & -1 \end{pmatrix}$$

the matrix of the change of basis, we compute explicitly,

$$\begin{pmatrix} 1 & 1 \\ 1 & -1 \end{pmatrix}^{-1} \begin{pmatrix} 0 & 1 \\ 1 & 0 \end{pmatrix} \begin{pmatrix} 1 & 1 \\ 1 & -1 \end{pmatrix} = \begin{pmatrix} 1 & 0 \\ 0 & -1 \end{pmatrix},$$

that is $P^{-1} M_{\phi}^{\mathcal{E},\mathcal{E}} P = M_{\phi}^{\mathcal{B},\mathcal{B}}$ (see the Proposition 9.1.3).

Not any endomorphism is simple as the following exercise shows.

Exercise 9.2.10 The endomorphism in $\mathbb{R}^2$ defined as $\phi((x, y)) = (-y, x)$ is not simple. For $v = (a, b)$ to be an eigenvector, $\phi((a, b)) = \lambda(a, b)$ it would be equivalent to $(-b, a) = \lambda(a, b)$, leading to $\lambda^2 = -1$. The only solution in $\mathbb{R}$ is then $a = b = 0$, showing that ϕ is not simple.

Proposition 9.2.11 *Let V be a real vector space with $\phi \in \text{End}(V)$. If λ_1, λ_2 are distinct eigenvalues, any two corresponding eigenvectors, $0 \neq v_1 \in V_{\lambda_1}$ and $0 \neq v_2 \in V_{\lambda_2}$, are linearly independent. Also, the sum $V_{\lambda_1} + V_{\lambda_2}$ is direct.*

Proof Let us assume that $v_2 = \alpha v_1$, with $\mathbb{R} \ni \alpha \neq 0$. By applying the linear map ϕ to both members, we have $\phi(v_2) = \alpha\phi(v_1)$. Since v_1 and v_2 are eigenvectors with eigenvalues λ_1 and λ_2,

$$\begin{aligned} \phi(v_1) &= \lambda_1 v_1 \\ \phi(v_2) &= \lambda_2 v_2 \end{aligned}$$

and the relation $\phi(v_2) = \alpha\phi(v_1)$, using $v_2 = \alpha v_1$ become

$$\lambda_2 v_2 = \alpha(\lambda_1 v_1) = \lambda_1(\alpha v_1) = \lambda_1 v_2,$$

that is

$$(\lambda_2 - \lambda_1) v_2 = 0_V.$$

Since $\lambda_2 \neq \lambda_1$, this would lead to the contradiction $v_2 = 0_V$. We therefore conclude that v_1 and v_2 are linearly independent.

For the last claim we use the Proposition 2.2.13 and show that $V_{\lambda_1} \cap V_{\lambda_2} = \{0_V\}$. If $v \in V_{\lambda_1} \cap V_{\lambda_2}$, we could write both $\phi(v) = \lambda_1 v$ (since $v \in V_{\lambda_1}$) and $\phi(v) = \lambda_2 v$ (since $v \in V_{\lambda_2}$): it would then be $\lambda_1 v = \lambda_2 v$, that is $(\lambda_1 - \lambda_2)v = 0_V$. From the hypothesis $\lambda_1 \neq \lambda_2$, we would get $v = 0_V$. □

The following proposition is proven along the same lines.

Proposition 9.2.12 *Let V be a real vector space, with $\phi \in \mathrm{End}(V)$. Let $\lambda_1, \dots, \lambda_s \in \mathbb{R}$ be distinct eigenvalues of ϕ with $0_V \neq v_j \in V_{\lambda_j}$, $j = 1, \dots, s$ corresponding eigenvectors. The set $\{v_1, \dots, v_s\}$ is free, and the sum $V_{\lambda_1} + \cdots + V_{\lambda_s}$ is direct.*

Corollary 9.2.13 *If ϕ is an endomorphism of the real vector space V, with $\dim(V) = n$, then ϕ has at most n distinct eigenvalues.*

Proof If ϕ had $s > n$ distinct eigenvalues, there would exist a set $v_1, \dots, v_s$ of non zero corresponding eigenvectors. From the proposition above, such a system should be free, thus contradicting the fact that the dimension of V is n. □

Remark 9.2.14 Let ϕ and ψ be two commuting endomorphisms, that is they are such that $\phi(\psi(w)) = \psi(\phi(w))$ for any $v \in V$. If $v \in V_\lambda$ is an eigenvector for ϕ corresponding to λ, it follows that

$$\phi(\psi(v)) = \psi(\phi(v)) = \lambda\, \psi(v).$$

Thus the endomorphism ψ maps any eigenspace V_λ of ϕ into itself, and analogously ϕ preserves any eigenspace $V'_{\lambda'}$ of ψ.

Finding the eigenspaces of an endomorphism amounts to compute suitable kernels. Let $f : V \to W$ be a linear map between real vector spaces with bases $\mathcal{B}$ and $\mathcal{C}$. We recall (see Proposition 7.5.1) that if $A = M_f^{\mathcal{C},\mathcal{B}}$ and $\Sigma : AX = 0$ is the linear system associated to A, the map $S_\Sigma \to \ker(f)$ given by

$$(x_1, \dots, x_n) \mapsto (x_1, \dots, x_n)_{\mathcal{B}}$$

is an isomorphism of vector spaces.

Lemma 9.2.15 *If V is a real vector space with basis $\mathcal{B}$, let $\phi \in \mathrm{End}(V)$ and $\lambda \in \mathbb{R}$. Then*

$$V_\lambda = \ker(\phi - \lambda\,\mathrm{id}_V) \cong S_{\Sigma_\lambda},$$

where S_{Σ_λ} is the space of the solutions of the linear homogeneous system

$$S_{\Sigma_\lambda} : \left(M_\phi^{\mathcal{B},\mathcal{B}} - \lambda I_n\right) X = 0.$$

Proof From the Definition 9.2.4 we write

$$\begin{aligned} V_\lambda &= \{v \in V : \phi(v) = \lambda v\} \\ &= \{v \in V : \phi(v) - \lambda v = 0_V\} \\ &= \ker(\phi - \lambda\,\mathrm{id}_V). \end{aligned}$$

Such a kernel is isomorphic (as recalled above) to the space of solutions of the linear system given by the matrix $M_{\phi-\lambda\mathrm{id}_V}^{\mathcal{B},\mathcal{B}}$, where $\mathcal{B}$ is an arbitrary basis of V. We conclude by noticing that $M_{\phi-\lambda\mathrm{id}_V}^{\mathcal{B},\mathcal{B}} = M_\phi^{\mathcal{B},\mathcal{B}} - \lambda I_n$. □

Proposition 9.2.16 *Let $\phi \in \mathrm{End}(V)$ be an endomorphism of the real vector space V, with $\dim(V) = n$, and let $\lambda \in \mathbb{R}$. The following are equivalent:*

(i) λ *is an eigenvalue for* ϕ,
(ii) $\dim(V_\lambda) \geq 1$,
(iii) $\det(M_\phi^{\mathcal{B},\mathcal{B}} - \lambda I_n) = 0$ *for any basis* $\mathcal{B}$ *in* V.

Proof (i) $\Leftrightarrow$ (ii) is the content of the Remark 9.2.6;
(ii) $\Leftrightarrow$ (iii). Let $\mathcal{B}$ be an arbitrary basis of V, and consider the linear system $S_{\Sigma_\lambda} : \left(M_\phi^{\mathcal{B},\mathcal{B}} - \lambda I_n\right)$
$X = 0$. We have

$$\dim(V_\lambda) = \dim(S_{\Sigma_\lambda}) = n - \mathrm{rk}\left(M_\phi^{\mathcal{B},\mathcal{B}} - \lambda I_n\right);$$

the first and the second equality follow from Definition 6.2.1 and Theorem 6.4.3 respectively. From Proposition 5.3.1 we finally write

$$\dim(V_\lambda) \geq 1 \quad \Leftrightarrow \quad \mathrm{rk}\left(M_\phi^{\mathcal{B},\mathcal{B}} - \lambda I_n\right) < n \quad \Leftrightarrow \quad \det\left(M_\phi^{\mathcal{B},\mathcal{B}} - \lambda I_n\right) = 0,$$

which concludes the proof. □

This proposition shows that the computation of an eigenspace reduces to finding the kernel of a linear map, a computation which has been described in the Proposition 7.5.1.

9.3 The Characteristic Polynomial of an Endomorphism

In this section we describe how to compute the eigenvalues of an endomorphism. These will be the roots of a canonical polynomial associate with the endomorphism.

Definition 9.3.1 Given a square matrix $A \in \mathbb{R}^{n,n}$, the expression

$$p_A(T) = \det(A - T I_n)$$

is a polynomial of order n in T with real coefficients. Such a polynomial is called the *characteristic polynomial* of the matrix A.

Exercise 9.3.2 If $A = \begin{pmatrix} a_{11} & a_{12} \\ a_{21} & a_{22} \end{pmatrix}$ is a square 2×2 matrix, then

$$\begin{aligned} p_A(T) &= \begin{vmatrix} a_{11} - T & a_{12} \\ a_{21} & a_{22} - T \end{vmatrix} \\ &= T^2 - (a_{11} + a_{22})\,T + (a_{11}a_{22} - a_{12}a_{21}) \\ &= T^2 - (\mathrm{tr}(A))\,T + (\det(A)). \end{aligned}$$

If λ_1 and λ_2 are the zeros (the roots) of the polynomial $p_A(T)$, with elementary algebra we write

$$p_A(T) = T^2 - (\lambda_1 + \lambda_2)\, T + \lambda_1\lambda_2$$

thus obtaining

$$\lambda_1 + \lambda_2 = a_{11} + a_{22} = \mathrm{tr}(A), \qquad \lambda_1\lambda_2 = (a_{11}a_{22} - a_{12}a_{21}) = \det(A).$$

Proposition 9.3.3 *Let V be a real vector space with $\dim(V) = n$, and let $\phi \in \mathrm{End}(V)$. For any choice of bases $\mathcal{B}$ and $\mathcal{C}$ in V, with corresponding matrices $A = M_\phi^{\mathcal{B},\mathcal{B}}$ and $B = M_\phi^{\mathcal{C},\mathcal{C}}$, it is*

$$p_A(T) = p_B(T).$$

Proof We know that $B = P^{-1}AP$, with $P = M^{\mathcal{B},\mathcal{C}}$ the matrix of change of basis. So we write

$$B - TI_n = P^{-1}AP - P^{-1}(TI_n)P = P^{-1}(A - TI_n)P.$$

From the Binet Theorem 5.1.16 we have then

$$\begin{aligned}\det(B - TI_n) &= \det(P^{-1}(A - TI_n)P) = \det(P^{-1})\det(A - TI_n)\det(P) \\ &= \det(A - TI_n),\end{aligned}$$

which yields a proof of the claim, since $\det(P^{-1})\det(P) = \det(I_n) = 1$. □

Given a matrix $A \in \mathbb{R}^{n,n}$, an explicit computation of $\det(A - TI_n)$ shows that

$$p_A(T) = (-1)^n T^n + (-1)^{n-1}\mathrm{tr}(A)\, T^{n-1} + \cdots + \det(A).$$

The case $n = 2$ is the Exercise 9.3.2.

Given $\phi \in \mathrm{End}(V)$, the Proposition 9.3.3 shows that the characteristic polynomial of the matrix associated to ϕ does not depend on the given basis of V.

Definition 9.3.4 For any matrix A associated to the endomorphism $\phi \in \mathrm{End}(V)$, the polynomial $p_\phi(T) = p_A(T)$ is called the *characteristic polynomial* of ϕ.

From the Proposition 9.2.16 and the Definition 9.3.4 we have the following result.

Corollary 9.3.5 *The eigenvalues of the endomorphism $\phi \in \mathrm{End}(V)$ (the spectrum of ϕ) are the real roots of the characteristic polynomial $p_\phi(T)$.*

Exercise 9.3.6 Let $\phi \in \mathrm{End}(\mathbb{R}^2)$ be associated to the matrix

$$M_\phi^{\mathcal{E},\mathcal{E}} = \begin{pmatrix} 0 & 1 \\ -1 & 0 \end{pmatrix}.$$

Since $p_\phi(T) = T^2 + 1$, the endomorphism has no (real) eigenvalues.

Definition 9.3.7 Let $p(X)$ be a polynomial with real coefficients, and let α be one of its real root. From the fundamental theorem of algebra (see the Proposition A.5.7) we know that then $(X - \alpha)$ is a divisor for $p(X)$, and that we have the decomposition

$$p(X) = (X - \alpha)^{m}(\alpha) \cdot q(X)$$

where $q(X)$ is not divisible by $(X - \alpha)$ and $1 \leq m(\alpha)$ is an integer depending on α. Such an integer is called the multiplicity of α.

Exercise 9.3.8 Let $p(X) = (X - 2)(X - 3)(X^2 + 1)$. Its real roots are 2 (with multiplicity $m(2) = 1$, since $(X - 3)(X^2 + 1)$ cannot be divided by 2) and 3 (with multiplicity $m(3) = 1$). Clearly the polynomial $p(X)$ has also two imaginary roots, given by $\pm\mathrm{i}$.

Proposition 9.3.9 *Let V be a real vector space with $\phi \in \mathrm{End}(V)$. If λ is an eigenvalue for ϕ with multiplicity $m(\lambda)$ and eigenspace V_λ, it holds that*

$$1 \leq \dim(V_\lambda) \leq m(\lambda).$$

Proof Let $r = \dim(V_\lambda)$ and $\mathcal{C}$ be a basis of V_λ. We complete $\mathcal{C}$ to a basis $\mathcal{B}$ for V. We then have $\mathcal{B} = (v_1, \ldots, v_r, v_{r+1}, \ldots, v_n)$, where the first elements $v_1, \ldots, v_r \in V_\lambda$ are eigenvectors for λ. The matrix $M_\phi^{\mathcal{B},\mathcal{B}}$ has the following block form,

$$A = M_\phi^{\mathcal{B},\mathcal{B}} = \begin{pmatrix} \lambda & 0 & \ldots & 0 & a_{1,r+1} & \ldots & a_{1,n} \\ 0 & \lambda & \ldots & 0 & a_{2,r+1} & \ldots & a_{2,n} \\ \vdots & \vdots & & \vdots & \vdots & & \vdots \\ 0 & 0 & \ldots & \lambda & a_{r,r+1} & \ldots & a_{r,n} \\ 0 & 0 & \ldots & 0 & a_{r+1,r+1} & \ldots & a_{r+1,n} \\ 0 & 0 & \ldots & 0 & a_{r+2,r+1} & \ldots & a_{r+2,n} \\ \vdots & \vdots & & \vdots & \vdots & & \vdots \\ 0 & 0 & \ldots & 0 & a_{n,r+1} & \ldots & a_{n,n} \end{pmatrix}.$$

If $\det(A - TI_n)$ is computed by the Laplace theorem (with respect to the first row, say), we have

$$p_\phi(T) = \det(A - T\,I_n) = (\lambda - T)^r\, g(T),$$

where $g(T)$ is the characteristic polynomial of the lower diagonal $(n - r) \times (n - r)$ square block of A. We can then conclude that $r \leq m(\lambda)$. □

Definition 9.3.10 The integer $\dim(V_\lambda)$ is called the *geometric multiplicity* of the eigenvalue λ, while $m(\lambda)$ is called the *algebraic multiplicity* of the eigenvalue λ.

Remark 9.3.11 Let $\phi \in \mathrm{End}(V)$.

(a) If $\lambda = 0$ is an eigenvalue for ϕ, the corresponding eigenspace V_0 is $\ker(\phi)$.

(b) If $\lambda \neq 0$ is an eigenvalue for ϕ, then $V_\lambda \subseteq \mathrm{Im}(\phi)$:
let us indeed consider $0_V \neq v \in V_\lambda$ with $\phi(v) = \lambda v$. Since $\lambda \neq 0$, we divide by λ and write
$$v = \lambda^{-1}\phi(v) = \phi(\lambda^{-1}v) \in \mathrm{Im}(\phi).$$

(c) If $\lambda_1 \neq \lambda_2 \neq \cdots \neq \lambda_s$ are distinct non zero eigenvalues for ϕ, from the Proposition 9.2.12 we have the direct sum of corresponding eigenspaces and
$$V_{\lambda_1} \oplus \cdots \oplus V_{\lambda_s} \subseteq \mathrm{Im}(\phi).$$

Exercise 9.3.12 Let $\phi \in \mathrm{End}(\mathbb{R}^4)$ be given by
$$\phi((x, y, z, t)) = (2x + 4y, x + 2y, -z - 2t, z + t).$$

The corresponding matrix with respect to the canonical basis $\mathcal{E}_4$ is
$$A = M_\phi^{\mathcal{E},\mathcal{E}} = \begin{pmatrix} 2 & 4 & 0 & 0 \\ 1 & 2 & 0 & 0 \\ 0 & 0 & -1 & -2 \\ 0 & 0 & 1 & 1 \end{pmatrix}.$$

Its characteristic polynomial reads
$$\begin{aligned} p_\phi(T) = p_A(T) &= \det(A - T I_4) \\ &= \begin{vmatrix} 2-T & 4 & 0 & 0 \\ 1 & 2-T & 0 & 0 \\ 0 & 0 & -1-T & -2 \\ 0 & 0 & 1 & 1-T \end{vmatrix} \\ &= \begin{vmatrix} 2-T & 4 \\ 1 & 2-T \end{vmatrix} \begin{vmatrix} -1-T & -2 \\ 1 & 1-T \end{vmatrix} \\ &= T(T-4)(T^2+1). \end{aligned}$$

The eigenvalues (the real roots of such a polynomial) of ϕ are $\lambda = 0, 4$. It is easy to compute that
$$\begin{aligned} V_0 &= \ker(\phi) = \mathcal{L}((-2, 1, 0, 0)), \\ V_4 &= \ker(\phi - 4I_4) = \mathcal{L}((2, 1, 0, 0)). \end{aligned}$$

This shows that V_4 is the only eigenspace corresponding to a non zero eigenvalue for ϕ.

From the Theorem 7.6.4 we know that $\dim \mathrm{Im}(\phi) = 4 - \dim \ker(\phi) = 3$, with a basis of the image of ϕ given by 3 linearly independent columns in A. It is immediate to notice that the second column is a multiple of the first one, so we have

$$\mathrm{Im}(\phi) = \mathcal{L}((2, 1, 0, 0), (0, 0, -1, 1), (0, 0, -2, 1)).$$

It is evident that $V_4 \subset \mathrm{Im}(\phi)$, as shown in general in the Remark 9.3.11.

Exercise 9.3.13 We consider the endomorphism in $\mathbb{R}^4$ given by

$$\phi((x, y, z, t)) = (2x + 4y, x + 2y, -z, z + t),$$

whose corresponding matrix with respect to the canonical basis $\mathcal{E}_4$ is

$$A = M_{\phi}^{\mathcal{E},\mathcal{E}} = \begin{pmatrix} 2 & 4 & 0 & 0 \\ 1 & 2 & 0 & 0 \\ 0 & 0 & -1 & 0 \\ 0 & 0 & 1 & 1 \end{pmatrix}.$$

The characteristic polynomial reads

$$\begin{aligned} p_{\phi}(T) = p_A(T) &= \det(A - T I_4) \\ &= \begin{vmatrix} 2-T & 4 & 0 & 0 \\ 1 & 2-T & 0 & 0 \\ 0 & 0 & -1-T & 0 \\ 0 & 0 & 1 & 1-T \end{vmatrix} \\ &= T(T-4)(T+1)(T-1). \end{aligned}$$

The eigenvalues are given by $\lambda = 0, 4, -1, 1$. The corresponding eigenspaces are

$$\begin{aligned} V_0 &= \ker(\phi) = \mathcal{L}((-2, 1, 0, 0)), \\ V_4 &= \ker(\phi - 4I_4) = \mathcal{L}((2, 1, 0, 0)), \\ V_{-1} &= \ker(\phi + I_4) = \mathcal{L}((0, 0, -2, 1)), \\ V_1 &= \ker(\phi - I_4) = \mathcal{L}((0, 0, 0, 1)), \end{aligned}$$

with

$$\mathrm{Im}(\phi) = V_{-1} \oplus V_1 \oplus V_4.$$

The characteristic polynomial $p_{\phi}(T)$ of an endomorphism over a real vector space has real coefficients. If $\lambda_1, \ldots, \lambda_s$ are its non zero real distinct roots (that is, the eigenvalues of ϕ), we can write

$$p_{\phi}(T) = (T - \lambda_1)^{m_1} \cdots (T - \lambda_p)^{m_s} \cdot q(T),$$

where m_j, $j = 1, \dots, s$ are the algebraic multiplicities and $q(T)$ has no real roots. We have then

$$\deg(p_\phi(T)) \geq m_1 + \cdots + m_s.$$

This proves the following proposition.

Proposition 9.3.14 *Let V be a real vector space with* $\dim(V) = n$, *and let* $\phi \in$ $\text{End}(V)$. *By denoting $\lambda_1, \dots, \lambda_s$ the distinct eigenvalues of ϕ with corresponding algebraic multiplicities $m_1, \dots, m_s$, one has*

$$m_1 + \cdots + m_s \leq n,$$

with the equality holding if and only if every root in $p_\phi(T)$ is real. □

9.4 Diagonalisation of an Endomorphism

In this section we describe conditions under which an endomorphism is simple. As we have seen, this problem is equivalent to study conditions under which a square matrix is diagonalisable. The first theorem we prove characterises simple endomorphims.

Theorem 9.4.1 *Let V be a real n-dimensional vector space, with* $\phi \in \text{End}(V)$. *If $\lambda_1, \dots, \lambda_s$ are the different roots of $p_\phi(T)$ with multiplicities $m_1, \dots, m_s$, the following claims are equivalent:*

(a) ϕ is a simple endomorphism,
(b) V has a basis of eigenvectors for ϕ,
(c) $\lambda_i \in \mathbb{R}$ for any $i = 1, \dots, s$, with $V = V_{\lambda_1} \oplus \cdots \oplus V_{\lambda_s}$,
(d) $\lambda_i \in \mathbb{R}$ and $m_i =$ $\dim(V_{\lambda_i})$ for any $i = 1, \dots, s$.

When ϕ is simple, each basis of V of eigenvectors for ϕ contains m_i eigenvectors for each distinct eigenvalues λ_i, for $i = 1, \dots, s$.

Proof • (a) $\Leftrightarrow$ (b): this has been shown in the Remark 9.2.3.

• (b) $\Rightarrow$ (c): let $\mathcal{B} = (v_1, \dots, v_n)$ be a basis of V of eigenvectors for ϕ. Any vector v_i belongs to one of the eigenspaces, so we can write

$$V = \mathcal{L}(v_1, \dots, v_n) \subseteq V_{\lambda_1} + \cdots + V_{\lambda_s},$$

while the opposite inclusion is obvious. Since the sum of eigenspaces corresponding to distinct eigenvalues is direct (see the Proposition 9.2.12), we have $V = V_{\lambda_1} \oplus \cdots \oplus V_{\lambda_s}$.

• (c) $\Rightarrow$ (b): let $\mathcal{B}_i$ be a basis of V_{λ_i} for any i. Since V is the direct sum of all the eigenspaces V_{λ_i}, the set $\mathcal{B} = \mathcal{B}_1 \cup \ldots \cup \mathcal{B}_s$ is a basis of V made by eigenvectors for ϕ.

- (c) $\Rightarrow$ (d): from the Grassmann Theorem 2.5.8, we have

$$\begin{aligned} n = \dim(V) &= \dim(V_{\lambda_1} \oplus \cdots \oplus V_{\lambda_s}) \\ &= \dim(V_{\lambda_1}) + \cdots + \dim(V_{\lambda_s}) \\ &\leq m_1 + \cdots + m_s \\ &\leq n, \end{aligned}$$

  where the inequalities follow from the Propositions 9.3.9 and 9.3.14. We can then conclude that $\dim(V_{\lambda_i}) = m(\lambda_i)$ for any i.
- (d) $\Rightarrow$ (c): from the hypothesis $m_i = \dim(V_{\lambda_i})$ for any $i = 1, \dots, s$, and the Proposition 9.3.14 we have

$$n = m_1 + \cdots + m_s = \dim(V_{\lambda_1}) + \cdots + \dim(V_{\lambda_s}).$$

  We have then $n = \dim(V_{\lambda_1} \oplus \cdots \oplus V_{\lambda_s})$ and this equality amounts to prove the claim, since $V_{\lambda_1} \oplus \cdots \oplus V_{\lambda_s}$ has dimension n and therefore coincides with V. □

Corollary 9.4.2 *If $\lambda_i \in \mathbb{R}$ and $m(\lambda_i) = 1$ for any $i = 1, \dots, n$, then is ϕ simple.*

Proof It is immediate, by recalling the Proposition 9.3.9 and (d) in the Theorem 9.4.1. □

Exercise 9.4.3 Let ϕ be the endomorphism in $\mathbb{R}^2$ whose corresponding matrix with respect to the canonical basis is the matrix

$$A = \begin{pmatrix} 1 & 1 \\ 0 & 1 \end{pmatrix}.$$

It is $p_A(T) = (1 - T)^2$: such a polynomial has only one root $\lambda = 1$ with algebraic multiplicity $m = 2$. It is indeed easy to compute that $V_1 = \mathcal{L}((1, 0))$, so the geometric multiplicity is 1. This proves that the matrix A is not diagonalisable, the corresponding endomorphism is not simple.

Proposition 9.4.4 *Let $\phi \in \mathrm{End}(V)$ be a simple endomorphism and $\mathcal{C}$ be a basis of V such that $\Delta = M_\phi^{\mathcal{C},\mathcal{C}}$. Then,*

(a) the eigenvalues $\lambda_1, \dots, \lambda_s$ for ϕ, counted with their multiplicities $m(\lambda_1), \dots, m(\lambda_s)$, are the diagonal elements for Δ;
(b) the diagonal matrix Δ is uniquely determined up to permutations of the eigenvalues (such a permutation corresponds to a permutation in the ordering of the basis elements in $\mathcal{C}$).

Proof (a) From the Remark 9.2.1 we know that the diagonal elements in $\Delta = M_\phi^{\mathcal{C},\mathcal{C}} \in \mathbb{R}^{n,n}$ are given by the eigenvalues $\lambda_1, \dots, \lambda_s$: each eigenvalue λ_i must be counted as many times as the geometric multiplicity of the eigenvector v_i, since $\mathcal{C}$ is a basis of eigenvectors. From the claim (d) in the Theorem 9.4.1, the geometric multiplicity of each eigenvalue coincides with its algebraic multiplicity.

(a) This is obvious. □

Proposition 9.4.5 *Let ϕ be a simple endomorphism on V, with $\mathcal{B}$ an arbitrary basis of V. By setting $A = M_{\phi}^{\mathcal{B},\mathcal{B}}$, let P be a matrix such that*

$$P^{-1}\,A\,P = \Delta.$$

Then the columns in P are the components, with respect to $\mathcal{B}$, of a basis of V made by eigenvectors for ϕ.

Proof Let $\mathcal{C}$ be a basis of V such that $\Delta = M_{\phi}^{\mathcal{C},\mathcal{C}}$. From the Remark 9.2.3 the basis $\mathcal{C}$ is made by eigenvectors for ϕ. The claim follows by setting $P = M^{\mathcal{B},\mathcal{C}}$, that is the matrix of the change of basis. □

Definition 9.4.6 Given a matrix $A \in \mathbb{R}^{n,n}$, its *diagonalisation* consists of determining, (if they exist) a diagonal matrix $\Delta \sim A$ and an invertible matrix $P \in \mathrm{GL}(n)$ such that $P^{-1}\,A\,P = \Delta$.

The following remark gives a resumé of the steps needed for the diagonalisation of a given matrix.

Remark 9.4.7 (**An algorithm for the diagonalisation**) Let $A \in \mathbb{R}^{n,n}$ be a square matrix. In order to diagonalise it:

(1) Write the characteristic polynomial $p_A(T)$ of A and find its roots $\lambda_1, \ldots, \lambda_s$ with the corresponding algebraic multiplicities $m_1, \ldots, m_s$.
(2) If one of the roots $\lambda_i \notin \mathbb{R}$, then A is *not* diagonalisable.
(3) If $\lambda_i \in \mathbb{R}$ for any $i = 1, \ldots, s$, compute the geometric multiplicities

$$\dim(V_{\lambda_i}) = n - \mathrm{rk}(A - \lambda_i I_n).$$

If there is an eigenvalue λ_i such that $m_i \neq \dim(V_{\lambda_i})$, then A is *not* diagonalisable.
(4) if $\lambda_i \in \mathbb{R}$ and $m(\lambda)_i = \dim(V_{\lambda_i})$ for any $i = 1, \ldots, s$, then A *is* diagonalisable. In such a case, A is similar to a diagonal matrix Δ: the eigenvalues λ_i, counted with their multiplicities, give the diagonal elements for Δ.
(5) it is $\Delta = M_{\phi}^{\mathcal{B},\mathcal{B}}$, where $\mathcal{B}$ is a basis of V given by eigenvectors for the endomorphism corresponding to the matrix A. By defining $P = M^{\mathcal{E},\mathcal{B}}$, it is $\Delta = P^{-1}AP$. Since V is the direct sum of the eigenspaces for A (see Theorem 9.4.1), it follows that $\mathcal{B} = \mathcal{B}_1 \cup \cdots \cup \mathcal{B}_s$, with $\mathcal{B}_i$ a basis of V_{λ_i} for any $i = 1, \ldots, s$. (The spaces V_{λ_i} can be obtained explicitly as in the Lemma 9.2.15.)

Exercise 9.4.8 We study whether the matrix

$$A = \begin{pmatrix} 3 & 1 & 1 \\ 1 & 0 & 2 \\ 1 & 2 & 0 \end{pmatrix}$$

is diagonalisable. Its characteristic polynomial is

$$\begin{aligned} p_A(T) &= \det(A - TI_3) \\ &= \begin{vmatrix} 3-T & 1 & 1 \\ 1 & -T & 2 \\ 1 & 2 & -T \end{vmatrix} \\ &= -T^3 + 3T^2 + 6T - 8 = (T-1)(T-4)(T+2). \end{aligned}$$

Its eigenvalues are found to be $\lambda_1 = 1, \lambda_2 = 4, \lambda_3 = -2$. Since each root of the characteristic polynomial has algebraic multiplicity $m = 1$, from the Corollary 9.4.2 the matrix A is diagonalisable, and indeed similar to

$$\Delta = \begin{pmatrix} 1 & 0 & 0 \\ 0 & 4 & 0 \\ 0 & 0 & -2 \end{pmatrix}.$$

We compute a basis $\mathcal{B}$ for $\mathbb{R}^3$ of eigenvectors for A. We know that $V_1 = \ker(A - I_3)$, so V_1 is the space of the solutions of the homogeneous linear system $(A - I_3)X = 0$ associated to the matrix

$$A - I_3 = \begin{pmatrix} 2 & 1 & 1 \\ 1 & -1 & 2 \\ 1 & 2 & -1 \end{pmatrix},$$

which is reduced to

$$\begin{pmatrix} 2 & 1 & 1 \\ 3 & 0 & 3 \\ 0 & 0 & 0 \end{pmatrix}.$$

The solution of such a linear system are given by $(x, y, z) = (x, -x, -x)$, thus $V_1 = \mathcal{L}((-1, 1, 1))$. Along the same lines we compute

$$\begin{aligned} V_4 &= \ker(A - 4I_3) = \mathcal{L}((2, 1, 1)), \\ V_{-2} &= \ker(A + 2I_3) = \mathcal{L}((0, -1, 1)). \end{aligned}$$

We have then $\mathcal{B} = ((-1, 1, 1), (2, 1, 1), (0, -1, 1))$ and

$$P = M^{\mathcal{E},\mathcal{B}} = \begin{pmatrix} -1 & 2 & 0 \\ 1 & 1 & -1 \\ 1 & 1 & 1 \end{pmatrix}.$$

It is easy to compute that $P^{-1} A P = \Delta$.

Proposition 9.4.9 *Let $A \in \mathbb{R}^{n,n}$ be diagonalisable, with eigenvalues $\lambda_1, \dots, \lambda_s$ and corresponding multiplicities $m_1, \dots, m_s$. Then*

$$\det(A) = \lambda_1^{m_1} \cdot \lambda_2^{m_2} \cdot \dots \cdot \lambda_s^{m_s},$$

$$\mathrm{tr}(A) = m_1\lambda_1 + m_2\lambda_2 + \dots + m_s\lambda_s.$$

Proof Since A is diagonalisable, there exists an invertible n-dimensional matrix P such that $\Delta = P^{-1} A P$. The matrix Δ is diagonal, and its diagonal elements are (see the Proposition 9.4.4) the eigenvalues of A counted with their multiplicities. Then, from the Proposition 9.1.5 on has,

$$\det(A) = \det(P^{-1}AP) = \det(\Delta) = \lambda_1^{m_1} \cdot \lambda_2^{m_2} \cdot \dots \cdot \lambda_s^{m_s}$$

and

$$\mathrm{tr}(A) = \mathrm{tr}(P^{-1}AP) = \mathrm{tr}(\Delta) = m_1\lambda_1 + m_2\lambda_2 + \dots + m_s\lambda_s.$$

□

9.5 The Jordan Normal Form

In this section we briefly describe the notion of *Jordan normal form* of a matrix. As we have described before in this chapter, a square matrix is not necessarily diagonalisable, that is it is not necessarily similar to a diagonal matrix. It is nonetheless possible to prove that any square matrix is similar to a triangular matrix J which is not far from being diagonal. Such a matrix J is diagonal if and only if A is diagonalisable; if not it has a 'standard' block structure.

An example of a so called *Jordan block* is the non diagonalisable matrix A in Exercise 9.4.3. We denote it by

$$J_2(1) = \begin{pmatrix} 1 & 1 \\ 0 & 1 \end{pmatrix}.$$

A Jordan block of order k is a k-dimensional upper triangular square matrix of the form

$$J_k(\lambda) = \begin{pmatrix} \lambda & 1 & 0 & \cdots & 0 \\ 0 & \lambda & 1 & \cdots & 0 \\ \vdots & \vdots & \vdots & & \vdots \\ \vdots & \vdots & \vdots & \cdots & 1 \\ 0 & 0 & 0 & \cdots & \lambda \end{pmatrix},$$

where the diagonal terms are given by a scalar $\lambda \in \mathbb{R}$, the $(J_k(\lambda))_{j,j+1}$ entries are 1 and the remaining entries are zero. It is immediate to show that the characteristic polynomial of such a matrix is given by

$$p_{J_k(\lambda)}(T) = (T - \lambda)^k,$$

and the parameter λ is the unique eigenvalue with algebraic multiplicity $m_\lambda = k$. The corresponding eigenspace is

$$V_\lambda = \ker(J_k(\lambda) - \lambda I_n) = \mathcal{L}((1, 0, \ldots, 0)),$$

with geometric multiplicity $\dim(V_\lambda) = 1$. Thus, if $k > 1$, a Jordan block is not diagonalisable.

A matrix J is said to be in (*canonical* or *normal*) *Jordan form* if it has a block diagonal form

$$J = \begin{pmatrix} J_{k_1}(\lambda_1) & \mathbf{0} & \ldots & \mathbf{0} \\ \mathbf{0} & J_{k_2}(\lambda_2) & \ldots & \mathbf{0} \\ \vdots & \vdots & \ddots & \vdots \\ \mathbf{0} & \mathbf{0} & \ldots & J_{k_s}(\lambda_s) \end{pmatrix},$$

where each $J_{k_j}(\lambda_j)$ is a Jordan block of order k_j and eigenvalue λ_j, for $j = 1, \ldots, s$.

Notice that nothing prevents from having the same eigenvalue in different Jordan blocks, that is $\lambda_j = \lambda_l$ even with $k_j \neq k_l$. Since each Jordan block $J_{k_j}(\lambda_j)$ provides a one dimensional eigenspace for λ_j, the geometric multiplicity of λ_j coincides with the number of Jordan blocks with eigenvalue λ_j. The algebraic multiplicity of λ_j coincides indeed with the sum of the orders of the Jordan blocks having the same eigenvalue λ_j.

Theorem 9.5.1 (Jordan) *Let $A \in \mathbb{R}^{n,n}$ such that its characteristic polynomial has only real roots (such roots are all the eigenvalues for A). Then,*

(i) the matrix A is similar to a Jordan matrix,
(ii) two Jordan matrices J and J' are similar if and only if one is mapped into the other under a block permutation.

We omit a complete proof of this theorem, and we limit ourselves to briefly introduce the notion of *generalised* eigenvector of a matrix A. We recall that, when A is not diagonalisable, the set of eigenvectors for A is not enough for a basis of $\mathbb{R}^n$. The columns of the invertible matrix P that realises the similarity between A and the Jordan form J (such that $P^{-1}AP = J$) are the components with respect to the canonical basis $\mathcal{E}_n$ of the so called generalised eigenvectors for A.

Given an eigenvalue λ for A with algebraic multiplicity $m_\lambda \geq 1$, a corresponding generalised eigenvector is a non zero vector v that solves the linear homogeneous system

$$(A - \lambda I_n)^m v = 0_{\mathbb{R}^n}.$$

It is possible to show that such a system has m solutions v_j (with $v_j = 1, \ldots, m$) which can be obtained by recursion,

$$\begin{aligned} (A - \lambda I_n)v_1 &= 0_{\mathbb{R}^n}, \\ (A - \lambda I_n)v_k &= v_{k-1}, \quad k = 2, \ldots m. \end{aligned}$$

The elements v_j span the *generalised eigenspace* V_λ for A corresponding to the eigenvalue λ. The generalised eigenvectors satisfy the condition

$$(A - \lambda I_n)^k v_k = 0_{\mathbb{R}^n} \qquad \text{for any} \quad k = 1, 2, \ldots m.$$

Since the characteristic polynomial of A has in general complex roots, we end by noticing that a more natural version of the Jordan theorem is valid on $\mathbb{C}$.

Exercise 9.5.2 We consider the matrix

$$A = \begin{pmatrix} 5 & 4 & 2 & 1 \\ 0 & 1 & -1 & -1 \\ -1 & -1 & 3 & 0 \\ 1 & 1 & -1 & 2 \end{pmatrix}.$$

Its characteristic polynomial is computed to be $p_A(T) = (T-1)(T-2)(T-4)^2$, so its eigenvalues are $\lambda = 1, 2, 4, 4$. Since the algebraic multiplicity of the eigenvalues $\lambda = 1$ and $\lambda = 2$ is 1, their geometric multiplicity is also 1. An explicit computation shows that

$$\dim(\ker(A - 4 I_4)) = 1.$$

We have then that A is not diagonalisable, and that the eigenvalue $\lambda = 4$ corresponds to a Jordan block. A canonical form for the matrix A is then given by

$$J = \begin{pmatrix} 1 & 0 & 0 & 0 \\ 0 & 2 & 0 & 0 \\ 0 & 0 & 4 & 1 \\ 0 & 0 & 0 & 4 \end{pmatrix}.$$

Exercise 9.5.3 The matrices

$$J = \begin{pmatrix} 3 & 1 & 0 & 0 \\ 0 & 3 & 0 & 0 \\ 0 & 0 & 3 & 0 \\ 0 & 0 & 0 & 3 \end{pmatrix}, \qquad J' = \begin{pmatrix} 3 & 1 & 0 & 0 \\ 0 & 3 & 0 & 0 \\ 0 & 0 & 3 & 1 \\ 0 & 0 & 0 & 3 \end{pmatrix}$$

have the same characteristic polynomial, the same determinant, and the same trace. They are however *not* similar, since they are in Jordan form, and there is no block permutation under which J is mapped into J'.

Chapter 10
Spectral Theorems on Euclidean Spaces

In Chap. 7 we studied the operation of changing a basis for a real vector space. In particular, in the Theorem 7.9.6 and the Remark 7.9.7 there, we showed that any matrix giving a change of basis for the vector space $\mathbb{R}^n$ is an invertible $n \times n$ matrix, and noticed that any $n \times n$ invertible yields a change of basis for $\mathbb{R}^n$.

In this chapter we shall consider the endomorphisms of the euclidean space $E^n = (\mathbb{R}^n, \cdot)$, where the symbol $\cdot$ denotes the euclidean scalar product, that we have described in Chap. 3.

10.1 Orthogonal Matrices and Isometries

As we noticed, the natural notion of basis for a euclidean space is that of orthonormal one. This restricts the focus to matrices which gives a change of basis between orthonormal bases for E^n.

Definition 10.1.1 A square matrix $A \in \mathbb{R}^{n,n}$ is called *orthogonal* if its columns form an orthonormal basis $\mathcal{B}$ for E^n. In such a case $A = M^{\mathcal{E},\mathcal{B}}$, that is A is the matrix giving the change of basis from the canonical basis $\mathcal{E}$ to the basis $\mathcal{B}$.

It follow from this definition that an orthogonal matrix is invertible.

Exercise 10.1.2 The identity matrix I_n is clearly orthogonal for each E^n. Since the vectors

$$v_1 = \tfrac{1}{\sqrt{2}}(1, 1), \qquad v_2 = \tfrac{1}{\sqrt{2}}(1, -1)$$

G. Landi and A. Zampini, *Linear Algebra and Analytic Geometry for Physical Sciences*, Undergraduate Lecture Notes in Physics,
https://doi.org/10.1007/978-3-319-78361-1_10

form an orthonormal basis for E^2, the matrix

$$A = \frac{1}{\sqrt{2}} \begin{pmatrix} 1 & 1 \\ 1 & -1 \end{pmatrix}$$

is orthogonal.

Proposition 10.1.3 *A matrix A is orthogonal if and only if*

$$^{t}A\,A = I_n,$$

that is if and only if $A^{-1} = {}^{t}A$.

Proof With $(v_1, \ldots, v_n)$ a system of vectors in E^n, we denote by $A = (v_1 \cdots v_n)$ the matrix with columns given by the given vectors, and by

$$^{t}A = \begin{pmatrix} {}^{t}v_1 \\ \vdots \\ {}^{t}v_n \end{pmatrix}$$

its transpose. We have the following equivalences. The matrix A is orthogonal (by definition) if and only if $(v_1, \ldots, v_n)$ is an orthonormal basis for E^n, that is if and only if $v_i \cdot v_j = \delta_{ij}$ for any i, j. Recalling the representation of the row by column product of matrices, one has $v_i \cdot v_j = \delta_{ij}$ if and only if $({}^{t}A A)_{ij} = \delta_{ij}$ for any i, j, which amounts to say that ${}^{t}A A = I_n$. □

Exercise 10.1.4 For the matrix A considered in the Exercise 10.1.2 one has easily compute that $A = {}^{t}A$ and $A^2 = I_2$.

Exercise 10.1.5 The matrix

$$A = \begin{pmatrix} 1 & 0 \\ 1 & 1 \end{pmatrix}$$

is not orthogonal, since

$$^{t}AA = \begin{pmatrix} 1 & 1 \\ 0 & 1 \end{pmatrix} \begin{pmatrix} 1 & 0 \\ 1 & 1 \end{pmatrix} = \begin{pmatrix} 2 & 1 \\ 1 & 1 \end{pmatrix} \neq I_2.$$

Proposition 10.1.6 *If A is orthogonal, then* $\det(A) = \pm 1$.

Proof This statement easily follows from the Binet Theorem 5.1.16: with ${}^{t}AA = I_n$, one has

$$\det({}^{t}A)\det(A) = \det(I_n) = 1,$$

and the Corollary 5.1.12, that is $\det({}^{t}A) = \det(A)$, which then implies $(\det(A))^2 = 1$. □

Remark 10.1.7 The converse to this statement does not hold. The matrix A from the Exercise 10.1.5 is not orthogonal, while $\det(A) = 1$.

Definition 10.1.8 An orthogonal matrix A with $\det(A) = 1$ is called *special orthogonal.*

Proposition 10.1.9 *The set* $\mathrm{O}(n)$ *of orthogonal matrices in* $\mathbb{R}^{n,n}$ *is a group, with respect to the usual matrix product. Its subset* $\mathrm{SO}(n) = \{A \in \mathrm{O}(n) : \det(A) = 1\}$ *is a subgroup of* $\mathrm{O}(n)$ *with respect to the same product.*

Proof We prove that $\mathrm{O}(n)$ is stable under the matrix product, has an identity element, and the inverse of an orthogonal matrix is orthogonal as well.

- The identity matrix I_n is orthogonal, as we already mentioned.
- If A and B are orthogonal, then we can write

$$\begin{aligned} {}^t(AB)AB &= {}^tB\,{}^tAAB \\ &= {}^tB\, I_n\, B \\ &= {}^tBB \;=\; I_n, \end{aligned}$$

  that is, AB is orthogonal.
- If A is orthogonal, ${}^tAA = I_n$, then

$${}^t(A^{-1})A^{-1} \;=\; (A\,{}^tA)^{-1} \;=\; I_n,$$

  that proves that A^{-1} is orthogonal.

From the Binet theorem it easily follows that the set of special orthogonal matrices is stable under the product, and the inverse of a special orthogonal matrix is special orthogonal. □

Definition 10.1.10 The group $\mathrm{O}(n)$ is called the *orthogonal group* of order n, its subset $\mathrm{SO}(n)$ is called the *special orthogonal group* of order n.

We know from the Definition 10.1.1 that a matrix is orthogonal if and only if it is the matrix of the change of basis between the canonical basis $\mathcal{E}$ (which is orthonormal) and a second orthonormal basis $\mathcal{B}$. A matrix A is then orthogonal if and only if $A^{-1} = {}^tA$ (Proposition 10.1.3).

The next theorem shows that we do not need the canonical basis. If one defines a matrix A to be orthogonal by the condition $A^{-1} = {}^tA$, then A is the matrix for a change between two orthonormal bases and viceversa, any matrix A giving the change between orthonormal bases satisfies the condition $A^{-1} = {}^tA$.

Theorem 10.1.11 *Let* $\mathcal{C}$ *be an orthonormal basis for the euclidean vector space* E^n*, with* $\mathcal{B}$ *another (arbitrary) basis for it. The matrix* $M^{\mathcal{C},\mathcal{B}}$ *of the change of basis from* $\mathcal{C}$ *to* $\mathcal{B}$ *is orthogonal if and only if also the basis* $\mathcal{B}$ *is orthonormal.*

Proof We start by noticing that, since $\mathcal{C}$ is an orthonormal basis, the matrix $M^{\mathcal{E},\mathcal{C}}$ giving the change of basis between the canonical basis $\mathcal{E}$ and $\mathcal{C}$ is orthogonal by the Definition 10.1.1. It follows that, being $\mathrm{O}(n)$ a group, the inverse $M^{\mathcal{C},\mathcal{E}} = (M^{\mathcal{E},\mathcal{C}})^{-1}$ is orthogonal. With $\mathcal{B}$ an arbitrary basis, from the Theorem 7.9.9 we can write

$$\begin{aligned} M^{\mathcal{C},\mathcal{B}} &= M^{\mathcal{C},\mathcal{E}}\, M^{\mathcal{E},\mathcal{E}}\, M^{\mathcal{E},\mathcal{B}} \\ &= M^{\mathcal{C},\mathcal{E}}\, I_n\, M^{\mathcal{E},\mathcal{B}} = M^{\mathcal{C},\mathcal{E}} M^{\mathcal{E},\mathcal{B}}. \end{aligned}$$

Firstly, let us assume $\mathcal{B}$ to be orthonormal. We have then that $M^{\mathcal{E},\mathcal{B}}$ is orthogonal; thus $M^{\mathcal{C},\mathcal{B}}$ is orthogonal since it is the product of orthogonal matrices.

Next, let us assume that $M^{\mathcal{C},\mathcal{B}}$ is orthogonal; from the chain relations displayed above we have

$$M^{\mathcal{E},\mathcal{B}} = (M^{\mathcal{C},\mathcal{E}})^{-1} M^{\mathcal{C},\mathcal{B}} = M^{\mathcal{E},\mathcal{C}} M^{\mathcal{C},\mathcal{B}}.$$

This matrix $M^{\mathcal{E},\mathcal{B}}$ is then orthogonal (being the product of orthogonal matrices), and therefore $\mathcal{B}$ is an orthonormal basis. □

We pass to endomorphisms corresponding to orthogonal matrices. We start by recalling, from the Definition 3.1.4, that a scalar product has a 'canonical' form when it is given with respect to orthonormal bases.

Remark 10.1.12 Let $\mathcal{C}$ be an orthonormal basis for the euclidean space E^n. If $v, w \in E^n$ are given by $v = (x_1, \ldots, x_n)_{\mathcal{C}}$ and $w = (y_1, \ldots, y_n)_{\mathcal{C}}$, one has that $v \cdot w = x_1 y_1 + \cdots + x_n y_n$. By denoting X and Y the one-column matrices whose entries are the components of v, w with respect to $\mathcal{C}$, that is

$$X = \begin{pmatrix} x_1 \\ \vdots \\ x_n \end{pmatrix}, \quad Y = \begin{pmatrix} y_1 \\ \vdots \\ y_n \end{pmatrix},$$

we can write

$$v \cdot w = x_1 y_1 + \cdots + x_n y_n = \begin{pmatrix} x_1 & \ldots & x_n \end{pmatrix} \begin{pmatrix} y_1 \\ \vdots \\ y_n \end{pmatrix} = {}^t X Y.$$

Theorem 10.1.13 *Let $\phi \in \mathrm{End}(E^n)$, with $\mathcal{E}$ the canonical basis of E^n. The following statements are equivalent:*

(i) The matrix $A = M_{\phi}^{\mathcal{E},\mathcal{E}}$ is orthogonal.
(ii) It holds that $\phi(v) \cdot \phi(w) = v \cdot w$ for any $v, w \in E^n$.
(iii) If $\mathcal{B} = (b_1, \ldots, b_n)$ is an orthonormal basis for E^n, then the set $\mathcal{B}' = (\phi(b_1), \ldots, \phi(b_n))$ is such.

Proof (i) $\Rightarrow$ (ii): by denoting $X = {}^t v$ and $Y = {}^t w$ we can write

$$v \cdot w = {}^t XY, \qquad \phi(v) \cdot \phi(w) = {}^t(AX)(AY) = {}^t X({}^t AA)Y,$$

and since A is orthogonal, ${}^t AA = I_n$, we conclude that $\phi(v) \cdot \phi(w) = v \cdot w$ for any $v, w \in E^n$.

(ii) $\Rightarrow$ (iii): let $A = M_\phi^{\mathcal{C},\mathcal{C}}$ be the matrix of the endomorphism ϕ with respect to the basis $\mathcal{C}$. We start by proving that A is invertible. By adopting the notation used above, we can represent the condition $\phi(v) \cdot \phi(w) = v \cdot w$ as ${}^t(AX)(AY) = {}^t XY$ for any $X, Y \in E^n$. It follows that ${}^t AA = I_n$, that is A is orthogonal, and then invertible. This means (see Theorem 7.8.4) that ϕ is an isomorphism, so it maps a basis for E^n into a basis for E^n. If $\mathcal{B}$ is an orthonormal basis, then we can write

$$\phi(b_i) \cdot \phi(b_j) = b_i \cdot b_j = \delta_{ij}$$

which proves that $\mathcal{B}'$ is an orthonormal basis.

(iii) $\Rightarrow$ (i): since $\mathcal{E}$, the canonical basis for E^n, is orthonormal, then $(\phi(e_1), \ldots, \phi(e_n))$ is orthonormal. Recall the Remark 7.1.10: the components with respect to $\mathcal{E}$ of the elements $\phi(e_i)$ are the column vectors of the matrix $M_\phi^{\mathcal{E},\mathcal{E}}$, thus $M_\phi^{\mathcal{E},\mathcal{E}}$ is orthogonal. □

We have seen that, if the action of $\phi \in \mathrm{End}(E^n)$ is represented with respect to the canonical basis by an orthogonal matrix, then ϕ is an isomorphism and preserves the scalar product, that is, for any $v, w \in E^n$ one has that,

$$v \cdot w = \phi(v) \cdot \phi(w).$$

The next result is therefore evident.

Corollary 10.1.14 *If $\phi \in \mathrm{End}(E^n)$ is an endomorphism of the euclidean space E^n whose corresponding matrix with respect to the canonical basis is orthogonal then ϕ preserves the norms, that is, for any $v \in E^n$ one has*

$$\|\phi(v)\| = \|v\|.$$

This is the reason why such an endomorphism is also called an isometry.

The analysis we developed so far allows us to introduce the following definition, which will be more extensively scrutinised when dealing with rotations maps.

Definition 10.1.15 If $\phi \in \mathrm{End}(E^n)$ takes the orthonormal basis $\mathcal{B} = (b_1, \ldots, b_n)$ to the orthonormal basis $\mathcal{B}' = (b'_1 = \phi(b_1), \ldots, b'_n = \phi(b_n))$ in E^n, we say that $\mathcal{B}$ and $\mathcal{B}'$ have the same *orientation* if the matrix representing the endomorphism ϕ is special orthogonal.

Remark 10.1.16 It is evident that this definition provides an equivalence relation within the collection of all orthonormal bases for E^n. The corresponding quotient can be labelled by the values of the determinant of the orthogonal map giving the change of basis, that is $\det \phi = \{\pm 1\}$. This is usually referred to by saying that the euclidean space E^n has two orientations.

10.2 Self-adjoint Endomorphisms

We need to introduce an important class of endomorphisms.

Definition 10.2.1 An endomorphism ϕ of the euclidean vector space E^n is called *self-adjoint* if

$$\phi(v) \cdot w = v \cdot \phi(w) \qquad \forall\, v, w \in E.$$

From the Proposition 9.2.11 we know that eigenvectors corresponding to distinct eigenvalues are linearly independent. When dealing with self-adjoint endomorphisms, a stronger property holds.

Proposition 10.2.2 *Let ϕ be a self-adjoint endomorphism of E^n, with $\lambda_1, \lambda_2 \in \mathbb{R}$ different eigenvalues for it. Any two corresponding eigenvectors, $0 \neq v_1 \in V_{\lambda_1}$ and $0 \neq v_2 \in V_{\lambda_2}$, are orthogonal.*

Proof Since ϕ is self-adjoint, one has $\phi(v_1) \cdot v_2 = v_1 \cdot \phi(v_2)$ while, v_1 and v_2 being eigenvectors, one has $\phi(v_i) = \lambda_i v_i$ for $i = 1, 2$. We can then write

$$(\lambda_1 v_1) \cdot v_2 = v_1 \cdot (\lambda_2 v_2)$$

which reads

$$\lambda_1 (v_1 \cdot v_2) = \lambda_2 (v_1 \cdot v_2) \qquad \Rightarrow \qquad (\lambda_2 - \lambda_1)(v_1 \cdot v_2) = 0.$$

The assumption that the eigenvalues are different allows one to conclude that $v_1 \cdot v_2 = 0$, that is v_1 is orthogonal to v_2. □

The self-adjointness of an endomorphism can be characterised in terms of properties of the matrices representing its action on E^n. We recall from the Definition 4.1.21 that a matrix $A = (a_{ij}) \in \mathbb{R}^{n,n}$ is called symmetric if ${}^tA = A$, that is if one has $a_{ij} = a_{ji}$, for any i, j.

Theorem 10.2.3 *Let $\phi \in \mathrm{End}(E^n)$ and $\mathcal{B}$ an orthonormal basis for E^n. The endomorphism ϕ is self-adjoint if and only if $M_{\phi}^{\mathcal{B},\mathcal{B}}$ is symmetric.*

Proof Using the usual notation, we set $A = (a_{ij}) = M_{\phi}^{\mathcal{B},\mathcal{B}}$ and X, Y be the columns giving the components with respect to $\mathcal{B}$ of the vectors v, w in E^n. From the Remark 10.1.12 we write

$$\begin{aligned} \phi(v)\cdot w &= {}^t(AX)Y = ({}^tX{}^tA)Y = {}^tX{}^tAY \\ \text{and} \quad v\cdot\phi(w) &= {}^tX(AY) = {}^tXAY. \end{aligned}$$

Let us assume A to be symmetric. From the relations above we conclude that $\phi(v)\cdot w = v\cdot\phi(w)$ for any $v, w \in E^n$, that is ϕ is self-adjoint.

If we assume ϕ to be self-adjoint, then we can equate

$${}^tX{}^tAY = {}^tXAY$$

for any X, Y in $\mathbb{R}^n$. If we let X and Y to range on the elements of the canonical basis $\mathcal{E} = (e_1, \dots, e_n)$ in $\mathbb{R}^n$, such a condition is just the fact that $a_{ij} = a_{ji}$ for any i, j, that is A is symmetric. □

Exercise 10.2.4 The following matrix is symmetric:

$$A = \begin{pmatrix} 2 & -1 \\ -1 & 3 \end{pmatrix}.$$

Then the endomorphism $\phi \in \mathrm{End}(E^2)$ corresponding to A with respect to the canonical basis is self-adjoint. This can also be shown by a direct calculation: $\phi((x, y)) = (2x - y, -x + 3y)$; then

$$\begin{aligned} (a, b)\cdot\phi((x, y)) &= a(2x - y) + b(-x + 3y) \\ &= (2a - b)x + (-a + 3b)y \\ &= \phi((a, b))\cdot(x, y). \end{aligned}$$

Exercise 10.2.5 The following matrix is not symmetric

$$B = \begin{pmatrix} 1 & 1 \\ -1 & 0 \end{pmatrix}.$$

The corresponding (with respect to the canonical basis) endomorphism $\phi \in \mathrm{End}(E^2)$ is indeed not self-adjoint since for instance,

$$\begin{aligned} \phi(e_1)\cdot e_2 &= (1, -1)\cdot(0, 1) = -1, \\ e_1\cdot\phi(e_2) &= (1, 0)\cdot(1, 0) = 1. \end{aligned}$$

An important family of self-adjoint endomorphisms is illustrated in the following exercise.

Exercise 10.2.6 We know from Sect. 8.2 that, if $\mathcal{B} = (e_1, \dots, e_n)$ is an orthonormal basis for E^n, then the action of an endomorphism ϕ whose associated matrix is $\Phi = M_\phi^{\mathcal{B},\mathcal{B}}$ can be written with the Dirac's notation as

$$\phi = \sum_{a,b=1}^{n} \Phi_{ab} |e_a\rangle\langle e_b|,$$

with $\Phi_{ab} = \langle e_a|\phi(e_b)\rangle$. Then, the endomorphism ϕ is self-adjoint if and only if $\Phi_{ab} = \Phi_{ba}$. Consider vectors $u = (u_1, \dots, u_n)_\mathcal{B}$, $v = (v_1, \dots, v_n)_\mathcal{B}$ in E^n, and define the operator $L = |u\rangle\langle v|$. We have

$$\begin{aligned}\langle e_a|Le_b\rangle &= \langle e_a|u\rangle\langle v|e_b\rangle = u_a v_b, \\ \langle e_b|Le_a\rangle &= \langle e_b|u\rangle\langle v|e_a\rangle = u_b v_a,\end{aligned}$$

so we conclude that the operator $L = |u\rangle\langle v|$ is self-adjoint if and only if $u = v$.

Exercise 10.2.7 Let ϕ be a self-adjoint endomorphism of the euclidean space E^n, and let the basis $\mathcal{B} = (e_1, \dots, e_n)$ made of orthonormal eigenvectors for ϕ with corresponding eigenvalues $(\lambda_1, \dots, \lambda_n)$ (not necessarily all distinct). A direct computation shows that, in the Dirac's notation, the action of ϕ can be written as

$$\phi = \lambda_1|e_1\rangle\langle e_1| + \cdots + \lambda_n|e_n\rangle\langle e_n|,$$

so that, for any $v \in E^n$, one writes

$$\phi(v) = \lambda_1|e_1\rangle\langle e_1|v\rangle + \cdots + \lambda_n|e_n\rangle\langle e_n|v\rangle.$$

10.3 Orthogonal Projections

As we saw in Chap. 3, given any vector subspace $W \subset E^n$, with orthogonal complement $W^\perp$ we have a direct sum decomposition $E^n = W \oplus W^\perp$, so for any vector $v \in E^n$ we have (see the Proposition 3.2.5) a unique decomposition $v = v_W + v_{W^\perp}$. This suggests the following definition.

Definition 10.3.1 Given the (canonical) euclidean space E^n with $W \subset E^n$ a vector subspace and the orthogonal sum decomposition $v = v_W + v_{W^\perp}$, the map

$$P_W : E^n \to E^n, \qquad v \mapsto u_W$$

is linear, and it is called the *orthogonal projection onto the subspace* W. The dimension of W is called the *rank* of the orthogonal projection P_W.

If $W \subset E^n$ it is easy to see that $\mathrm{Im}(P_W) = W$ while $\ker(P_W) = W^\perp$. Moreover, since P_W acts as an identity operator on its range W, one also has $P_W^2 = P_W$. If

u, v are vectors in E^n, with orthogonal sum decomposition $u = u_W + u_{W^\perp}$ and $v = v_W + v_{W^\perp}$, we can explicitly compute

$$\begin{aligned} P_W(u) \cdot v &= u_W \cdot (v_W + v_{W^\perp}) \\ &= u_W \cdot v_W \qquad\qquad \text{and} \\ u \cdot P_W(v) &= (u_W + u_{W^\perp}) \cdot v_W \\ &= u_W \cdot v_W. \end{aligned}$$

This shows that orthogonal projectors are self-adjoint endomorphisms. To which extent can one reverse these computations, that is can one characterise, within all self-adjoint endomorphisms, the collection of orthogonal projectors? This is the content of the next proposition.

Proposition 10.3.2 *Given the euclidean vector space E^n, an endomorphism $\phi \in \mathrm{End}(E^n)$ is an orthogonal projection if and only if it is self-adjoint and satisfies the condition $\phi^2 = \phi$.*

Proof We have already shown that the conditions are necessary for an endomorphism to be an orthogonal projection in E^n. Let us now assume that ϕ is a self-adjoint endomorphism fulfilling $\phi^2 = \phi$. For any choice of $u, v \in E^n$ we have

$$\begin{aligned} ((1-\phi)(u)) \cdot \phi(v) &= u \cdot \phi(v) - \phi(u) \cdot \phi(v) \\ &= u \cdot \phi(v) - u \cdot \phi^2(v) \\ &= u \cdot \phi(v) - u \cdot \phi(v) = 0 \end{aligned}$$

with the second line coming from the self-adjointness of ϕ and the third line from the condition $\phi^2 = \phi$. This shows that the vector subspace $\mathrm{Im}(1-\phi)$ is orthogonal to the vector subspace $\mathrm{Im}(\phi)$. We can then decompose any vector $y \in E^n$ as an orthogonal sum $y = y_{\mathrm{Im}(1-\phi)} + y_{\mathrm{Im}\phi} + \xi$, where ξ is an element in the vector subspace orthogonal to the sum $\mathrm{Im}(1-\phi) \oplus \mathrm{Im}(\phi)$. For any $u \in E^n$ and any such vector ξ we have

$$\phi(u) \cdot \xi = 0, \qquad ((1-\phi)(u)) \cdot \xi = 0.$$

These conditions give that $u \cdot \xi = 0$ for any $u \in E^n$, so we can conclude that $\xi = 0$. Thus we have the orthogonal vector space decomposition

$$E^n = \mathrm{Im}(1-\phi) \oplus \mathrm{Im}(\phi).$$

We show next that $\ker(\phi) = \mathrm{Im}(1-\phi)$. If $u \in \mathrm{Im}(1-\phi)$, we have $u = (1-\phi)v$ with $v \in E^n$, thus $\phi(u) = \phi(1-\phi)v = 0$, that is $\mathrm{Im}(1-\phi) \subseteq \ker(\phi)$. Conversely, if $u \in \ker(\phi)$, then $\phi(u) \cdot v = 0$ for any $v \in E^n$, and $u \cdot \phi(v) = 0$, since ϕ is self-adjoint, which gives $\ker(\phi) \subseteq (\mathrm{Im}(\phi))^\perp$ and $\ker(\phi) \subseteq \mathrm{Im}(1-\phi)$, from the decomposition of E^n above.

If $w \in \mathrm{Im}(\phi)$, then $w = \phi(x)$ for a given $x \in E^n$, thus $\phi(w) = \phi^2(x) = \phi(x) = w$. We have shown that we can identify $\phi = P_{\mathrm{Im}(\phi)}$. This concludes the proof. □

Exercise 10.3.3 Consider the three dimensional euclidean space E^3 with canonical basis and take $W = \mathcal{L}((1, 1, 1))$. Its orthogonal subspace is given by the vectors (x, y, z) whose components solve the linear equation $\Sigma : x + y + z = 0$, so we get $S_\Sigma = W^\perp = \mathcal{L}((1, -1, 0), (1, 0, -1))$. The vectors of the canonical basis when expressed with respect to the vectors $u_1 = (1, 1, 1)$ spanning W and $u_2 = (1, -1, 0)$, $u_3 = (1, 0, -1)$ spanning $W^\perp$, are written as

$$\begin{aligned} e_1 &= \frac{1}{3}\,(u_1 + u_2 + u_3), \\ e_2 &= \frac{1}{3}\,(u_1 - 2u_2 + u_3), \\ e_3 &= \frac{1}{3}\,(u_1 + u_2 - 2u_3). \end{aligned}$$

Therefore,

$$P_W(e_1) = \frac{1}{3}\,u_1, \qquad P_W(e_2) = \frac{1}{3}\,u_1, \qquad P_W(e_3) = \frac{1}{3}\,u_1,$$

and

$$P_{W^\perp}(e_1) = \frac{1}{3}\,(u_2 + u_3), \qquad P_{W^\perp}(e_2) = \frac{1}{3}\,(-2u_2 + u_3), \qquad P_{W^\perp}(e_3) = \frac{1}{3}\,(u_2 - 2u_3).$$

Remark 10.3.4 Given an orthogonal space decomposition $E^n = W \oplus W^\perp$, the union of the basis $\mathcal{B}_W$ and $\mathcal{B}_{W^\perp}$ of W and $W^\perp$, is a basis $\mathcal{B}$ for W. It is easy to see that the matrix associated to the orthogonal projection operator P_W with respect to such a basis $\mathcal{B}$ has a block diagonal structure

$$M_{P_W}^{\mathcal{B},\mathcal{B}} = \begin{pmatrix} 1 & \cdots & 0 & 0 & \cdots & 0 \\ \vdots & & \vdots & \vdots & & \vdots \\ 0 & \cdots & 1 & 0 & \cdots & 0 \\ 0 & \cdots & 0 & 0 & \cdots & 0 \\ \vdots & & \vdots & \vdots & & \vdots \\ 0 & \cdots & 0 & 0 & \cdots & 0 \end{pmatrix},$$

where the order of the diagonal identity block is the dimension of $W = \mathrm{Im}(P_W)$. This makes it evident that an orthogonal projection operator is diagonalisable: its spectrum contains the real eigenvalue $\lambda = 1$ with multiplicity equal to $m_{\lambda=1} = \dim(W)$ and the real eigenvalue $\lambda = 0$ with multiplicity equal to $m_{\lambda=0} = \dim(W^\perp)$.

It is clear that the rank of P_W (the dimension of W) is given by the trace $\mathrm{tr}(M_{P_W}^{\mathcal{B},\mathcal{B}})$ irrespectively of the basis chosen to represent the projection (see the Proposition 9.1.5) since as usual, for a change of basis with matrix $M^{\mathcal{B},\mathcal{C}}$, one has that $M^{\mathcal{C},\mathcal{C}}_{P_W} = M^{\mathcal{C},\mathcal{B}} M^{\mathcal{B},\mathcal{B}}_{P_W} M^{\mathcal{B},\mathcal{C}}$, with $M^{\mathcal{B},\mathcal{C}} = (M^{\mathcal{C},\mathcal{B}})^{-1}$.

Exercise 10.3.5 The matrix

$$M = \begin{pmatrix} a & \sqrt{a-a^2} \\ \sqrt{a-a^2} & 1-a \end{pmatrix}$$

is symmetric and satisfies $M^2 = M$ for any $a \in (0, 1]$. With respect to an orthonormal basis (e_1, e_2) for E^2, it is then associated to an orthogonal projection with rank given by $\mathrm{tr}(M) = 1$. In order to determine its range, we diagonalise M. Its characteristic polynomial is

$$p_M(T) = \det(M - T I_2) = T^2 - T$$

and the eigenvalues are then $\lambda = 0$ and $\lambda = 1$. Since they are both simple, the matrix M is diagonalisable. The eigenspace $V_{\lambda=1}$ corresponding to the range of the orthogonal projection is one dimensional and given as the solution (x, y) of the system

$$\begin{pmatrix} a-1 & \sqrt{a-a^2} \\ \sqrt{a-a^2} & -a \end{pmatrix} \begin{pmatrix} x \\ y \end{pmatrix} = \begin{pmatrix} 0 \\ 0 \end{pmatrix},$$

that is $(x, \sqrt{\frac{1-a}{a}}x)$ with $x \in \mathbb{R}$. This means that the range of the projection is given by $\mathcal{L}((1, \sqrt{\frac{1-a}{a}}))$.

We leave as an exercise to show that M is the most general rank 1 orthogonal projection in E^2.

Exercise 10.3.6 We know from Exercise 10.2.6 that the operator $L = |u\rangle\langle u|$ is self-adjoint. We compute

$$L^2 = |u\rangle\langle u|u\rangle\langle u| = \|u\|^2 L.$$

Thus such an operator L is an orthogonal projection if and only if $\|u\| = 1$. It is then the rank one orthogonal projection $L = |u\rangle\langle u| = P_{\mathcal{L}(u)}$.

Let us assume that W_1 and W_2 are two orthogonal subspaces (to be definite we take $W_2 \subseteq W_1^{\perp}$). By using for instance the Remark 10.3.4 it is not difficult to show that $P_{W_1} P_{W_2} = P_{W_2} P_{W_1} = 0$. As a consequence,

$$(P_{W_1} + P_{W_2})(P_{W_1} + P_{W_2}) = P^2_{W_1} + P^2_{W_2} + P_{W_1} P_{W_2} + P_{W_2} P_{W_1} = (P_{W_1} + P_{W_2}).$$

Since the sum of two self-adjoint endomorphisms is self-adjoint, we can conclude (from Proposition 10.3.2) that the sum $P_{W_1} + P_{W_2}$ is an orthogonal projector, with $P_{W_1} + P_{W_2} = P_{W_1 \oplus W_2}$. This means that with two orthogonal subspaces, the sum of the corresponding orthogonal projectors is the orthogonal projection onto the direct sum of the given subspaces.

These results can be extended. If the euclidean space has a finer orthogonal decomposition, that is there are mutually orthogonal subspaces $\{W_a\}_{a=1,\dots,k}$ with $E^n = W_1 \oplus \cdots \oplus W_k$, then we have a corresponding set of orthogonal projectors P_{W_a}. We omit the proof of the following proposition, which we shall use later on in the chapter.

Proposition 10.3.7 *If $E^n = W_1 \oplus \cdots \oplus W_k$ with mutually orthogonal subspaces W_a, $a = 1, \dots, k$, then the following hold:*

(a) For any $a, b = 1, \dots, k$, one has

$$P_{W_a} P_{W_b} = \delta_{ab} P_{W_a}.$$

(b) If $\widetilde{W} = W_{a_1} \oplus \cdots \oplus W_{a_s}$ is the vector subspace given by the direct sum of the orthogonal subspaces $\{W_{a_j}\}$ with a_j any subset of $(1, \dots, k)$ without repetition, then the sum $\widetilde{P} = P_{W_{a_1}} + \dots + P_{W_{a_s}}$ is the orthogonal projection operator $\widetilde{P} = P_{\widetilde{W}}$.

(c) For any $v \in E^n$, one has

$$v = (P_{W_1} + \cdots + P_{W_k})(v).$$

Notice that point (c) shows that the identity operator acting on E^n can be *decomposed* as the sum of *all* the orthogonal projectors corresponding to *any* orthogonal subspace decomposition of E^n.

Remark 10.3.8 All we have described for the euclidean space E^n can be naturally extended to the hermitian space $(\mathbb{C}^n, \cdot)$ introduced in Sect. 3.4. If for example $(e_1, \dots, e_n)$ gives a hermitian orthonormal basis for H^n, the orthogonal projection onto $W_a = \mathcal{L}(e_a)$ can be written in the Dirac's notation (see the Exercise 10.3.6) as

$$P_{W_a} = |e_a\rangle\langle e_a|,$$

while the orthogonal projection onto $\widetilde{W} = W_{a_1} \oplus \cdots \oplus W_{a_s}$ (point b) of the Proposition 10.3.7) as

$$P_{\widetilde{W}} = |e_{a_1}\rangle\langle e_{a_1}| + \cdots + |e_{a_s}\rangle\langle e_{a_s}|.$$

The decomposition of the identity operator can be now written as

$$\mathrm{id}_{H^n} = |e_1\rangle\langle e_1| + \cdots + |e_n\rangle\langle e_n|.$$

Thus, any vector $v \in H^n$ can be decomposed as

$$v = |v\rangle = |e_1\rangle\langle e_1|v\rangle + \cdots + |e_n\rangle\langle e_n|v\rangle.$$

10.4 The Diagonalization of Self-adjoint Endomorphisms

The following theorem is a central result for the diagonalization of real symmetric matrices.

Theorem 10.4.1 *Let $A \in \mathbb{R}^{n,n}$ be symmetric, ${}^tA = A$. Then, any root of its characteristic polynomial $p_A(T)$ is real.*

Proof Let us assume λ to be a root of $p_A(T)$. Since $p_A(T)$ has real coefficients, its roots are in general complex (see the fundamental theorem of algebra, Theorem A.5.7). We therefore think of A as the matrix associate to an endomorphism

$$\phi : \mathbb{C}^n \longrightarrow \mathbb{C}^n,$$

with $M_{\phi}^{\mathcal{E},\mathcal{E}} = A$ with respect to the canonical basis $\mathcal{E}$ for $\mathbb{C}^n$ as a complex vector space. Let v be a non zero eigenvector for ϕ, that is

$$\phi(v) = \lambda v.$$

By denoting with X the column of the components of $v = (x_1, \ldots, x_n)$ with respect to $\mathcal{E}$, we write

$${}^tX = {}^t(x_1, \ldots, x_n), \qquad AX = \lambda X.$$

Under complex conjugation, with $\bar{A} = A$ since A has real entries, we get

$${}^t\bar{X} = {}^t(\bar{x}_1, \ldots, \bar{x}_n), \qquad A\bar{X} = \bar{\lambda}\bar{X}.$$

From these relations we can write the scalar ${}^t\bar{X}AX$ in the following two ways,

$$\begin{aligned} &{}^t\bar{X}AX = {}^t\bar{X}(AX) = {}^t\bar{X}(\lambda X) = \lambda\,({}^t\bar{X}X) \\ \text{and} \quad &{}^t\bar{X}AX = ({}^t\bar{X}A)X = {}^t(A\bar{X})X = {}^t(\bar{\lambda}\bar{X})X = \bar{\lambda}\,({}^t\bar{X}X). \end{aligned}$$

By equating them, we have

$$(\lambda - \bar{\lambda})\,({}^t\bar{X}X) = 0.$$

The quantity ${}^t\bar{X}X = \bar{x}_1x_1 + \bar{x}_2x_2 + \cdots + \bar{x}_nx_n$ is a positive real number, since $v \neq 0_{\mathbb{C}^n}$; we can then conclude $\lambda = \bar{\lambda}$, that is $\lambda \in \mathbb{R}$. □

Example 10.4.2 The aim of this example is threefold, namely

- it provides an *ad hoc* proof of the Theorem 10.4.1 for symmetric 2×2 matrices;
- it provides a direct proof for the Proposition 10.2.2 for symmetric 2×2 matrices;
- it shows that, if ϕ is a self-adjoint endomorphism in E^2, then E^2 has an orthonormal basis made up of eigenvectors for ϕ. This result anticipates the general result which will be proven in the Theorem 10.4.5.

We consider then a symmetric matrix $A \in \mathbb{R}^{2,2}$,

$$A = \begin{pmatrix} a_{11} & a_{12} \\ a_{12} & a_{22} \end{pmatrix}.$$

Its characteristic polynomial $p_A(T) = \det(A - T I_2)$ is then

$$p_A(T) = T^2 - (a_{11} + a_{22})T + a_{11}a_{22} - a_{12}^2.$$

The discriminant of this degree 2 characteristic polynomial $p_A(T)$ is not negative:

$$\begin{aligned} \Delta &= (a_{11} + a_{22})^2 - 4(a_{11}a_{22} - a_{12}^2) \\ &= (a_{11} - a_{22})^2 + 4a_{12}^2 \geq 0 \end{aligned}$$

being the sum of two square terms; therefore the roots λ_1, λ_2 of $p_A(T)$ are both real.

We prove next that A is diagonalisable, and that the matrix P giving the change of basis is orthogonal. We consider the endomorphism ϕ corresponding to $A = M_\phi^{\mathcal{E},\mathcal{E}}$, for the canonical basis $\mathcal{E}$ for E^2, and compute the eigenspaces V_{λ_1} and V_{λ_2}.

- If $\Delta = 0$, then $a_{11} = a_{22}$ and $a_{12} = 0$. The matrix A is already diagonal, so we may take $P = I_2$. There is only one eigenvalue $\lambda_1 = a_{11} = a_{22}$. Its algebraic multiplicity is 2 and its geometric multiplicity is 2, with corresponding eigenspace $V_{\lambda_1} = E^2$.
- If $\Delta > 0$ the characteristic polynomial has two simple roots $\lambda_1 \neq \lambda_2$ with corresponding one dimensional orthogonal (from the Proposition 10.2.2) eigenspaces V_{λ_1} and V_{λ_2}. The change of basis matrix P, whose columns are the normalised eigenvectors

$$\frac{v_1}{\|v_1\|} \quad \text{and} \quad \frac{v_2}{\|v_2\|},$$

  is then orthogonal by construction. We notice that P can be always chosen to be an element in SO(2), since a permutation of its columns changes the sign of its determinant, and is compatible with the permutation of the eigenvalue λ_1, λ_2 in the diagonal matrix.

In order to explicitly compute the matrix P we see that the eigenspace V_{λ_i} for any $i = 1, 2$ is given by the solutions of the linear homogeneous system associated to the matrix

$$A - \lambda_i I_2 = \begin{pmatrix} a_{11} - \lambda_i & a_{12} \\ a_{12} & a_{22} - \lambda_i \end{pmatrix}.$$

Since we already know that $\dim(V_{\lambda_i}) = 1$, such a linear system is equivalent to a single linear equation. We can write

$$\begin{aligned} V_{\lambda_i} &= \{(x, y) : (a_{11} - \lambda_i)x + a_{12}y = 0\} \\ &= \mathcal{L}((-a_{12}, a_{11} - \lambda_i)) = \mathcal{L}(v_i), \end{aligned}$$

where we set

$$v_1 = (-a_{12}, a_{11} - \lambda_1), \qquad v_2 = (-a_{12}, a_{11} - \lambda_2).$$

For the scalar product,

$$v_1 \cdot v_2 = a_{12}^2 + a_{11}^2 - (\lambda_1 + \lambda_2)a_{11} + \lambda_1\lambda_2 = 0$$

since one has

$$\lambda_1 + \lambda_2 = a_{11} + a_{22}, \qquad \lambda_1\lambda_2 = a_{11}a_{22} - a_{12}^2.$$

Exercise 10.4.3 We consider again the symmetric matrix

$$A = \begin{pmatrix} 2 & -1 \\ -1 & 3 \end{pmatrix}$$

from the Exercise 10.2.4. Its characteristic polynomial is

$$p_A(T) = \det(A - TI_2) = p_A(T) = T^2 - 5T + 5,$$

with roots

$$\lambda_{\pm} = \frac{1}{2}(5 \pm \sqrt{5}).$$

The corresponding eigenspaces $V_{\pm}$ are the solutions of the homogeneous linear systems associated to the matrices

$$A - \lambda_{\pm} I_2 = \frac{1}{2}\begin{pmatrix} (-1 \mp \sqrt{5}) & -2 \\ -2 & (1 \pm \sqrt{5}) \end{pmatrix}.$$

one has $\dim(V_{\pm}) = 1$, so each system is equivalent to a single linear equation, that is

$$V_{\pm} = \mathcal{L}((-2, 1 \pm \sqrt{5}) = \mathcal{L}(v_{\pm}),$$

where

$$v_+ = (-2, 1 + \sqrt{5}), \qquad v_- = (-2, 1 - \sqrt{5}),$$

and one computes that

$$v_+ \cdot v_- = 4 - 4 = 0,$$

that is the eigenspaces are orthogonal. The elements

$$u_1 = \frac{v_+}{\|v_+\|} \quad \text{and} \quad u_2 = \frac{v_-}{\|v_-\|}$$

form an orthonormal basis for E^2 of eigenvectors for the endomorphism ϕ_A.

We present now the fundamental result of this chapter, that is the *spectral theorem* for self-adjoint endomorphisms and for symmetric matrices. Towards this, it is worth mentioning that the whole theory, presented in this chapter for the euclidean space E^n, can be naturally formulated for any finite dimensional real vector space equipped with a scalar product (see Chap. 3).

Definition 10.4.4 Let $\phi : V \to V$ be an endomorphism of the real vector space V, and let $\widetilde{V} \subset V$ be a vector subspace in V. If the image of $\widetilde{V}$ for ϕ is a subset of the same $\widetilde{V}$ (that is, $\phi(\widetilde{V}) \subseteq \widetilde{V}$), there is a well defined endomorphism $\phi_{\widetilde{V}} : \widetilde{V} \to \widetilde{V}$ given by

$$\phi_{\widetilde{V}}(v) = \phi(v), \qquad \text{for all} \quad v \in \widetilde{V}$$

(clearly a linear map). The endomorphism $\phi_{\widetilde{V}}$ acts in the same way as the endomorphism ϕ, but on a restricted domain. This is why $\phi_{\widetilde{V}}$ is called the *restriction* to $\widetilde{V}$ of ϕ.

Proposition 10.4.5 (Spectral theorem for endomorphisms) *Let* $(V, \cdot)$ *be a real vector space equipped with a scalar product, and let* $\phi \in \mathrm{End}(V)$. *The endomorphism* ϕ *is self-adjoint if and only if* V *has an orthonormal basis of eigenvectors for* ϕ.

Proof Let us assume the orthonormal basis $\mathcal{C}$ for V is made of eigenvectors for ϕ. This implies that $M_\phi^{\mathcal{C},\mathcal{C}}$ is diagonal and therefore symmetric. From the Theorem 10.2.3 we conclude that ϕ is self-adjoint.

The proof of the converse is by induction on $n = \dim(V)$. For $n = 2$ the statement is true, as we explicitly proved in the Example 10.4.2. Let us then assume it to be true for any $(n-1)$-dimensional vector space. Then, let us consider a real n-dimensional vector space $(V, \cdot)$ equipped with a scalar product, and let ϕ be a self-adjoint endomorphism on V. With $\mathcal{B}$ an orthonormal basis for V (remember from the Theorem 3.3.9 that such a basis always exists V finite dimensional), the matrix $A = M_\phi^{\mathcal{B},\mathcal{B}}$ is symmetric (from the Theorem 10.2.3) and thus any root of the characteristic polynomial $p_A(T)$ is real. Denote by λ one such an eigenvalue for ϕ, with v_1 a corresponding eigenvector that we can assume of norm 1.

Then, let us consider the orthogonal complement to the vector line spanned by v_1,

$$\widetilde{V} = (\mathcal{L}(v_1))^{\perp}.$$

In order to meaningfully define the restriction to $\widetilde{V}$ of ϕ, we have to verify that for any $v \in \widetilde{V}$ one has $\phi(v) \in \widetilde{V}$, that is, we have to prove the implication

$$v \cdot v_1 = 0 \quad \Rightarrow \quad \phi(v) \cdot v_1 = 0.$$

By recalling that ϕ is self-adjoint and $\phi(v_1) = \lambda v_1$ we can write

$$\begin{aligned}\phi(v) \cdot v_1 &= v \cdot \phi(v_1) = v \cdot (\lambda v_1) \\ &= \lambda\,(v \cdot v_1) = 0.\end{aligned}$$

This proves that ϕ can be restricted to a $\phi_{\widetilde{V}} : \widetilde{V} \to \widetilde{V}$, clearly self-adjoint. Since $\dim(\widetilde{V}) = n - 1$, by the inductive assumption there exist $n - 1$ elements $(v_2, \ldots, v_n)$ of eigenvectors for $\phi_{\widetilde{V}}$ making up an orthonormal basis for $\widetilde{V}$. Since $\phi_{\widetilde{V}}$ is a restriction of ϕ, the elements $(v_2, \ldots, v_n)$ are eigenvectors for ϕ as well, and orthogonal to v_1 as they all belong to $\widetilde{V}$. Then the elements $(v_1, v_2, \ldots, v_n)$ are orthonormal and eigenvectors for ϕ. Being $n = \dim(V)$, they are an orthonormal basis for V. □

10.5 The Diagonalization of Symmetric Matrices

There is a counterpart of Proposition 10.4.5 for symmetric matrices.

Proposition 10.5.1 (Spectral theorem for symmetric matrices) *Let $A \in \mathbb{R}^{n,n}$ be symmetric. There exists an orthogonal matrix P such that ${}^tP\,A\,P$ is diagonal. This result is often referred to by saying that* symmetric matrices *are* orthogonally diagonalisable.

Proof Let us consider the endomorphism $\phi = f_A^{\mathcal{E},\mathcal{E}} : E^n \to E^n$, which is self-adjoint since A is symmetric and $\mathcal{E}$ is the canonical basis (see the Theorem 10.2.3). From the Proposition 10.4.5, the space E^n has an orthonormal basis $\mathcal{C}$ of eigenvectors for ϕ, so the matrix $M_\phi^{\mathcal{C},\mathcal{C}}$ is diagonal. From Theorem 7.9.9 we can write

$$M_\phi^{\mathcal{C},\mathcal{C}} = M^{\mathcal{C},\mathcal{E}}\, M_\phi^{\mathcal{E},\mathcal{E}}\, M^{\mathcal{E},\mathcal{C}}.$$

Since $M_\phi^{\mathcal{E},\mathcal{E}} = A$, by setting $P = M^{\mathcal{C},\mathcal{E}}$ we have that $P^{-1}AP$ is diagonal. The columns of the matrix P are given by the components with respect to $\mathcal{E}$ of the elements in $\mathcal{C}$, so P is orthogonal since $\mathcal{C}$ is orthonormal. □

Remark 10.5.2 The orthogonal matrix P can always be chosen in $\mathrm{SO}(n)$, since, as already mentioned, the sign of its determinant changes under a permutation of two columns.

Exercise 10.5.3 Consider $\phi \in \mathrm{End}(\mathbb{R}^4)$ given by

$$\phi((x, y, z, t)) = (x + y, x + y, -z + t, z - t).$$

Its corresponding matrix with respect to the canonical basis $\mathcal{E}$ in $\mathbb{R}^4$ is given by

$$A = M_{\phi}^{\mathcal{E},\mathcal{E}} = \begin{pmatrix} 1 & 1 & 0 & 0 \\ 1 & 1 & 0 & 0 \\ 0 & 0 & -1 & 1 \\ 0 & 0 & 1 & -1 \end{pmatrix}.$$

Being A symmetric and $\mathcal{E}$ orthonormal, than ϕ is self-adjoint. Its characteristic polynomial is

$$\begin{aligned} p_{\phi}(T) = p_A(T) &= \det(A - T\, I_4) \\ &= \begin{vmatrix} 1-T & 1 \\ 1 & 1-T \end{vmatrix} \begin{vmatrix} -1-T & 1 \\ 1 & -1-T \end{vmatrix} \\ &= T^2(T-2)(T+2). \end{aligned}$$

The eigenvalues are then $\lambda_1 = 0$ with (algebraic) multiplicity $m(0) = 2$, $\lambda_2 = -2$ with $m(-2) = 1$ and $\lambda_2 = 2$ with $m(2) = 1$. The corresponding eigenspaces are computed to be

$$\begin{aligned} V_0 &= \ker(\phi) = \mathcal{L}((1, -1, 0, 0), (0, 0, 1, 1)), \\ V_{-2} &= \ker(\phi - 2I_4) = \mathcal{L}((1, 1, 0, 0)), \\ V_2 &= \ker(\phi - I_4) = \mathcal{L}((0, 0, 1, -1)) \end{aligned}$$

and as we expect, these three eigenspaces are mutually orthogonal, with the two basis vectors spanning V_0 orthogonal as well. In order to write the matrix P which diagonalises A one just needs to normalise such a system of four basis eigenvectors. We have

$$P = \frac{1}{\sqrt{2}} \begin{pmatrix} 1 & 0 & 1 & 0 \\ -1 & 0 & 1 & 0 \\ 0 & 1 & 0 & 1 \\ 0 & 1 & 0 & -1 \end{pmatrix}, \qquad {}^t P A P = \begin{pmatrix} 0 & 0 & 0 & 0 \\ 0 & 0 & 0 & 0 \\ 0 & 0 & -2 & 0 \\ 0 & 0 & 0 & 2 \end{pmatrix},$$

where we have chosen an ordering for the eigenvalues that gives $\det(P) = 1$.

Corollary 10.5.4 *Let $\phi \in \mathrm{End}(E^n)$. If the endomorphism ϕ is self-adjoint then it is simple.*

Proof The proof is immediate. From the Proposition 10.4.5 we know that the self-adjointness of ϕ implies that E^n has an orthonormal basis of eigenvectors for ϕ. From the Remark 9.2.3 we conclude that ϕ is simple. □

Exercise 10.5.5 The converse of the previous corollary does not hold in general. Consider for example the endomorphism ϕ in E^2 whose matrix with respect to the canonical basis $\mathcal{E}$ is

$$A = \begin{pmatrix} 1 & 1 \\ 0 & -1 \end{pmatrix}.$$

An easy calculation gives for the eigenvalues $\lambda_1 = 1$ e $\lambda_2 = -1$ and ϕ is (see the Corollary 9.4.2) therefore simple. But ϕ is not self-adjoint, since

$$\begin{aligned} \phi(e_1) \cdot e_2 &= (1, 0) \cdot (0, 1) = 0, \\ e_1 \cdot \phi(e_2) &= (1, 0) \cdot (1, -1) = 1, \end{aligned}$$

or simply because A is not symmetric. The eigenspaces are given by

$$V_1 = \mathcal{L}((1, 0)), \qquad V_{-1} = \mathcal{L}((1, -2)),$$

and they are not orthogonal. As a further remark, notice that the diagonalising matrix

$$P = \begin{pmatrix} 1 & 1 \\ 0 & -2 \end{pmatrix}$$

is not orthogonal.

What we have shown in the previous exercise is a general property characterising self-adjoint endomorphisms within the class of simple endomorphisms, as the next theorem shows.

Theorem 10.5.6 *Let $\phi \in \mathrm{End}(E^n)$ be simple, with $V_{\lambda_1}, \ldots, V_{\lambda_s}$ the corresponding eigenspaces. Then ϕ is self-adjoint if and only if $V_{\lambda_i} \perp V_{\lambda_j}$ for any $i \neq j$.*

Proof That the eigenspaces corresponding to distinct eigenvalues are orthogonal for a self-adjoint endomorphism comes directly from the Proposition 10.2.2.

Conversely, let us assume that ϕ is simple, so that $E^n = V_{\lambda_1} \oplus \cdots \oplus V_{\lambda_s}$. The union of the bases given by applying the Gram-Schmidt orthogonalisation procedure to an arbitrary basis for each V_{λ_j}, yield an orthonormal basis for E^n, which is clearly made of eigenvectors for ϕ. The statement then follows from the Proposition 10.4.5. □

Exercise 10.5.7 The aim of this exercise is to define (if possible) a self-adjoint endomorphism $\phi : E^3 \to E^3$ such that $\ker(\phi) = \mathcal{L}((1, 2, 1))$ and $\lambda_1 = 1$, $\lambda_2 = 2$ are eigenvalues of ϕ.

Since $\ker(\phi) \neq \{(0, 0, 0)\}$, then $\lambda_3 = 0$ is the third eigenvalue for ϕ, with $\ker(\phi) = V_0$. Thus ϕ is simple since it has three distinct eigenvalues, with $E^3 = V_1 \oplus V_2 \oplus V_0$. In order for ϕ to be self-adjoint, we have to impose that $V_{\lambda_i} \perp V_{\lambda_j}$, for all $i \neq j$. In particular, one has

$$(\ker(\phi))^\perp = (V_0)^\perp = V_1 \oplus V_2.$$

We compute

$$\begin{aligned}(\ker(\phi))^{\perp} &= (\mathcal{L}((1,2,1))^{\perp} \\ &= \{(\alpha, \beta, -\alpha - 2\beta) \ : \ \alpha, \beta \in \mathbb{R}\} \\ &= \mathcal{L}((1,0,-1), (a,b,c))\end{aligned}$$

where we impose that (a, b, c) belongs to $\mathcal{L}((1, 2, 1))^{\perp}$ and is orthogonal to $(1, 0, -1)$. By setting

$$\begin{cases} (1,2,1) \cdot (a,b,c) = 0 \\ (1,0,-1) \cdot (a,b,c) = 0 \end{cases},$$

we have $(a, b, c) = (1, -1, 1)$, so we select

$$V_1 = \mathcal{L}((1,0,-1)), \qquad V_2 = \mathcal{L}((1,-1,1)).$$

Having a simple ϕ with mutually orthogonal eigenspaces, the endomorphism ϕ self-adjoint. To get a matrix representing ϕ we can choose the basis in E^3

$$\mathcal{B} = ((1,0,-1), (1,-1,1), (1,2,1)),$$

thus obtaining

$$M_{\phi}^{\mathcal{B},\mathcal{B}} = \begin{pmatrix} 1 & 0 & 0 \\ 0 & 2 & 0 \\ 0 & 0 & 0 \end{pmatrix}.$$

By defining $e_1 = (1, 0, -1)$, $e_2 = (1, -1, 1)$ we can write, in the Dirac's notation,

$$\phi = |e_1\rangle\langle e_1| + 2|e_2\rangle\langle e_2|.$$

Exercise 10.5.8 This exercise defines a simple, but not self-adjoint, endomorphism $\phi : E^3 \to E^3$ such that $\ker(\phi) = \mathcal{L}((1, -1, 1))$ and $\mathrm{Im}(\phi) = (\ker(\phi))^{\perp}$.

We know that ϕ has the eigenvalue $\lambda_1 = 0$ with $V_0 = \ker(\phi)$. For ϕ to be simple, the algebraic multiplicity of the eigenvalue λ_1 must be 1, and there have to be two additional eigenvalues λ_2 and λ_3 with either $\lambda_2 = \lambda_3$ or $\lambda_2 \neq \lambda_3$. If $\lambda_2 = \lambda_3$, one has then

$$V_{\lambda_2} = \mathrm{Im}(f) = (\ker(f))^{\perp} = (V_{\lambda_1})^{\perp}.$$

In such a case, ϕ would be a simple endomorphism with mutually orthogonal eigenspaces for distinct eigenvalues. This would imply ϕ to be self-adjoint. Thus to satisfy the conditions we require for ϕ we need $\lambda_2 \neq \lambda_3$. In such a case, one has $V_{\lambda_2} \oplus V_{\lambda_3} = \mathrm{Im}(\phi)$ and also clearly $V_{\lambda_i} \perp V_0$ for $i = 2, 3$. In order for ϕ to be simple but not self-adjoint, we select the eigenspaces V_{λ_2} and V_{λ_3} to be not mutually orthogonal subspaces in $\mathrm{Im}(f)$. Since

$$\begin{aligned}\mathrm{Im}(f) &= (\mathcal{L}((1,-1,1)))^{\perp} \\ &= \{(x,y,z) \,:\, x-y+z=0\} \\ &= \mathcal{L}((1,1,0),(0,1,1))\end{aligned}$$

we can choose

$$V_{\lambda_2} = \mathcal{L}((1,1,0)), \qquad V_{\lambda_3} = \mathcal{L}((0,1,1)).$$

If we set $\mathcal{B} = \big((1,-1,1),(1,1,0),(0,1,1)\big)$ (clearly not an orthonormal basis for E^3), we have

$$M_{\phi}^{\mathcal{B},\mathcal{B}} = \begin{pmatrix} 0 & 0 & 0 \\ 0 & \lambda_2 & 0 \\ 0 & 0 & \lambda_3 \end{pmatrix}.$$

Exercise 10.5.9 Consider the endomorphism $\phi : E^3 \to E^3$ whose corresponding matrix with respect to the basis $\mathcal{B} = \big(v_1 = (1,1,0), v_2 = (1,-1,0), v_3 = (0,0,-1)\big)$ is

$$M_{\phi}^{\mathcal{B},\mathcal{B}} = \begin{pmatrix} 1 & 0 & 0 \\ 0 & 2 & 0 \\ 0 & 0 & 3 \end{pmatrix}.$$

With $\mathcal{E}$ as usual the canonical basis for E^3, in this exercise we would like to determine:

(1) an orthonormal basis $\mathcal{C}$ for E^3 given by eigenvectors for ϕ,
(2) the orthogonal matrix $M^{\mathcal{E},\mathcal{C}}$,
(3) the matrix $M^{\mathcal{C},\mathcal{E}}$,
(4) the matrix $M_{\phi}^{\mathcal{E},\mathcal{E}}$,
(5) the eigenvalues of ϕ with their corresponding multiplicities.

(1) We start by noticing that, since $M_{\phi}^{\mathcal{B},\mathcal{B}}$ is diagonal, the basis $\mathcal{B}$ is given by eigenvectors of ϕ, as the action of ϕ on the basis vectors in $\mathcal{B}$ can be clearly written as $\phi(v_1) = v_1$, $\phi(v_2) = 2v_2$, $\phi(v_3) = 3v_3$. The basis $\mathcal{B}$ is indeed orthogonal, but not orthonormal, and for an orthonormal basis $\mathcal{C}$ of eigenvectors for ϕ we just need to normalize, that is to consider

$$u_1 = \frac{v_1}{\|v_1\|}, \qquad u_2 = \frac{v_2}{\|v_2\|}, \qquad u_3 = \frac{v_3}{\|v_3\|}$$

just obtaining $\mathcal{C} = (\frac{1}{\sqrt{2}}(1,1,0), \frac{1}{\sqrt{2}}(1,-1,0), (0,0,-1))$. While the existence of such a basis $\mathcal{C}$ implies that ϕ is self-adjoint, the self-adjointness of ϕ could not be derived from the matrix $M_{\phi}^{\mathcal{B},\mathcal{B}}$, which is symmetric with respect to a basis $\mathcal{B}$ which is not orthonormal.

(2) From its definition, the columns of $M^{\mathcal{E},\mathcal{C}}$ are given by the components with respect to $\mathcal{E}$ of the vectors in $\mathcal{C}$. We then have

$$M^{\mathcal{E},\mathcal{C}} = \frac{1}{\sqrt{2}} \begin{pmatrix} 1 & 1 & 0 \\ 1 & -1 & 0 \\ 0 & 0 & -\sqrt{2} \end{pmatrix}.$$

(3) We know that $M^{\mathcal{C},\mathcal{E}} = \left(M^{\mathcal{E},\mathcal{C}}\right)^{-1}$. Since the matrix above is orthogonal, we have

$$M^{\mathcal{C},\mathcal{E}} = {}^{t}\left(M^{\mathcal{E},\mathcal{C}}\right) = \frac{1}{\sqrt{2}} \begin{pmatrix} 1 & 1 & 0 \\ 1 & -1 & 0 \\ 0 & 0 & -\sqrt{2} \end{pmatrix}.$$

(4) From the Theorem 7.9.9 we have

$$M_{\phi}^{\mathcal{E},\mathcal{E}} = M^{\mathcal{E},\mathcal{C}} \, M_{\phi}^{\mathcal{C},\mathcal{C}} \, M^{\mathcal{C},\mathcal{E}}.$$

Since $M_{\phi}^{\mathcal{C},\mathcal{C}} = M_{\phi}^{\mathcal{B},\mathcal{B}}$, the matrix $M_{\phi}^{\mathcal{E},\mathcal{E}}$ can be now directly computed.

(5) Clearly, from $M_{\phi}^{\mathcal{B},\mathcal{B}}$ the eigenvalues for ϕ are all simple and given by $\lambda = 1, 2, 3$.

Chapter 11
Rotations

The notion of *rotation* appears naturally in physics, and is geometrically formulated in terms of a euclidean structure as a suitable linear map on a real vector space. The aim of this chapter is to analyse the main properties of rotations using the spectral theory previously developed, as well as to recover known results from classical mechanics, using the geometric language we are describing.

11.1 Skew-Adjoint Endomorphisms

In analogy to the Definition 10.2.1 of a self-adjoint endomorphism, we have the following.

Definition 11.1.1 An endomorphism ϕ of the euclidean vector space E^n is called *skew-adjoint* if

$$\phi(v) \cdot w = -\, v \cdot \phi(w), \qquad \text{for all} \quad v, w \in E^n.$$

From the Definition 4.1.7 we call a matrix $A = (a_{ij}) \in \mathbb{R}^{n,n}$ *skew-symmetric* (or *anti-symmetric*) if ${}^tA = -A$, that is if $a_{ij} = -a_{ji}$, for any i, j. Notice that the skew-symmetry condition for A clearly implies for its diagonal elements that $a_{ii} = 0$. The following result is an analogous of the Theorem 10.2.3 and can be established in a similar manner.

Theorem 11.1.2 *Let $\phi \in \mathrm{End}(E^n)$ and $\mathcal{B}$ an orthonormal basis for E^n. The endomorphism ϕ is skew-adjoint if and only if $M_\phi^{\mathcal{B},\mathcal{B}}$ is skew-symmetric.*

G. Landi and A. Zampini, *Linear Algebra and Analytic Geometry for Physical Sciences*, Undergraduate Lecture Notes in Physics,
https://doi.org/10.1007/978-3-319-78361-1_11

Proposition 11.1.3 *Let $\phi \in \mathrm{End}(E^n)$ be skew-adjoint. It holds that*

(a) the euclidean vector space E^n has an orthogonal decomposition

$$E^n = \mathrm{Im}(\phi) \oplus \ker(\phi),$$

(b) the rank of ϕ is even.

Proof (a) Let $u \in E^n$ and $v \in \ker(\phi)$. We can write

$$0 = u \cdot \phi(v) = -\phi(u) \cdot v.$$

Since this is valid for any $u \in E^n$, the element $\phi(u)$ ranges over the whole space $\mathrm{Im}(\phi)$, so we have that $\ker(\phi) = (\mathrm{Im}(\phi))^{\perp}$.

(b) From ${}^tM_{\phi}^{\mathcal{B},\mathcal{B}} = -M_{\phi}^{\mathcal{B},\mathcal{B}}$, it follows $\det(M_{\phi}^{\mathcal{B},\mathcal{B}}) = (-1)^n \det(M_{\phi}^{\mathcal{B},\mathcal{B}})$. Thus a skew-adjoint endomorphism on an odd dimensional euclidean space is singular (that is it is not invertible). From the orthogonal decomposition for E^n of point (a) we conclude that the restriction $\widetilde{\phi}_{\mathrm{Im}(\phi)} : \mathrm{Im}(\phi) \to \mathrm{Im}(\widetilde{\phi})$ is regular (that is it is invertible). Since such a restriction is skew-adjoint, we have that $\dim(\mathrm{Im}(\phi)) = \dim(\mathrm{Im}(\widetilde{\phi})) = \mathrm{rk}(\phi)$ is even. □

A skew-adjoint endomorphism ϕ on E^n can have only the zero as (real) eigenvalue, so it is not diagonalisable. Indeed, if λ is an eigenvalue for ϕ, that is $\phi(v) = \lambda v$ for $v \neq 0_{E^n} \in E^n$, from the skew-symmetry condition we have that $0 = v \cdot \phi(v) = \lambda\, v \cdot v$, which implies $\lambda = 0$. Also, since its characteristic polynomial has non real roots, it does not have a Jordan form (see Theorem 9.5.1).

Although not diagonalisable, a skew-adjoint endomorphism has nonetheless a canonical form.

Proposition 11.1.4 *Given a skew-adjoint invertible endomorphism $\phi : E^{2p} \to E^{2p}$, there exists an orthonormal basis $\mathcal{B}$ for E^{2p} with respect to which the representing matrix for ϕ is of the form,*

$$M_{\phi}^{\mathcal{B},\mathcal{B}} = \begin{pmatrix} 0 & \mu_1 & & & & \\ -\mu_1 & 0 & & & & \\ & & \ddots & & & \\ & & & \ddots & & \\ & & & & 0 & \mu_p \\ & & & & -\mu_p & 0 \end{pmatrix}$$

with $\mu_j \in \mathbb{R}$ for $j = 1, \ldots, p$.

Proof The map $S = \phi^2 = \phi \circ \phi$ is a self-adjoint endomorphism on E^{2p}, so there exists an orthonormal basis of eigenvectors for S. Given $S(w_j) = \lambda_j w_j$ with $\lambda_j \in \mathbb{R}$, each eigenvalue λ_j has even multiplicity, since the identity

$$S(\phi(w_i)) = \phi(S(w_i)) = \lambda_i \phi(w_i)$$

shows that w_i and $\phi(w_i)$ are eigenvectors of S with the same eigenvalue λ_i. We label then the spectrum of S by $(\lambda_1, \ldots, \lambda_k)$ and the basis $\mathcal{C} = \big(w_1, \phi(w_1), \ldots, w_k, \phi(w_k)\big)$. We also have

$$\lambda_i = w_i \cdot S(w_i) = w_i \cdot \phi^2(w_i) = -\phi(w_i) \cdot \phi(w_i) = -\|\phi(w_i)\|^2$$

and, since we took ϕ to be invertible, we have $\lambda_i < 0$. Define the set $\mathcal{B} = (e_1, \ldots, e_{2p})$ of vectors as

$$e_{2j-1} = w_j, \qquad e_{2j} = \frac{1}{\sqrt{|\lambda_j|}}\,\phi(w_j)$$

for $j = 1, \ldots, p$. A direct computation shows that $e_j \cdot e_k = \delta_{jk}$ with $j, k = 1, \ldots, 2p$ and

$$\phi(e_{2j-1}) = \sqrt{|\lambda_j|}\, e_{2j}, \qquad \phi(e_{2j}) = -\sqrt{|\lambda_j|}\, e_{2j-1}.$$

Thus $\mathcal{B}$ is an orthonormal basis with respect to which the matrix representing the endomorphism ϕ has the form above, with $\mu_j = \sqrt{|\lambda_j|}$. □

Corollary 11.1.5 *If ϕ is a skew-adjoint endomorphism on E^n, then there exists an orthonormal basis $\mathcal{B}$ for E^n with respect to which the associated matrix of ϕ has the form*

$$M_\phi^{\mathcal{B},\mathcal{B}} = \begin{pmatrix} 0 & \mu_1 & & & & & & & \\ -\mu_1 & 0 & & & & & & & \\ & & \ddots & & & & & & \\ & & & \ddots & & & & & \\ & & & & 0 & \mu_p & & & \\ & & & & -\mu_p & 0 & & & \\ & & & & & & 0 & & \\ & & & & & & & \ddots & \\ & & & & & & & & 0 \end{pmatrix},$$

with $\mu_j \in \mathbb{R}$, $j = 1, \cdots, p$, and $2p \le n$.

The study of antisymmetric matrices makes it natural to introduce the notion of Lie algebra.

Definition 11.1.6 Given $A, B \in \mathbb{R}^{n,n}$, one defines the map $[\ ,\ ] : \mathbb{R}^{n,n} \times \mathbb{R}^{n,n} \to \mathbb{R}^{n,n}$,

$$[A, B] = AB - BA$$

as the *commutator* of A and B. Using the properties of the matrix product is it easy to prove that the following hold, for any $A, B, C \in \mathbb{R}^{n,n}$ and any $\alpha \in \mathbb{R}$:

(1) $[A, B] = -[B, A]$, $\quad [\alpha A, B] = \alpha[A, B]$, $\quad [A + B, C] = [A, C] + [B, C]$, that is the commutator is bilinear and antisymmetric,
(2) $[AB, C] = A[B, C] + [A, C]B$,
(3) $[A, [B, C]] + [B, [C, A]] + [C, [A, B]] = 0$; this is called the *Jacoby identity*.

Definition 11.1.7 If $W \subseteq \mathbb{R}^{n,n}$ is a vector subspace such that the commutator maps $W \times W$ into W, we say that W is a *(matrix) Lie algebra*. Its rank is the dimension of W as a vector space.

Excercise 11.1.8 The collection of all antisymmetric matrices $W_A \subset \mathbb{R}^{n,n}$ is a matrix Lie algebra since, if ${}^tA = -A$ and ${}^tB = -B$ it is

$${}^t([A, B]) = {}^tB\,{}^tA - {}^tA\,{}^tB = BA - AB.$$

As a Lie algebra, it is denoted $\mathfrak{so}(n)$ and one easily computed its dimension to be $n(n-1)/2$. As we shall see, this Lie algebra has a deep relation with the orthogonal group $\mathrm{SO}(n)$.

Remark 11.1.9 It is worth noticing that the vector space $W_S \subset \mathbb{R}^{n,n}$ of symmetric matrices is *not* a matrix Lie algebra, since the commutator of two symmetric matrices is an antisymmetric matrix.

Excercise 11.1.10 It is clear that the matrices

$$L_1 = \begin{pmatrix} 0 & 0 & 0 \\ 0 & 0 & -1 \\ 0 & 1 & 0 \end{pmatrix}, \qquad L_2 = \begin{pmatrix} 0 & 0 & 1 \\ 0 & 0 & 0 \\ -1 & 0 & 0 \end{pmatrix}, \qquad L_3 = \begin{pmatrix} 0 & -1 & 0 \\ 1 & 0 & 0 \\ 0 & 0 & 0 \end{pmatrix}$$

provide a basis for the three dimensional real vector space of antisymmetric matrices $W_A \subset \mathbb{R}^{3,3}$. As the matrix Lie algebra $\mathfrak{so}(3)$, one computes the commutators:

$$[L_1, L_2] = L_3, \qquad [L_2, L_3] = L_1, \qquad [L_3, L_1] = L_2.$$

Excercise 11.1.11 We consider the most general skew-adjoint endomorphism ϕ on E^3. With respect to the canonical orthonormal basis $\mathcal{E} = (e_1, e_2, e_3)$ it has associated matrix of the form

$$M_\phi^{\mathcal{E},\mathcal{E}} = \begin{pmatrix} 0 & -\gamma & \beta \\ \gamma & 0 & -\alpha \\ -\beta & \alpha & 0 \end{pmatrix} = \alpha L_1 + \beta L_2 + \gamma L_3.$$

with $\alpha, \beta, \gamma \in \mathbb{R}$. Any vector (x, y, z) in its kernel is a solution of the system

$$\begin{pmatrix} 0 & -\gamma & \beta \\ \gamma & 0 & -\alpha \\ -\beta & \alpha & 0 \end{pmatrix} \begin{pmatrix} x \\ y \\ z \end{pmatrix} = \begin{pmatrix} 0 \\ 0 \\ 0 \end{pmatrix}.$$

It is easy to show that the kernel is one-dimensional with $\ker(\phi) = \mathcal{L}((\alpha, \beta, \gamma))$. Since ϕ is defined on a three dimensional space and has a one-dimensional kernel, from the Proposition 11.1.4 the spectrum of the map $S = \phi^2$ is made of the simple eigenvalue $\lambda_0 = 0$ and a multiplicity 2 eigenvalue $\lambda < 0$, which is such that $2\lambda = \text{tr}(M_S^{\mathcal{E},\mathcal{E}})$ with

$$M_S^{\mathcal{E},\mathcal{E}} = \begin{pmatrix} 0 & -\gamma & \beta \\ \gamma & 0 & -\alpha \\ -\beta & \alpha & 0 \end{pmatrix} \begin{pmatrix} 0 & -\gamma & \beta \\ \gamma & 0 & -\alpha \\ -\beta & \alpha & 0 \end{pmatrix} = \begin{pmatrix} -\gamma^2 - \beta^2 & \alpha\beta & \alpha\gamma \\ \alpha\beta & -\gamma^2 - \alpha^2 & \beta\gamma \\ \alpha\gamma & \beta\gamma & -\beta^2 - \alpha^2 \end{pmatrix};$$

thus $\lambda = -(\alpha^2 + \beta^2 + \gamma^2)$. For the corresponding eigenspace $V_\lambda \ni (x, y, x)$ one has

$$\begin{pmatrix} \alpha^2 & \alpha\beta & \alpha\gamma \\ \alpha\beta & \beta^2 & \beta\gamma \\ \alpha\gamma & \beta\gamma & \gamma^2 \end{pmatrix} \begin{pmatrix} x \\ y \\ z \end{pmatrix} = \begin{pmatrix} 0 \\ 0 \\ 0 \end{pmatrix} \quad \Leftrightarrow \quad \begin{cases} \alpha(\alpha x + \beta y + \gamma z) = 0 \\ \beta(\alpha x + \beta y + \gamma z) = 0 \\ \gamma(\alpha x + \beta y + \gamma z) = 0 \end{cases}.$$

Such a linear system is equivalent to the single equation $(\alpha x + \beta y + \gamma z) = 0$, which shows that $\ker(S)$ is orthogonal to $\text{Im}(S)$. To be definite, assume $\alpha \neq 0$, and fix as basis for V_λ

$$\begin{aligned} w_1 &= (-\gamma, 0, \alpha), \\ w_2 &= \phi(w_1) = (-\alpha\beta, \alpha^2 + \gamma^2, -\beta\gamma), \end{aligned}$$

with $w_1 \cdot \phi(w_1) = 0$. With the appropriate normalization, we define

$$\begin{aligned} u_1 &= \frac{w_1}{\|w_1\|}, \\ u_2 &= \frac{\phi(w_1)}{\|\phi(w_1)\|}, \\ u_3 &= \frac{1}{\sqrt{\alpha^2 + \beta^2 + \gamma^2}} (\alpha, \beta, \gamma) \end{aligned}$$

and verify that $\mathcal{C} = (u_1, u_2, u_3)$ is an orthonormal basis for E^3. With $M^{\mathcal{C},\mathcal{E}}$ the orthogonal matrix of change of bases (see the Theorem 7.9.9), this leads to

$$M^{\mathcal{C},\mathcal{E}} M_\phi^{\mathcal{E},\mathcal{E}} M^{\mathcal{E},\mathcal{C}} = M_\phi^{\mathcal{C},\mathcal{C}} = \begin{pmatrix} 0 & -\rho & 0 \\ \rho & 0 & 0 \\ 0 & 0 & 0 \end{pmatrix}, \qquad \rho = |\lambda| = \alpha^2 + \beta^2 + \gamma^2,$$

an example indeed of Corollary 11.1.5.

11.2 The Exponential of a Matrix

In Sect. 10.1 we studied the properties of the orthogonal group $\mathrm{O}(n)$ in E^n. Before studying the spectral properties of orthogonal matrices we recall some general results.

Definition 11.2.1 Given a matrix $A \in \mathbb{R}^{n,n}$, its *exponential* is the matrix e^A defined by

$$e^A = \sum_{k=0}^{\infty} \frac{1}{k!} A^k$$

where the sum is defined component-wise, that is $(e^A)_{jl} = \sum_{k=0}^{\infty} \frac{1}{k!} (A^k)_{jl}$.

We omit the proof that such a limit exists (that is each series converges) for every matrix A, and we omit as well the proof of the following proposition, which lists several properties of the exponential maps on matrices.

Proposition 11.2.2 *Given matrices $A, B \in \mathbb{R}^{n,n}$ and an invertible matrix $P \in \mathrm{GL}(n)$, the following identities hold:*

(a) $e^A \in \mathrm{GL}(n)$, that is the matrix e^A is invertible, with $(e^A)^{-1} = e^{-A}$ and $\det(e^A) = e^{\mathrm{tr}\, A}$,
(b) if $A = \mathrm{diag}(a_{11}, \dots, a_{nn})$, then $e^A = \mathrm{diag}(e^{a_{11}}, \dots, e^{a_{nn}})$,
(c) $e^{PAP^{-1}} = Pe^A P^{-1}$,
(d) if $AB = BA$, that is $[A, B] = 0$, then $e^A e^B = e^B e^A = e^{A+B}$,
(e) it is $e^{{}^tA} = {}^t(e^A)$,
(f) if $W \subset \mathbb{R}^{n,n}$ is a matrix Lie algebra, the elements e^M with $M \in W$ form a group with respect to the matrix product.

Excercise 11.2.3 Let as determine the exponential e^Q of the symmetric matrix

$$Q = \begin{pmatrix} 0 & a \\ a & 0 \end{pmatrix}, \quad a \in \mathbb{R}.$$

We can proceed in two ways. On the one hand, it is easy to see that

$$Q^{2k} = \begin{pmatrix} a^{2k} & 0 \\ 0 & a^{2k} \end{pmatrix}, \qquad Q^{2k+1} = \begin{pmatrix} 0 & a^{2k+1} \\ a^{2k+1} & 0 \end{pmatrix}.$$

Thus, by using the definition we compute

$$e^Q = \begin{pmatrix} \sum_{k=0}^{\infty} \frac{a^{2k}}{(2k)!} & \sum_{k=0}^{\infty} \frac{a^{2k+1}}{(2k+1)!} \\ \sum_{k=0}^{\infty} \frac{a^{2k+1}}{(2k+1)!} & \sum_{k=0}^{\infty} \frac{a^{2k}}{(2k)!} \end{pmatrix} = \begin{pmatrix} \cosh a & \sinh a \\ \sinh a & \cosh a \end{pmatrix}.$$

Alternatively, we can use the identities (c) and (b) in the previous proposition, once Q has been diagonalised. It is easy to compute the eigenvalues of Q to be $\lambda_{\pm} = \pm a$, with diagonalising orthogonal matrix $P = \frac{1}{\sqrt{2}}\begin{pmatrix} 1 & 1 \\ -1 & 1 \end{pmatrix}$. That is, $P\Delta_Q P^{-1} = Q$ with with $\Delta_Q = \mathrm{diag}(-a, a)$,

$$\tfrac{1}{2}\begin{pmatrix} 1 & 1 \\ -1 & 1 \end{pmatrix}\begin{pmatrix} -a & 0 \\ 0 & a \end{pmatrix}\begin{pmatrix} 1 & -1 \\ 1 & 1 \end{pmatrix} = \begin{pmatrix} 0 & a \\ a & 0 \end{pmatrix}.$$

We then compute

$$\begin{aligned} e^Q &= e^{P\Delta_Q P^{-1}} = Pe^{\Delta_Q}T^{-1} \\ &= \tfrac{1}{2}\begin{pmatrix} 1 & 1 \\ -1 & 1 \end{pmatrix}\begin{pmatrix} e^{-a} & 0 \\ 0 & e^{a} \end{pmatrix}\begin{pmatrix} 1 & -1 \\ 1 & 1 \end{pmatrix} = \begin{pmatrix} \cosh a & \sinh a \\ \sinh a & \cosh a \end{pmatrix}. \end{aligned}$$

Notice that $\det(e^Q) = \cosh^2 a - \sinh^2 a = 1 = e^{\mathrm{tr}\, Q}$.

Excercise 11.2.4 Let us determine the exponential e^M of the anti-symmetric matrix

$$M = \begin{pmatrix} 0 & a \\ -a & 0 \end{pmatrix}, \quad a \in \mathbb{R}.$$

Since M is not diagonalisable, we explicitly compute e^M as we did in the previous exercise, finding

$$M^{2k} = (-1)^k \begin{pmatrix} a^{2k} & 0 \\ 0 & a^{2k} \end{pmatrix}, \qquad M^{2k+1} = (-1)^k \begin{pmatrix} 0 & a^{2k+1} \\ -a^{2k+1} & 0 \end{pmatrix}.$$

By putting together all terms, one finds

$$e^M = \begin{pmatrix} \sum_{k=0}^{\infty}(-1)^k \frac{a^{2k}}{(2k)!} & \sum_{k=0}^{\infty}(-1)^k \frac{a^{2k+1}}{(2k+1)!} \\ -\sum_{k=0}^{\infty}(-1)^k \frac{a^{2k+1}}{(2k+1)!} & \sum_{k=0}^{\infty}(-1)^k \frac{a^{2k}}{(2k)!} \end{pmatrix} = \begin{pmatrix} \cos a & \sin a \\ -\sin a & \cos a \end{pmatrix}.$$

We see that if M is a 2×2 anti-symmetric matrix, the matrix e^M is special orthogonal. This is an example for the point (f) in the Proposition 11.2.2.

Excercise 11.2.5 In order to further explore the relations between anti-symmetric matrices and special orthogonal matrices, consider the matrix

$$M = \begin{pmatrix} 0 & a & 0 \\ -a & 0 & 0 \\ 0 & 0 & 0 \end{pmatrix} \quad a \in \mathbb{R}.$$

In parallel with the computations from the previous exercise, it is immediate to see that

$$e^M = \begin{pmatrix} \cos a & \sin a & 0 \\ -\sin a & \cos a & 0 \\ 0 & 0 & 1 \end{pmatrix}.$$

This hints to the conclusion that $e^M \in \mathrm{SO}(3)$ if $M \in \mathbb{R}^{3,3}$ is anti-symmetric.

The following proposition generalises the results of the exercises above, and provides a further example for the claim (f) from the Proposition 11.2.2, since the set $W_A \subset \mathbb{R}^{n,n}$ of antisymmetric matrices is the (matrix) Lie algebra $\mathfrak{so}(n)$, as shown in the exercise 11.1.8.

Proposition 11.2.6 *If $M \in \mathbb{R}^{n,n}$ is anti-symmetric, then e^M is special orthogonal. The restriction of the exponential map to the Lie algebra $\mathfrak{so}(n)$ of anti-symmetric matrices is surjective onto* $\mathrm{SO}(n)$.

Proof We focus on the first claim which follows from point (a) of Proposition 11.2.2. If $M \in \mathbb{R}^{n,n}$ is anti-symmetric, ${}^tM = -M$ and $\mathrm{tr}(M) = 0$. Thus ${}^t(e^M) = e^{{}^tM} = e^{-M} = (e^M)^{-1}$ and $\det(e^M) = e^{\mathrm{tr}(M)} = e^0 = 1$. □

Remark 11.2.7 As the Exercise 11.2.5 directly shows, the restriction of the exponential map to the Lie algebra $\mathfrak{so}(n)$ of anti-symmetric matrices is *not* injective into $\mathrm{SO}(n)$.

In the Example 11.3.1 below, we shall sees explicitly that the exponential map, when restricted to 2-dimensional anti-symmetric matrices, is indeed *surjective* onto the group $\mathrm{SO}(2)$.

11.3 Rotations in Two Dimensions

We study now spectral properties of orthogonal matrices. We start with the orthogonal group $\mathrm{O}(2)$.

Example 11.3.1 Let $A = \begin{pmatrix} a_{11} & a_{12} \\ a_{21} & a_{22} \end{pmatrix} \in \mathbb{R}^{2,2}$. The condition ${}^tAA = A\,{}^tA = I_2$ is equivalent to the conditions for its entries given by

$$\begin{aligned} a_{11}^2 + a_{12}^2 &= 1 \\ a_{21}^2 + a_{22}^2 &= 1 \\ a_{11}a_{21} + a_{12}a_{22} &= 0. \end{aligned}$$

To solve these equations, let us assume $a_{11} \neq 0$ (the case $a_{22} \neq 0$ is analogous). We have then $a_{21} = -(a_{12}a_{22})/a_{11}$ from the third equation while, from the others, we have

$$a_{22}^2 \left(\frac{a_{12}^2}{a_{11}^2} + 1 \right) = 1 \qquad \Rightarrow \qquad a_{22}^2 = a_{11}^2 .$$

There are two possibilities.

- If $a_{11} = a_{22}$, it follows that $a_{12} + a_{21} = 0$, so the matrix A can be written as

$$A_+ = \begin{pmatrix} a & b \\ -b & a \end{pmatrix} \qquad \text{with } a^2 + b^2 = 1,$$

  and $\det(A_+) = a^2 + b^2 = 1$. One can write $a = \cos\varphi$, $b = \sin\varphi$, for $\varphi \in \mathbb{R}$, so to get

$$A_+ = \begin{pmatrix} \cos\varphi & \sin\varphi \\ -\sin\varphi & \cos\varphi \end{pmatrix}.$$

- If $a_{11} = -a_{22}$, it follows that $a_{12} = a_{21}$, so the matrix A can be written as

$$A_- = \begin{pmatrix} a & b \\ b & -a \end{pmatrix} \qquad \text{with } a^2 + b^2 = 1,$$

  and we can write

$$A_- = \begin{pmatrix} \cos\varphi & \sin\varphi \\ \sin\varphi & -\cos\varphi \end{pmatrix}$$

  with $\det(A_-) = -a^2 + b^2 = -1$.

Finally, it is easy to see that $a_{11} = 0$ would imply $a_{22} = 0$ and $a_{12}^2 = a_{21}^2 = 1$. These four cases correspond to $\varphi = \pm\frac{\pi}{2}$ for A_+ or A_-, according to wether $a_{12} = -a_{21}$ or $a_{12} = a_{21}$ respectively.

We see that A_+ makes up the special orthogonal group SO(2), while A_- the orthogonal transformations in E^2 which in physics are usually called *improper* rotations.

Given the 2π-periodicity of the trigonometric functions, we see that any element in the special orthogonal group SO(2) corresponds bijectively to an angle $\varphi \in [0, 2\pi)$.

On the other hand, any improper orthogonal transformation can be factorised as the product of a SO(2) matrix times the matrix $Q = \mathrm{diag}(1, -1)$,

$$\begin{pmatrix} \cos\varphi & \sin\varphi \\ \sin\varphi & -\cos\varphi \end{pmatrix} = \begin{pmatrix} 1 & 0 \\ 0 & -1 \end{pmatrix} \begin{pmatrix} \cos\varphi & \sin\varphi \\ -\sin\varphi & \cos\varphi \end{pmatrix}.$$

Thus, an improper orthogonal transformation 'reverses' one of the axis of any given orthogonal basis for E^2 and so changes its orientation.

Remark 11.3.2 Being O(2) a group, the product of two improper orthogonal transformations is a special orthogonal transformation. We indeed compute

$$\begin{pmatrix} \cos\varphi & \sin\varphi \\ \sin\varphi & -\cos\varphi \end{pmatrix} \begin{pmatrix} \cos\varphi' & \sin\varphi' \\ \sin\varphi' & -\cos\varphi' \end{pmatrix} = \begin{pmatrix} \cos(\varphi'-\varphi) & \sin(\varphi'-\varphi) \\ -\sin(\varphi'-\varphi) & \cos(\varphi'-\varphi) \end{pmatrix} \in \mathrm{SO}(2).$$

Proposition 11.3.3 *A matrix $A \in \mathrm{SO}(2)$ is diagonalisable if and only if $A = \pm I_2$. An orthogonal matrix A with $\det(A) = -1$ is diagonalisable, with spectrum given by $\lambda = \pm 1$.*

Proof From the previous example we have:

(a) The eigenvalues λ for a special orthogonal matrix are given by the solutions of the equation

$$p_{A_+}(T) = (\cos\varphi - T)^2 + \sin^2\varphi = T^2 - 2(\cos\varphi)\,T + 1 = 0,$$

which are $\lambda_\pm = \cos\varphi \pm \sqrt{\cos^2\varphi - 1}$. This shows that A_+ is diagonalisable if and only if $\cos^2 = 1$, that is $A_+ = \pm I_2$.

(b) Improper orthogonal matrices A_- turn to be diagonalisable since they are symmetric. The eigenvalue equation is

$$p_{A_-} = (T + \cos\varphi)(T - \cos\varphi) - \sin^2\varphi = T^2 - 1 = 0,$$

giving $\lambda_\pm = \pm 1$. □

11.4 Rotations in Three Dimensions

We move to the analysis of rotations in three dimensional spaces.

Excercise 11.4.1 From the Exercise 11.1.11 we know that the anti-symmetric matrices in $\mathbb{R}^{3,3}$ form a three dimensional vector space, thus any anti-symmetric matrix M is labelled by a triple (α, β, γ) of real parameters. The vector $a = (\alpha, \beta, \gamma)$ is the generator, with respect the canonical basis $\mathcal{E}$ of E^3, of the kernel of the endomorphism ϕ associated to M with respect to the basis $\mathcal{E}$, $M = M_\phi^{\mathcal{E},\mathcal{E}}$.

Moreover, from the same exercise we know that there exists an orthogonal matrix P which reduces M to its canonical form (see Corollary 11.1.5), that is $M = P\,\mathfrak{a}_M P^{-1}$ with

$$\mathfrak{a}_M = \begin{pmatrix} 0 & -\rho & 0 \\ \rho & 0 & 0 \\ 0 & 0 & 0 \end{pmatrix}, \qquad \text{and} \quad \rho^2 = \alpha^2 + \beta^2 + \gamma^2,$$

with respect to an orthonormal basis $\mathcal{C}$ for E^3 such that $P = M^{\mathcal{E},\mathcal{C}}$, the matrix of change of basis. From the Exercise 11.2.5 it is

$$\mathrm{SO}(3) \ni e^{\mathfrak{a}_M} = \begin{pmatrix} \cos\rho & -\sin\rho & 0 \\ \sin\rho & \cos\rho & 0 \\ 0 & 0 & 1 \end{pmatrix}, \tag{11.1}$$

and, if $R = e^M$, from the Proposition 11.2.2 one has $R = Pe^{\mathfrak{a}_M}P^{-1}$.

The only real eigenvalue of the orthogonal transformation $e^{\mathfrak{a}_M}$ is then $\lambda = 1$, corresponding to the 1-dimensional eigenspace spanned by the vector $a = (\alpha, \beta, \gamma)$. The vector line $\mathcal{L}(a)$ is therefore left unchanged by the isometry ϕ of E^3 corresponding to the matrix R, that is such that $M_\phi^{\mathcal{E},\mathcal{E}} = R$.

From the Proposition 11.2.6 we know that given $R \in \mathrm{SO}(3)$, there exists an antisymmetric matrix $M \in \mathbb{R}^{3,3}$ such that $R = e^M$. The previous exercise gives then the proof of the following theorem.

Theorem 11.4.2 *For any matrix $R \in \mathrm{SO}(3)$ with $R \neq I_3$ there exists an orthonormal basis $\mathcal{B}$ on E^3 with respect to which the matrix R has the form* (11.1).

This theorem, that is associated with the name of Euler, can also be stated as follow:

Theorem 11.4.3 *Any special orthogonal matrix $R \in \mathrm{SO}(3)$ has the eigenvalue* $+1$.

Those isometries $\phi \in \mathrm{End}(E^3)$ whose representing matrices $M_\phi^{\mathcal{E},\mathcal{E}}$ with respect to an orthonormal basis $\mathcal{E}$ are special orthogonal are also called 3-dimensional *rotation* endomorphisms or rotations tout court. With a language used for the euclidean affine spaces (Chap. 15), we then have:

- For each rotation R of E^3 there exists a unique vector line (a direction) which is left unchanged by the action of the rotation. Such a vector line is called the *rotation axis*.
- The width of the rotation around the rotation axis is given by an angle ρ obtained from (11.1), and implicitly given by

$$1 + 2\cos\rho = \mathrm{tr}\begin{pmatrix} \cos\rho & -\sin\rho & 0 \\ \sin\rho & \cos\rho & 0 \\ 0 & 0 & 1 \end{pmatrix} = \mathrm{tr}\, P^{-1}RP = \mathrm{tr} R, \tag{11.2}$$

from the cyclic property of the trace.

Excercise 11.4.4 Consider the rotation of E^3 whose matrix $\mathcal{E} = (e_1, e_2.e_3)$ is

$$R = \begin{pmatrix} \cos\alpha & \sin\alpha & 0 \\ -\sin\alpha & \cos\alpha & 0 \\ 0 & 0 & 1 \end{pmatrix} \begin{pmatrix} 1 & 0 & 0 \\ 0 & \cos\beta & \sin\beta \\ 0 & -\sin\beta & \cos\beta \end{pmatrix}$$

with respect to the canonical basis. Such a matrix is the product $R = R_1 R_2$ of two special orthogonal matrices. The matrix R_1 is a rotation by an angle α with rotation axis the vector line $\mathcal{L}(e_1)$ and angular width α, while R_2 is a rotation by the angle β with rotation axis $\mathcal{L}(e_3)$. We wish to determine the rotation axis for R with corresponding angle. A direct calculation yields

$$R = \begin{pmatrix} \cos\alpha & \sin\alpha\cos\beta & \sin\alpha\sin\beta \\ -\sin\alpha & \cos\alpha\cos\beta & \cos\alpha\sin\beta \\ 0 & -\sin\beta & \cos\beta \end{pmatrix}.$$

Since $R \neq I_3$ for $\alpha \neq 0$ and $\beta \neq 0$, the rotation axis is given by the eigenspace corresponding to the eigenvalue $\lambda = 1$. This eigenspace is found to be spanned by the vector v with

$$v = \big(\sin\alpha(1-\cos\beta),\ (\cos\alpha - 1)(1-\cos\beta),\ \sin\beta(1-\cos\alpha)\big) \quad \text{if } \alpha \neq 0,\ \beta \neq 0,$$
$$v = (1, 0, 0) \quad \text{if } \alpha = 0,$$
$$v = (0, 0, 1) \quad \text{if } \beta = 0.$$

The rotation angle ρ can be obtained (implicitly) from the Eq. (11.2) as

$$1 + 2\cos\rho = \mathrm{tr}(R) = \cos\alpha + \cos\beta + \cos\alpha\cos\beta.$$

Excercise 11.4.5 Since the special orthogonal group $\mathrm{SO}(n)$ is non abelian for $n > 2$, for the special orthogonal matrix given by $R' = R_2 R_1$ one has $R' \neq R$. The matrix R' can be written as

$$R' = \begin{pmatrix} \cos\alpha & \sin\alpha & 0 \\ -\sin\alpha\cos\beta & \cos\alpha\cos\beta & \sin\beta \\ \sin\alpha\sin\beta & -\sin\beta\cos\alpha & \cos\beta \end{pmatrix}.$$

One now computes that while the rotation angle is the same as in the previous exercise, the rotation axis is spanned by the vector v' with

$$v' = \big(\sin\alpha\sin\beta,\ (1-\cos\alpha)\sin\beta,\ (1-\cos\alpha)(1+\cos\beta)\big) \quad \text{if } \alpha \neq 0,\ \beta \neq 0,$$
$$v' = (1, 0, 0) \quad \text{if } \alpha = 0,$$
$$v' = (0, 0, 1) \quad \text{if } \beta = 0.$$

Excercise 11.4.6 Consider the matrix $R'' = Q_1 Q_2$ given by

$$R'' = \begin{pmatrix} \cos\alpha & \sin\alpha & 0 \\ \sin\alpha & -\cos\alpha & 0 \\ 0 & 0 & 1 \end{pmatrix} \begin{pmatrix} 1 & 0 & 0 \\ 0 & \cos\beta' & \sin\beta' \\ 0 & \sin\beta' & -\cos\beta' \end{pmatrix}.$$

Now neither Q_1 nor Q_2 are (proper) rotation matrix: both Q_1 and Q_2 are in O(3), but $\det(Q_1) = \det(Q_2) = -1$ (see the Example 11.3.1, where O(2) has been described, and the Remark 11.3.2). The matrix R'' is nonetheless special orthogonal since O(3) is a group and $\det(R'') = 1$.

One finds that the rotation axis is the vector line spanned by the vector v'' with

$$\begin{aligned}
v'' &= \big(\sin\alpha\,\sin\beta',\ (1-\cos\alpha)\sin\beta',\ (1-\cos\alpha)(1-\cos\beta')\big) \qquad \text{if } \alpha \neq 0,\ \beta' \neq 0,\\
v'' &= (1,0,0) \qquad \text{if } \alpha = 0,\\
v'' &= (0,0,1) \qquad \text{if } \beta' = 0.
\end{aligned}$$

One way to establish this result without doing explicit computation, is to observe that R'' is obtained from R' in Exercise 11.4.5 under a transposition and the identification $\beta' = \pi - \beta$.

Excercise 11.4.7 As an easy application of the Theorem 10.1.13 we know that, if $\mathcal{B} = (u_1, u_2, u_3)$ and $\mathcal{C} = (v_1, v_2, v_3)$ are orthonormal bases in E^3, then the orthogonal endomorphism ϕ mapping $v_k \mapsto u_k$ is represented by a matrix whose entry Φ_{ab} is given by the scalar product $u_b \cdot v_a$

$$M_\phi^{\mathcal{C},\mathcal{C}} = \Phi = \Big(\Phi_{ab} = u_b \cdot v_a\Big)_{a,b=1,2,3}.$$

It is easy indeed to see that the matrix element $({}^t\Phi\,\Phi)_{ks}$ is given by

$$\sum_{a=1}^{3} \Phi_{ak}\Phi_{as} = \sum_{a=1}^{3}(u_a \cdot v_k)(u_a \cdot v_s) = v_k \cdot v_s = \delta_{ks}$$

thus proving that Φ is orthogonal. Notice that $M_\phi^{\mathcal{B},\mathcal{B}} = {}^t\Phi = \Phi^{-1}$.

Excercise 11.4.8 Let $\mathcal{E} = (e_1, e_2, e_3)$ be an orthonormal basis for E^3. We compute the rotation matrix corresponding to the change of basis $\mathcal{E} \to \mathcal{B}$ with $\mathcal{B} = (u_1, u_2, u_3)$ for any given basis $\mathcal{B}$ with the same orientation (see the Definition 10.1.15) of $\mathcal{E}$.

Firstly, consider a vector u of norm 1. Since such a vector defines a point on a sphere of radius 1 in the three dimensional physical space $\mathcal{S}$, which can be identified by a *latitude* and a *longitude*, its components with respect to $\mathcal{E}$ are determined by two angles. With respect to Figure 11.1 we write them as

$$u = (\sin\varphi\,\sin\theta, -\cos\varphi\,\sin\theta, \cos\theta)$$

with $\theta \in (0, \pi)$ and $\varphi \in [0, 2\pi)$. Then, to complete u to an orthonormal basis for E^3 with $u'_3 = u$, one finds,

$$\begin{aligned}
u'_1 &= u_N = (\cos\varphi, \sin\varphi, 0),\\
u'_2 &= (-\sin\varphi\,\cos\theta, \cos\varphi\,\cos\theta, \sin\theta).
\end{aligned}$$

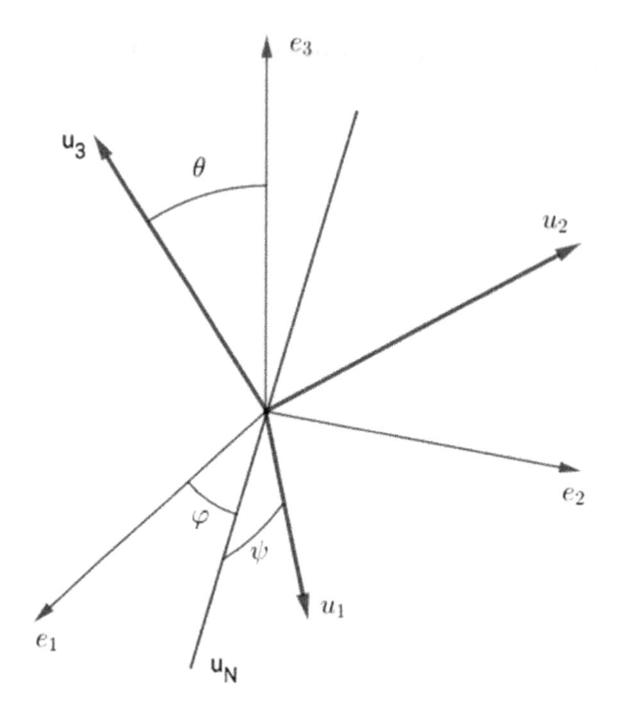

Fig. 11.1 The Euler angles

The rotation matrix (with respect to the basis $\mathcal{E}$) of the transformation $\mathcal{E} \to (u_1', u_2', u_3')$ is given by

$$R'(\theta, \varphi) = \begin{pmatrix} \cos\varphi & -\sin\varphi\cos\theta & \sin\varphi\sin\theta \\ \sin\varphi & \cos\varphi\cos\theta & -\cos\varphi\sin\theta \\ 0 & \sin\theta & \cos\theta \end{pmatrix}.$$

Since the choice of u_1', u_2' is unique up to a rotation around the orthogonal vector u, we see that the most general SO(3) rotation matrix mapping $e_3 \to u$ is given by

$$\begin{aligned} R(\theta, \varphi, \psi) &= R(\theta, \varphi) \begin{pmatrix} \cos\psi & -\sin\psi & 0 \\ \sin\psi & \cos\psi & 0 \\ 0 & 0 & 1 \end{pmatrix} \\ &= \begin{pmatrix} \cos\varphi\cos\psi - \sin\varphi\cos\theta\sin\psi & -\cos\varphi\sin\psi - \sin\varphi\cos\theta\cos\psi & \sin\varphi\cos\theta \\ \sin\varphi\cos\psi + \cos\varphi\cos\theta\sin\psi & -\sin\varphi\sin\psi + \cos\varphi\cos\theta\cos\psi & -\cos\varphi\sin\theta \\ \sin\theta\sin\psi & \sin\theta\cos\psi & \cos\theta \end{pmatrix} \end{aligned}$$

with $\psi \in [0, 2\pi)$. This shows that the proper 3-dimensional rotations, that is the group SO(3), can be parametrised by 3 angles. Such angles are usually called *Euler angles*, and clearly there exist several (consistent and equivalent) different choices for them.

Our result depends on the assumption that $\sin\theta \neq 0$, which means that $u_1 \neq \pm e_3$ (this corresponds to the case when u_1 is the north-south pole direction). The most general rotation matrix representing an orthogonal transformation with$e_1 \to u_1 = \pm e_3$ is given by

$$R(\psi) = \begin{pmatrix} 0 & \cos\psi & \mp\sin\psi \\ 0 & \sin\psi & \pm\cos\psi \\ \pm 1 & 0 & 0 \end{pmatrix}.$$

We finally remark that the rotation matrix $R(\theta, \varphi, \psi)$ can be written as the product

$$R(\theta, \varphi, \psi) = \begin{pmatrix} \cos\varphi & -\sin\varphi & 0 \\ \sin\varphi & \cos\varphi & 0 \\ 0 & 0 & 1 \end{pmatrix} \begin{pmatrix} 1 & 0 & 0 \\ 0 & \cos\theta & -\sin\theta \\ 0 & \sin\theta & \cos\theta \end{pmatrix} \begin{pmatrix} \cos\psi & -\sin\psi & 0 \\ \sin\psi & \cos\psi & 0 \\ 0 & 0 & 1 \end{pmatrix}.$$

This identity shows that we can write

$$R(\theta, \beta, \psi) = e^{\varphi L_3}\, e^{\theta L_1}\, e^{\psi L_3}.$$

where L_1 and L_3 are the matrices in Exercise 11.1.10. These matrices are the 'generators' of the rotations around the first and third axis, respectively.

In applications to the dynamics of a rigid body, with reference to the Figure 11.1, the angle φ parametrises the motion of *precession* of the axis u_3 around the axis e_3, the angle θ the motion of *nutation* of the axis u_3 and the angle φ the *intrinsic rotation* around the axis u_3. The unit vector u_N indicates the *line of nodes*, the intersection of the plane $(e_1 e_2)$ with the plane $(u_1 u_2)$.

We close this section by listing the most interesting properties of orthogonal endomorphisms in E^n with $n > 0$. Endomorphisms ϕ whose representing matrix $M_\phi^{\mathcal{E},\mathcal{E}}$ are special orthogonal, with respect to an orthonormal basis $\mathcal{E}$ for E^n, are called *rotations*. From the Proposition 11.2.6 we know that there exists an anti-symmetric matrix M such that $M_\phi^{\mathcal{E},\mathcal{E}} = e^M$. When $\mathrm{rk}(M) = 2k$, the matrix e^M depends on k angular variables.

From the Corollary 11.1.5 and a direct generalisation of the computations above, one can conclude that for each n-dimensional rotation:

- There exists a vector subspace $V \subset E^n$ which is left unchanged by the action of the rotation, with $\dim(V) = n - \mathrm{rk}(M)$.
- Since $\mathrm{rk}(M)$ is even, we have that, if n is odd, then V is odd dimensional as well, and at least one dimensional. If n is even and the matrix M is invertible, then V is the null space.

11.5 The Lie Algebra $\mathfrak{so}(3)$

We have a closer look at the Lie algebra $\mathfrak{so}(3)$ introduced in the Exercise 11.1.10. As mentioned, it is three dimensional and generated by the three matrices

$$L_1 = \begin{pmatrix} 0 & 0 & 0 \\ 0 & 0 & -1 \\ 0 & 1 & 0 \end{pmatrix}, \qquad L_2 = \begin{pmatrix} 0 & 0 & 1 \\ 0 & 0 & 0 \\ -1 & 0 & 0 \end{pmatrix}, \qquad L_3 = \begin{pmatrix} 0 & -1 & 0 \\ 1 & 0 & 0 \\ 0 & 0 & 0 \end{pmatrix},$$

which are closed under matrix commutator.

Consider the three dimensional euclidean *totally antisymmetric* Levi-Civita symbol $\varepsilon_{a_1a_2a_3}$ with indices $a_j = 1, 2, 3$ and defined by

$$\varepsilon_{a_1a_2a_3} = \begin{cases} +1 & \text{if } (a_1, a_2, a_3) \text{ is an even permutation of } (1, 2, 3) \\ -1 & \text{if } (a_1, a_2, a_3) \text{ is an odd permutation of } (1, 2, 3) \\ 0 & \text{if any two indices are equal} \end{cases}.$$

One has the identity $\sum_{a=1}^{3} \varepsilon_{abc}\epsilon_{aks} = (\delta_{bk}\delta_{cs} - \delta_{bs}\delta_{ck})$.

Excercise 11.5.1 Using the Levi-Civita symbol, it is easy to see that the generators L_a have components given by

$$(L_a)_{mn} = \varepsilon_{man},$$

while their commutators are written as

$$[L_m, L_n] = \sum_{a=1}^{3} \varepsilon_{mna} L_a.$$

There is an important subtlety when identifying 3×3 antisymmetric matrices with three dimensional vectors. The most general antisymmetric matrix in indeed characterised by three scalars,

$$A = \begin{pmatrix} 0 & -v_3 & v_2 \\ v_3 & 0 & -v_1 \\ -v_2 & v_1 & 0 \end{pmatrix} = \sum_{a=1}^{3} v_a L_a$$

For the time being, this only defines a triple of numbers (v_1, sv_2, v_3) in E^3. Whether this triple provides the components of a vector in the three dimensional euclidean space, will depend on how it transforms under an orthonormal transformation. Now, we may think of A as the matrix, with respect to the canonical orthonormal basis $\mathcal{E}$ of a skew-adjoint endomorphism ϕ on E^3: $A = M_{\phi}^{\mathcal{E},\mathcal{E}}$. When changing basis to an orthonormal basis $\mathcal{B}$ with matrix of change of basis $R = M^{\mathcal{E},\mathcal{B}} \in \mathrm{O}(3)$, the matrix A is transformed to

$$A' = RAR^{-1} = RA\,{}^tR,$$

since R is orthogonal and thus $R^{-1} = {}^tR$. Since A' is antisymmetric as well, it can be written as $A' = \sum_{a,b=1}^{3} v'_a L_a$ for some (v'_1, v'_2, v'_3). In order to establish the transformation rule from (v_1, v_2, v_3) to (v'_1, v'_2, v'_3), we need an additional result on orthogonal matrices.

Excercise 11.5.2 Using the expression in Sect. 5.3 for the inverse of an invertible matrix, the orthogonality condition for a matrix $R \in \mathrm{O}(3)$, that is $R_{ab} = ({}^tR)_{ba} = (R^{-1})_{ba}$, can be written as

$$R_{ab} = \frac{1}{\det R}(-1)^{a+b} \det(\widehat{R}_{ab}),$$

where $\widehat{R}_{ab}$ is the 2 dimensional matrix obtained by deleting the row a and the column b in the 3 dimensional matrix R. (Then $\det(\widehat{R}_{ab}$ is the minor of the element R_{ab}, see the Definition 5.1.7.) In terms of the Levi-Civita symbol this identity transform to

$$\sum_{j=1}^{3} \varepsilon_{mjn} R_{jq} = \frac{1}{\det R} \sum_{a,b=1}^{3} R_{ma} \varepsilon_{aqb} R_{nb}, \tag{11.3}$$

or, being ${}^tR$ orthogonal as well, with $\det R = \det {}^tR$,

$$\sum_{j=1}^{3} \varepsilon_{mjn} R_{qj} = \frac{1}{\det R} \sum_{a,b=1}^{3} R_{am} \varepsilon_{aqb} R_{bn}. \tag{11.4}$$

Going back to $A = \sum_{a=1}^{3} v_a L_a$ and $A' = \sum_{a,b=1}^{3} v'_a L_a$, we have for their components:

$$A_{mn} = \sum_{j=1}^{3} v_j\, \varepsilon_{mjn} \quad \text{and} \quad A'_{mn} = \sum_{j=1}^{3} v'_j\, \varepsilon_{mjn}.$$

We then compute, using the relation (11.3),

$$\begin{aligned}
A'_{mn} = (RA\,{}^tR)_{mn} &= \sum_{a,b=1}^{3} R_{ma} A_{ab} R_{nb} \\
&= \sum_{j=1}^{3} \sum_{a,b=1}^{3} v_j\, \varepsilon_{ajb}\, R_{ma} R_{nb} \\
&= (\det R) \sum_{j=1}^{3} \sum_{c=1}^{3} R_{cj} v_j\, \varepsilon_{mcn} = (\det R) \sum_{c=1}^{3} (Rv)_c\, \varepsilon_{mcn},
\end{aligned}$$

that is,

$$v'_j = (\det R)\,(Rv)_j = (\det R)\sum_{c=1}^{3} R_{cj} v_j .$$

This shows that, under an orthogonal transformation between different bases of E^3, the components of an antisymmetric matrix transforms as the components of a vector only if the orientation is preserved, that is only if the transformation is special orthogonal.

Using a terminology from physics, elements in E^3 whose components with respect to orthonormal basis transform as the general theory (see the Proposition 7.9.2) prescribes are called *polar* vectors (or vectors tout court), while elements in E^3 whose components transform as the components of an antisymmetric matrix are called *axial* (or *pseudo*) vectors.

An example of an axial vector is given by the vector product in E^3 of two (polar) vector, that we recall from the Chap. 1. To be definite, let us start with the canonical orthonormal basis $\mathcal{E}$. If $v = (v_1, v_2, v_3)$ and $w = (w_1, w_2, w_3)$, the Proposition 1.3.15 define the vector product of v and w as,

$$\tau(v, w) = v \wedge w = (v_2 w_3 - v_3 w_2,\ v_3 w_1 - v_1 w_3,\ v_1 w_2 - v_2 w_1).$$

Using the Levi-Civita symbol, the components are written as

$$(v \wedge w)_a = \sum_{b,c=1}^{3} \varepsilon_{abc}\, v_b\, w_c .$$

If $R = M^{\mathcal{E},\mathcal{B}} \in \mathrm{O}(3)$ is the change of basis to a new orthonormal basis $\mathcal{B}$ for E^3, on one hand we have $(v \wedge w)'_q = (R(v \wedge w))_q$ while the relation (11.4) yields,

$$\begin{aligned}(v' \wedge w')_q &= \sum_{k,j=1}^{3} \varepsilon_{qkj} v'_k w'_j = \sum_{k,j,b,s=1}^{3} \varepsilon_{qkj} R_{kb} R_{js} v_b w_s \\ &= (\det R) \sum_{a,b,s=1}^{3} R_{qa} \varepsilon_{abs}\, v_b w_s = (\det R)(v \wedge w)'_q .\end{aligned}$$

This shows that the components of a vector product transforms as an axial vector under an orthogonal transformation between different bases of E^3. In a similar manner one shows that the vector product of an axial vector with a polar vector, is a polar vector.

Excercise 11.5.3 For example, the change of basis from $\mathcal{B}$ to $\mathcal{B}' = (b'_1 = -b_1, b'_2 = -b_2, b'_3 = -b_3)$ is clearly represented by the matrix $M^{\mathcal{B},\mathcal{B}'} = M^{\mathcal{B}',\mathcal{B}} = -I_3$ which is orthogonal but not special orthogonal. It is immediate to see that we have

$v = (-v_1, -v_2, -v_3)_{\mathcal{B}'}$ and $w = (-w_1, -w_2, -w_3)_{\mathcal{B}'}$, but $v \wedge w = (v_2 w_3 - v_3 w_2, v_3 w_1 - v_1 w_3, v_1 w_2 - v_2 w_1)_{\mathcal{B}'}$.

From the Example 1.3.17 we see that, since the physical observables position, velocity, acceleration and force are described by polar vectors, both momenta and angular momenta for the dynamics of a point mass are axial vectors.

Excercise 11.5.4 We recall from Sect. 1.4 the action of the operator rot on a vector field $\mathbf{A}(\mathbf{x})$,

$$\text{rot}\,\mathbf{A} = \nabla \wedge \mathbf{A} = \sum_{i,j,k=1}^{3} (\varepsilon_{ijk} \partial_j A_k)\, e_i$$

with respect to an orthonormal basis $\mathcal{E} = (e_1, e_2, e_3)$ of E^3 which represents the physical space $\mathcal{S}$. This identity shows that, if $\mathbf{A}$ is a polar vector (field) then rot $\mathbf{A}$ is an axial vector (field).

Example 11.5.5 The (Lorentz) force $\mathbf{F}$ acting on a point electric charge q whose motion is given by $\mathbf{x}(t)$, in the presence of an electric field $\mathbf{E}(\mathbf{x})$ and a magnetic field $\mathbf{B}(\mathbf{x})$ is written as

$$\mathbf{F} = q(\mathbf{E} + \dot{\mathbf{x}} \wedge \mathbf{B}).$$

We conclude that $\mathbf{E}$ is a polar vector field, while $\mathbf{B}$ is an axial vector field. Indeed, the correct way to describe $\mathbf{B}$ is with an antisymmetric 3×3 matrix.

11.6 The Angular Velocity

When dealing with rotations in physics, an important notion is that of *angular velocity*. This and several related notions can be analysed in terms of the spectral properties of orthogonal matrices that we have illustrated above. It is worth recalling from Chap. 1 that euclidean vector spaces with orthonormal bases are the natural framework for the notion of cartesian orthogonal coordinate systems for the physical space $\mathcal{S}$ (*inertial reference frames*).

Example 11.6.1 Consider the motion $\mathbf{x}(t)$ in E^3 of a point mass such that its distance $\|\mathbf{x}(t)\|$ from the origin of the coordinate system is fixed. We then consider a fixed orthonormal basis $\mathcal{E} = (e_1, e_2, e_3)$, and a orthonormal basis $\mathcal{E}' = (e'_1(t), e'_2(t), e'_3(t))$ which *rotates* with respect to $\mathcal{E}$ in such a way that the components of $\mathbf{x}(t)$ along $\mathcal{E}'$ do not depend on time — the point mass is at *rest* with respect to $\mathcal{E}'$. We can write the position vector $\mathbf{x}(t)$ as

$$\mathbf{x}(t) = \sum_{a=1}^{3} x_a(t) e_a = \sum_{k=1}^{3} x'_k e'_k(t).$$

Since $\mathcal{E}'$ depends on time, the change of the basis is given by a time-dependent orthogonal matrix $M^{\mathcal{E},\mathcal{E}'(t)} = R(t) \in \mathrm{SO}(3)$ as

$$x_k(t) = \sum_{j=1}^{3} R_{kj}(t)x'_j.$$

By differentiating with respect to time t (recall that the dot means time derivative), with $\dot{x}'_j = 0$, the above relation gives,

$$\dot{x}_k = \sum_{a=1}^{3} \dot{R}_a x'_a = \sum_{a,b=1}^{3} \dot{R}_{ka}(R^{-1})_{ab}x_b = \sum_{a,b=1}^{3} \dot{R}_{ka}({}^tR)_{ab}x_b.$$

From the relation $R(t)\,{}^tR(t) = I_3$ it follows that, by differentiating with respect to t,

$$\begin{aligned} \dot{R}\,{}^tR + R\,{}^t\dot{R} = 0 \quad &\Rightarrow \\ &\dot{R}\,{}^tR = -R\,({}^t\dot{R}) = -\,{}^t(\dot{R}\,{}^tR) \end{aligned}$$

We see that the matrix $\dot{R}\,{}^tR$ is antisymmetric, so from the Exercise 11.1.11 there exist real scalars $(\omega_1(t), \omega_2(t), \omega_3(t))$ such that

$$\dot{R}\,{}^tR = \begin{pmatrix} 0 & -\omega_3(t) & \omega_2(t) \\ \omega_3(t) & 0 & -\omega_1(t) \\ -\omega_2(t) & \omega_1(t) & 0 \end{pmatrix}. \tag{11.5}$$

A comparison with the Example 1.3.17 than shows that the expression for the velocity,

$$\begin{pmatrix} \dot{x}_1 \\ \dot{x}_2 \\ \dot{x}_3 \end{pmatrix} = \dot{R}\,{}^tR \begin{pmatrix} x_1 \\ x_2 \\ x_3 \end{pmatrix} = \begin{pmatrix} 0 & -\omega_3(t) & \omega_2(t) \\ \omega_3(t) & 0 & -\omega_1(t) \\ -\omega_2(t) & \omega_1(t) & 0 \end{pmatrix} \begin{pmatrix} x_1 \\ x_2 \\ x_3 \end{pmatrix},$$

can be written as

$$\dot{\mathbf{x}}(t) = \omega(t) \wedge \mathbf{x}(t). \tag{11.6}$$

The triple $\omega(t) = \big(\omega_1(t), \omega_2(t), \omega_3(t)\big)$ is the *angular velocity* vector of the motion described by the rotation $R(t)$.

As we shall see in the Exercise 11.7.1, this relation also describes the rotation of a rigid body with a fixed point.

Excercise 11.6.2 The velocity corresponding to the motion in E^3 given by (here $r > 0$)

$$x(t) = \big(r\cos\alpha(t),\ r\sin\alpha(t), 0\big)$$

with respect to an orthonormal basis $\mathcal{E}$ is

$$\dot{x}(t) = \dot{\alpha}\big(- r \sin \alpha(t),\ r \cos \alpha(t), 0\big) = \omega(t) \wedge x(t)$$

with $\omega(t) = (0, 0, \dot{\alpha})$.

From the Sect. 11.5, we know that the angular velocity is an axial vector, so we write

$$\omega_a(t) \mapsto \omega'_b(t) = (\det P) \sum_{a=1}^{3} P_{ab}\omega_a(t).$$

for the transformation of the components under a change of basis in E^3 given by an orthogonal matrix $P \in \mathrm{O}(3)$. Notice that the relation (11.6) shows that the vector $\dot{\mathbf{x}}(t)$, although expressed via an axial vector, is a polar vector, since the vector product between an axial vector and a polar vector yields a polar vector. This is consistent with the formulation of $\dot{\mathbf{x}}(t)$ as the physical velocity of a point mass.

A different perspective on these notions and examples, allows one to study how the dynamics of a point mass is described with respect to different reference systems, in physicists' parlance.

Example 11.6.3 We describe the motion of a point mass with respect to an orthonormal basis $\mathcal{E} = (e_1, e_2, e_3)$ *and* with respect to an orthonormal basis $\mathcal{E}'(t) = (e'_1(t), e'_2(t), e'_3(t))$ that *rotates* with respect to $\mathcal{E}$. So we write

$$\mathbf{x}(t) = \sum_{a=1}^{3} x_a(t)\, e_a = \sum_{k=1}^{3} x'_k(t)\, e'_k(t).$$

Considering the time derivative of both sides, we have

$$\dot{\mathbf{x}}(t) = \sum_{a=1}^{3} \dot{x}_a(t)\, e_a = \sum_{k=1}^{3} \dot{x}'_k(t)\, e'_k(t) + \sum_{k=1}^{3} x'_k\, \dot{e}'_k(t).$$

Using the results of the Example 11.6.1, the second term can be written by means of an angular velocity $\omega(t)$ and thus we have

$$\dot{\mathbf{x}}(t) = \dot{\mathbf{x}}'(t) + \omega(t) \wedge \mathbf{x}'(t),$$

where $\mathbf{v} = \dot{\mathbf{x}}$ is the velocity of the point mass with respect to $\mathcal{E}$, while $\mathbf{v}' = \dot{\mathbf{x}}'$ is the velocity of the point mass with respect to $\mathcal{E}'(t)$.

With one step further along the same line, by taking a second time derivative results in

$$\begin{aligned} \ddot{\mathbf{x}}(t) &= \ddot{\mathbf{x}}'(t) + \omega(t) \wedge \dot{\mathbf{x}}'(t) + \omega(t) \wedge \big(\dot{\mathbf{x}}'(t) + \omega(t) \wedge \mathbf{x}'(t)\big) + \dot{\omega}(t) \wedge \mathbf{x}'(t) \\ &= \ddot{\mathbf{x}}'(t) + 2\,\omega(t) \wedge \dot{\mathbf{x}}'(t) + \omega(t) \wedge \big(\omega(t) \wedge \mathbf{x}'(t)\big) + \dot{\omega}(t) \wedge \mathbf{x}'(t). \end{aligned}$$

Using the language of physics, the term $\dot{\mathbf{x}'}(t)$ is the acceleration of the point mass with respect to the 'observer' at rest $\mathcal{E}$, while $\dot{\mathbf{x}}'(t)$ gives its acceleration with respect to the moving 'observer' $\mathcal{E}'(t)$.

With the rotation of $\mathcal{E}'(t)$ with respect to $\mathcal{E}$ given in terms of the angular velocity $\omega(t)$, the term

$$\mathbf{a}_{\mathbf{C}} = 2\,\omega(t) \wedge \dot{\mathbf{x}}'(t)$$

is called the *Coriolis* acceleration, the term

$$\mathbf{a}_{\mathbf{R}} = \omega(t) \wedge \big(\omega(t) \wedge \mathbf{x}'(t)\big)$$

is the *radial* (that is parallel to $\mathbf{x}'(t)$) acceleration, while the term

$$\mathbf{a}_{\mathbf{T}} = \dot{\omega}(t) \wedge \mathbf{x}'(t)$$

is the *tangential* (that is orthogonal to $\mathbf{x}'(t)$) one, and depending on the variation of the angular velocity.

11.7 Rigid Bodies and Inertia Matrix

Example 11.7.1 Consider a system of point masses $\{m_{(j)}\}_{j=1,\ldots,N}$ whose mutual distances in E^3 is constant, so that it can be considered as an example of a *rigid body*. The dynamics of each point mass is described by vectors $\mathbf{x}_{(j)}(t)$.

If we do not consider rigid translations, each motion $\mathbf{x}_{(j)}(t)$ is a rotation with the same angular velocity $\omega(t)$ around a fixed point. If we assume, with no loss of generality, that the fixed point coincides with the centre of mass of the system, and we set it to be the origin of E^3, then the total angular momentum of the system (the natural generalization of the angular momentum defined for a single point mass in the Example 1.3.17) is given by (using (11.6))

$$\mathbf{L}(t) = \sum_{j=1}^{N} m_{(j)}\mathbf{x}_{(j)}(t) \wedge \dot{\mathbf{x}}_{(j)}(t) = \sum_{j=1}^{N} m_{(j)}\mathbf{x}_{(j)}(t) \wedge \big(\omega(t) \wedge \mathbf{x}_{(j)}(t)\big).$$

With an orthonormal basis $\mathcal{E} = (e_1, e_2, e_3)$ for E^3, so that $\mathbf{x}_{(j)} = (x_{(j)1}, x_{(j)2}, x_{(j)3})$ and using the definition of vector product in terms of the Levi-Civita symbol, it is straightforward to compute that $\mathbf{L} = (L_1, L_2, L_3)$ is given by

$$L_k = \sum_{s=1}^{3} \Big\{ \sum_{i=1}^{N} m_{(j)}(\|\mathbf{x}_{(j)}\|^2 \delta_{ks} - x_{(j)k}x_{(j)s}) \Big\}\, \omega_s .$$

(In order to lighten notations, we drop for this example the explicit t dependence on the maps.) This expression can be written as

$$L_k = \sum_{s=1}^{3} I_{ks}\,\omega_s$$

where the quantities

$$I_{ks} = \sum_{j=1}^{N} m_{(j)}\big(\|\mathbf{x}_{(j)}\|^2\delta_{ks} - x_{(j)k}x_{(j)s}\big)$$

are the entries of the so called *inertia matrix* $\mathcal{I}$ (or inertia tensor) of the rigid body under analysis.

It is evident that the inertia matrix is symmetric, so from the Proposition 10.5.1, there exists an orthonormal basis for E^3 of eigenvectors for it. Moreover, if λ is an eigenvalue with eigenvector u, we have

$$\lambda\|u\|^2 = \sum_{k,s=1}^{3} I_{ks}u_k u_s = \sum_{j=1}^{N} m_{(j)}\big(\|u\|^2\|\mathbf{x}_{(j)}\|^2 - (u\cdot\mathbf{x}_{(j)})^2\big) \geq 0$$

where the last relation comes from the Schwarz inequality of Proposition 3.1.8. This means that $\mathcal{I}$ has no negative eigenvalues. If (u_1, u_2, u_3) is the orthonormal basis for which the inertia matrix is diagonal, and $(\lambda_1, \lambda_2, \lambda_3)$ are the corresponding eigenvalues, the vector lines $\mathcal{L}(u_a)$ are the so called *principal axes of inertia* for the rigid body, while the eigenvalues are the *moments of inertia*.

We give some basic examples for the inertia matrix of a rigid body.

Excercise 11.7.2 Consider a rigid body given by two point masses with $m_{(1)} = \alpha m_{(2)} = \alpha m$ with $\alpha > 0$, whose position is given in E^3 by the vectors $\mathbf{x}_{(1)} = (0, 0, r)$ and $\mathbf{x}_{(2)} = (0, 0, -\alpha r)$ with $r > 0$. The corresponding inertia matrix is found to be

$$\mathcal{I} = \alpha(1+\alpha)mr^2\begin{pmatrix}1&0&0\\0&1&0\\0&0&0\end{pmatrix}.$$

The principal axes of inertia coincide with the vector lines spanned by the orthonormal basis $\mathcal{E}$. The rigid body has two non zero momenta of inertia; the third momentum of inertia is zero since the rigid body is one dimensional.

Consider a rigid body given by three equal masses $m_{(j)} = m$ and

$$\mathbf{x}_{(1)} = (r, 0, 0), \qquad \mathbf{x}_{(2)} = \frac{1}{2}(-r, \sqrt{3}r, 0), \qquad \mathbf{x}_{(3)} = \frac{1}{2}(-r, -\sqrt{3}r, 0)$$

with $r > 0$, with respect to an orthonormal basis $\mathcal{E}$ in E^3. The inertia matrix is computed to be

$$\mathcal{I} = \frac{3mr^2}{2} \begin{pmatrix} 1 & 0 & 0 \\ 0 & 1 & 0 \\ 0 & 0 & 2 \end{pmatrix},$$

so the basis elements $\mathcal{E}$ provide the inertia principal axes.

Finally, consider a rigid body in E^3 consisting of four point masses with $m_{(j)} = m$ and

$$\mathbf{x}_{(1)} = (r, 0, 0), \qquad \mathbf{x}_{(2)} = (-r, 0, 0), \qquad \mathbf{x}_{(3)} = (0, r, 0), \qquad \mathbf{x}_{(4)} = (0, -r, 0)$$

with $r > 0$. The inertia matrix is already diagonal with respect to $\mathcal{E}$ whose basis elements give the principal axes of inertia for the rigid body, while the momenta of inertia is

$$\mathcal{I} = 2mr^2 \begin{pmatrix} 1 & 0 & 0 \\ 0 & 1 & 0 \\ 0 & 0 & 2 \end{pmatrix}.$$

Chapter 12
Spectral Theorems on Hermitian Spaces

In this chapter we shall extend to the complex case some of the notions and results of Chap. 10 on euclidean spaces, with emphasis on spectral theorems for a natural class of endomorphisms.

12.1 The Adjoint Endomorphism

Consider the vector space $\mathbb{C}^n$ and its dual space $\mathbb{C}^{n*}$, as defined in Sect. 8.1. The duality between $\mathbb{C}^n$ and $\mathbb{C}^{n*}$ allows one to define, for any endomorphism ϕ of $\mathbb{C}^n$, its adjoint.

Definition 12.1.1 Given $\phi : \mathbb{C}^n \to \mathbb{C}^n$, the map $\phi^\dagger : \omega \in \mathbb{C}^{n*} \mapsto \phi^\dagger(\omega) \in \mathbb{C}^{n*}$ defined by

$$(\phi^\dagger(\omega))(v) = \omega(\phi(v)) \qquad (12.1)$$

for any $\omega \in \mathbb{C}^{n*}$ and any $v \in \mathbb{C}^n$ is called the *adjoint* to ϕ.

Remark 12.1.2 From the linearity of ϕ and ω it follows that $\phi^\dagger$ is linear, so $\phi^\dagger \in \mathrm{End}(\mathbb{C}^{n*})$.

Example 12.1.3 Let $\mathcal{B} = (b_1, b_2)$ be a basis for $\mathbb{C}^2$, with $\mathcal{B}^* = (\beta_1, \beta_2)$ its dual basis for $\mathbb{C}^{2*}$. If ϕ is the endomorphism given by

$$\begin{aligned} \phi : b_1 &\mapsto kb_1 + b_2 \\ \phi : b_2 &\mapsto b_2, \end{aligned}$$

G. Landi and A. Zampini, *Linear Algebra and Analytic Geometry for Physical Sciences*, Undergraduate Lecture Notes in Physics,
https://doi.org/10.1007/978-3-319-78361-1_12

with $k \in \mathbb{C}$, we see from the definition of adjoint that

$$\begin{array}{rcl} \phi^{\dagger}(\beta_1) : b_1 & \mapsto & \beta_1(\phi(b_1)) = k \\ \phi^{\dagger}(\beta_1) : b_2 & \mapsto & \beta_1(\phi(b_2)) = 0 \\ \phi^{\dagger}(\beta_2) : b_1 & \mapsto & \beta_2(\phi(b_1)) = 1 \\ \phi^{\dagger}(\beta_2) : b_2 & \mapsto & \beta_2(\phi(b_2)) = 1. \end{array}$$

The (linear) action of the adjoint map $\phi^{\dagger}$ to ϕ is then

$$\begin{array}{rcl} \phi^{\dagger} : \beta_1 & \mapsto & k\beta_1 \\ \phi^{\dagger} : \beta_2 & \mapsto & \beta_1 + \beta_2. \end{array}$$

Consider now the canonical hermitian space $H^n = (\mathbb{C}^n, \cdot)$, that is the vector space $\mathbb{C}^n$ with the canonical hermitian product (see Sect. 3.4). As described in Sect. 8.2, the hermitian product allows one to identify $\mathbb{C}^{n*}$ with $\mathbb{C}^n$. Under such identification, the defining relation for $\phi^{\dagger}$ can be written as

$$(\phi^{\dagger}u) \cdot v = u \cdot (\phi\, v) \qquad \text{or equivalently} \qquad \langle \phi^{\dagger}(u) \mid v \rangle = \langle u | \phi(v) \rangle$$

for any $u, v \in \mathbb{C}^n$, so that $\phi^{\dagger}$ is an endomorphism of $H^n = (\mathbb{C}^n, \cdot)$.

Definition 12.1.4 Given a matrix $A = (a_{ij}) \in \mathbb{C}^{n,n}$, its adjoint $A^{\dagger} \in \mathbb{C}^{n,n}$ is the matrix whose entries are given by $(A^{\dagger})_{ab} = \overline{a_{ba}}$.

Thus, adjoining a matrix is the composition of two compatible involutions, the transposition and the complex conjugation.

Exercise 12.1.5 Clearly

$$A = \begin{pmatrix} 1 & \alpha \\ 0 & \beta \end{pmatrix} \quad \Rightarrow \quad A^{\dagger} = \begin{pmatrix} 1 & 0 \\ \bar{\alpha} & \bar{\beta} \end{pmatrix}.$$

Exercise 12.1.6 By using the matrix calculus we described in the previous chapters, it comes as no surprise that the following relations hold.

$$\begin{aligned} (A^{\dagger})^{\dagger} &= A, \\ (AB)^{\dagger} &= B^{\dagger}A^{\dagger}, \\ (A + \alpha B)^{\dagger} &= (A^{\dagger} + \bar{\alpha}B^{\dagger}) \end{aligned}$$

for any $A, B \in \mathbb{C}^{n,n}$ and $\alpha \in \mathbb{C}$. The second line indeed parallels the Remark 8.2.1. If we have two endomorphisms $\phi, \psi \in \mathrm{End}(H^n)$, one has

$$\langle (\phi\,\psi)^{\dagger}(u) | v \rangle = \langle u | \phi\,\psi(v) \rangle = \langle \phi^{\dagger}(u) | \psi(v) \rangle = \langle \psi^{\dagger}\phi^{\dagger}(u) | v \rangle,$$

for any $u, v \in H^n$. With $\alpha \in \mathbb{C}$, it is also

$$\langle(\phi + \alpha\,\psi)^\dagger u|v\rangle \;=\; \langle u|(\phi + \alpha\,\psi)v\rangle \;=\; \langle\phi(u)|v\rangle \;+\; \alpha\langle\psi(u)|v\rangle \;=\; \langle(\phi^\dagger + \bar{\alpha}\psi^\dagger)(u)|v\rangle.$$

Again using the properties of the hermitian product together with the definition of adjoint, it is

$$\langle(\phi^\dagger)^\dagger u|v\rangle \;=\; \langle u|\phi^\dagger(v)\rangle \;=\; \langle\phi(u)|v\rangle$$

The above lines establish the following identities

$$\begin{aligned}(\phi^\dagger)^\dagger &= \phi,\\ (\phi\,\psi)^\dagger &= \psi^\dagger\phi^\dagger,\\ (\phi + \alpha\,\psi)^\dagger &= \phi^\dagger + \bar{\alpha}\psi^\dagger\end{aligned}$$

which are the operator counterpart of the matrix identities described above.

Definition 12.1.7 An endomorphism ϕ on H^n is called

(a) *self-adjoint*, or *hermitian*, if

$$\phi = \phi^\dagger,$$

that is if $\langle\phi(u)|v\rangle = \langle u|\phi(v)\rangle$ for any $u, v \in H^n$,

(b) *unitary*, if

$$\phi\phi^\dagger = \phi^\dagger\phi = I_n,$$

that is if $\langle\phi(u)|\phi(v)\rangle = \langle u|v\rangle$ for any $u, v \in H^n$,

(c) *normal*, if $\phi\phi^\dagger = \phi^\dagger\phi$.

In parallel to these, a matrix $A \in \mathbb{C}^{n,n}$ is called

(a) *self-adjoint*, or *hermitian*, if $A^\dagger = A$,
(b) *unitary*, if $AA^\dagger = A^\dagger A = I_n$,
(c) *normal*, if $AA^\dagger = A^\dagger A$.

Remark 12.1.8 Clearly the condition of unitarity for ϕ is equivalent to the condition $\phi^\dagger = \phi^{-1}$. Also, both unitary and self-adjoint endomorphisms are normal. From the Remark 12.1.6 it follows that for any endomorphism ψ, the compositions $\psi\psi^\dagger$ and $\psi^\dagger\psi$ are self-adjoint.

Remark 12.1.9 The notion of adjoint of an endomorphism can be introduced also on euclidean spaces E^n, where it is identified, at a matrix level, by the transposition. Then, it is clear that the notion of self-adjointness in H^n generalises that in E^n, since if $A = {}^tA$ in E^n, then $A = A^\dagger$ in H^n, while orthogonal matrices in E^n are unitary matrices in H^n with real entries.

The following theorem is the natural generalisation for hermitian spaces of a similar result for euclidean spaces. Its proof, that we omit, mimics indeed that of the Theorem 10.1.11.

Theorem 12.1.10 *Let $\mathcal{C}$ be an orthonormal basis for the hermitian vector space H^n and let $\mathcal{B}$ be any other basis. The matrix $M^{\mathcal{C},\mathcal{B}}$ of the change of basis from $\mathcal{C}$ to $\mathcal{B}$ is unitary if and only if $\mathcal{B}$ is orthonormal.*

The following proposition, gives an *ex-post* motivation for the definitions above.

Proposition 12.1.11 *If $\mathcal{E}$ is the canonical basis for H^n, with $\phi \in \mathrm{End}(H^n)$, it holds that*

$$M_{\phi^\dagger}^{\mathcal{E},\mathcal{E}} = (M_{\phi}^{\mathcal{E},\mathcal{E}})^\dagger$$

Proof Let $\mathcal{E} = (e_1, \ldots, e_n)$ be the canonical basis for H^n. If $M_{\phi}^{\mathcal{E},\mathcal{E}} \in \mathbb{C}^{n,n}$ is the matrix that represents the action of ϕ on H^n with respect to the basis $\mathcal{E}$, its entries are given (see 8.7) by

$$(M_{\phi}^{\mathcal{E},\mathcal{E}})_{ab} = \langle e_a | \phi(e_b) \rangle .$$

By denoting $\phi_{ab} = (M_{\phi}^{\mathcal{E},\mathcal{E}})_{ab}$, the action of ϕ is given by $\phi(e_a) = \sum_{b=1}^{n} \phi_{ba} e_b$, so we can compute

$$(M_{\phi^\dagger}^{\mathcal{E},\mathcal{E}})_{ab} = \langle e_a | \phi^\dagger(e_b) \rangle = \langle \phi(e_a) | e_b \rangle = \sum_{c=1}^{n} \langle \phi_{ca} e_c | e_b \rangle = \overline{\phi_{ba}}$$

As an application of this proposition, the next proposition also generalises to hermitian spaces analogous results proven in Chap. 10 for euclidean spaces.

Proposition 12.1.12 *The endomorphism ϕ on H^n is self-adjoint (resp. unitary, resp. normal) if and only if there exists an orthonormal basis $\mathcal{B}$ for H^n with respect to which the matrix $M_{\phi}^{\mathcal{B},\mathcal{B}}$ is self-adjoint (resp. unitary, resp. normal).*

Exercise 12.1.13 Consider upper triangular matrices in $\mathbb{C}^{2,2}$,

$$M = \begin{pmatrix} a & b \\ 0 & c \end{pmatrix} \quad \Rightarrow \quad M^\dagger = \begin{pmatrix} \bar{a} & 0 \\ \bar{b} & \bar{c} \end{pmatrix} .$$

One explicitly computes

$$MM^\dagger = \begin{pmatrix} a\bar{a} + b\bar{b} & b\bar{c} \\ c\bar{b} & b\bar{b} + c\bar{c} \end{pmatrix}, \qquad M^\dagger M = \begin{pmatrix} a\bar{a} & b\bar{a} \\ a\bar{b} & b\bar{b} + c\bar{c} \end{pmatrix},$$

and the matrix M is normal, $MM^\dagger = M^\dagger M$, if and only if $b\bar{b} = 0 \Leftrightarrow b = 0$. Thus an upper triangular matrix in 2-dimension is normal if and only if it is diagonal. In such a case, the matrix is self-adjoint if the diagonal entries are real, and unitary if the diagonal entries have norm 1.

Exercise 12.1.14 We consider the following family of matrices in $\mathbb{C}^{2,2}$,

$$M = \begin{pmatrix} a & b \\ c & 0 \end{pmatrix}, \qquad M^{\dagger} = \begin{pmatrix} \bar{a} & \bar{c} \\ \bar{b} & 0 \end{pmatrix}.$$

It is

$$MM^{\dagger} = \begin{pmatrix} a\bar{a} + b\bar{b} & a\bar{c} \\ c\bar{a} & +c\bar{c} \end{pmatrix}, \qquad M^{\dagger}M = \begin{pmatrix} a\bar{a} + c\bar{c} & b\bar{a} \\ a\bar{b} & +b\bar{b} \end{pmatrix}.$$

The conditions for which M is normal are

$$b\bar{b} = c\bar{c}, \qquad a\bar{c} = b\bar{a}.$$

These are solved by $b = Re^{\mathrm{i}\beta}$, $c = Re^{\mathrm{i}\gamma}$, $A = |A|e^{\mathrm{i}\alpha}$ with $2\alpha = (\beta + \gamma) \bmod 2\pi$, where $R > 0$ and $|A| > 0$ are arbitrary moduli for complex numbers.

Exercise 12.1.15 With the Dirac's notation as in (8.6), an endomorphism ϕ and its adjoint are written as

$$\phi = \sum_{a,b=1}^{n} \phi_{ab} \, |e_a\rangle\langle e_b| \qquad \text{and} \qquad \phi^{\dagger} = \sum_{a,b=1}^{n} \overline{\phi_{ba}} \, |e_a\rangle\langle e_b|$$

with $\phi_{ab} = \langle e_a|\phi(e_b)\rangle = (M_{\phi}^{\mathcal{E},\mathcal{E}})_{ab}$ with respect to the orthonormal basis $\mathcal{E} = (e_1, \ldots, e_n)$.

With $u = (u_1, \cdots, u_n)$ and $v = (v_1, \ldots, v_n)$ vectors in H^n we have the endomorphism $P = |u\rangle\langle v|$. If we decompose the identity endomorphism (see the point (c) from the Proposition 10.3.7) as

$$\mathrm{id} = \sum_{s=1}^{n} |e_s\rangle\langle e_s|$$

we can write

$$P = |u\rangle\langle v| = \sum_{ab=1}^{n} |e_a\rangle\langle e_a|u\rangle\langle v|e_b\rangle\langle e_b| = \sum_{ab=1}^{n} P_{ab} \, |e_a\rangle\langle e_b|$$

with $P_{ab} = u_a\overline{v_b} = \langle e_a|P(e_b)\rangle$. Clearly then

$$P^{\dagger} = |v\rangle\langle u|.$$

Example 12.1.16 Let ϕ an endomorphism H^n with matrix $M_{\phi}^{\mathcal{E},\mathcal{E}}$ with respect to the canonical orthonormal basis, thus $(M_{\phi}^{\mathcal{E},\mathcal{E}})_{ab} = \langle e_a|\phi(e_b)\rangle$. If $\mathcal{B} = (b_1, \ldots, b_n)$ is a second orthonormal basis for H^n, we have two decompositions

$$\mathrm{id} = \sum_{k=1}^{n} |e_k\rangle\langle e_k| = \sum_{s=1}^{n} |e_s\rangle\langle e_s|.$$

Thus, by *inserting* these two expressions of the identity operators, we have

$$\langle e_a|\phi(e_b)\rangle = \sum_{k,s=1}^{n} \langle e_a|b_k\rangle\langle b_k|\phi(b_s)\rangle\langle b_s|e_b\rangle,$$

giving in components,

$$(M_{\phi}^{\mathcal{E},\mathcal{E}})_{ab} = \sum_{k,s=1}^{n} \langle e_a|b_k\rangle(M_{\phi}^{\mathcal{B},\mathcal{B}})_{ks}\langle b_s|e_b\rangle.$$

The matrix of the change of basis from $\mathcal{E}$ to $\mathcal{B}$ has entries $\langle e_a|b_k\rangle = (M^{\mathcal{E},\mathcal{B}})_{ak}$, with its inverse matrix entries given by $(M^{\mathcal{B},\mathcal{E}})_{sb} = \langle b_s|e_b\rangle$. From the previous examples we see that

$$(M^{\mathcal{B},\mathcal{E}\dagger})_{ak} = \overline{(M^{\mathcal{B},\mathcal{E}})_{ka}} = \overline{\langle b_k|e_a\rangle} = \langle e_a|b_k\rangle = (M^{\mathcal{E},\mathcal{B}})_{ak}$$

thus finding that the change of basis is given by a unitary matrix.

Proposition 12.1.17 *For any endomorphism ϕ in H^n, there is an orthogonal vector space decomposition*

$$H^n = \mathrm{Im}(\phi) \oplus \ker(\phi^{\dagger})$$

Proof If u is any vector in H^n, the vector $\phi(u)$ cover over all of $\mathrm{Im}(\phi)$, so the condition $\langle\phi(u)|w\rangle = 0$ characterises the elements $w \in (\mathrm{Im}(\phi))^{\perp}$. It is now easy to compute

$$0 = \langle\phi(u)|w\rangle = \langle u|\phi^{\dagger}(w)\rangle.$$

Since u is arbitrary and the hermitian product is not degenerate, we have $\ker(\phi^{\dagger}) = (\mathrm{Im}(\phi))^{\perp}$. □

12.2 Spectral Theory for Normal Endomorphisms

We prove a few results for normal endomorphisms which will be useful for spectral theorems.

Proposition 12.2.1 *Let ϕ be a normal endomorphism of H^n.*

(a) With $u \in H^n$, we can write

$$\|\phi(u)\|^2 = \langle\phi(u)|\phi(u)\rangle = \langle u|\phi^\dagger\phi(u)\rangle = \langle u|\phi\phi^\dagger(u)\rangle = \langle\phi^\dagger(u)|\phi^\dagger(u)\rangle = \|\phi^\dagger(u)\|^2.$$

Since the order of these computations can be reversed, we have the following characterisation.

$$\phi\phi^\dagger = \phi^\dagger\phi \quad \Leftrightarrow \quad \|\phi(u)\| = \|\phi^\dagger(u)\| \quad \textit{for all} \quad u \in H^n.$$

(b) From this it also follows that $\ker(\phi) = \ker(\phi^\dagger)$. *So from the Proposition 12.1.17, we have the following orthogonal decomposition,*

$$H^n = \mathrm{Im}(\phi) \oplus \ker(\phi).$$

(c) Clearly $(\phi - \lambda I)$ is a normal endomorphism if ϕ is such. This gives $\ker(\phi - \lambda I) = \ker(\phi^\dagger - \bar{\lambda} I)$, *meaning that if λ is an eigenvalue of a normal endomorphism ϕ, then $\bar{\lambda}$ is an eigenvalue for $\phi^\dagger$, with the same eigenspaces.*

(d) Let λ, μ be two distinct eigenvalues for ϕ, with $\phi(v) = \lambda v$ and $\phi(w) = \mu w$. Then we have

$$(\lambda - \mu)\langle v|w\rangle = \langle\bar{\lambda}v|w\rangle - \langle v|\mu w\rangle = \langle\phi^\dagger(v)|w\rangle - \langle v|\phi(w)\rangle = 0.$$

We can conclude that the eigenspaces corresponding to distinct eigenvalues for a normal endomorphism are mutually orthogonal. □

We are ready to characterise a normal operator in terms of its spectral properties. The proof of the following result generalises to hermitian spaces the proof of the Theorem 10.4.5 on the diagonalization of symmetric endomorphisms on euclidean spaces.

Theorem 12.2.2 *An endomorphism ϕ of H^n is normal if and only there exists an orthonormal basis for H^n made of eigenvectors for ϕ.*

Proof If $\mathcal{B} = (b_1, \ldots, b_n)$ is an orthonormal basis of eigenvectors for ϕ, with corresponding eigenvalues $(\lambda_1, \ldots, \lambda_n)$, we can write

$$\phi = \sum_{a=1}^{n} \lambda_a \, |b_a\rangle\langle b_a| \quad \text{and} \quad \phi^\dagger = \sum_{a=1}^{n} \overline{\lambda_a} \, |b_a\rangle\langle b_a|$$

which directly yields (see the Exercise 12.1.15)

$$\phi\phi^{\dagger} = \sum_{a=1}^{n} |\lambda_a|^2 \, |b_a\rangle\langle b_a| = \phi^{\dagger}\phi.$$

The converse, the less trivial part of the statement, is proven once again by induction.

Consider first a normal operator ϕ on the two dimensional hermitian space H^2. With respect to any basis, the characteristic polynomial $p_{\phi}(T)$ has two complex roots, from the fundamental theorem of algebra. A normal endomorphism of H^2 with only the zero eigenvalue, would be the null endomorphism. So we can assume there is a root $\lambda \neq 0$, with v a (normalised) eigenvectors, that is $\phi(v) = \lambda v$ with $\|v\| = 1$. If $\mathcal{C} = (v, w)$ is an orthonormal basis for H^2 that completes v, we have, from point (c) above,

$$\langle \phi(w)|v\rangle = \langle w|\phi^{\dagger}(v)\rangle = \langle w|v\rangle\bar{\lambda} = 0.$$

Being $\lambda \neq 0$, this shows that $\phi(w)$ is orthogonal to $\mathcal{L}(v)$, so that there must exists a scalar μ, such that $\phi(w) = \mu w$. In turn this shows that if ϕ is a normal endomorphism of H^2, then H^2 has an orthonormal basis of eigenvectors for ϕ.

Inductively, let us assume that the statement is valid when the dimension of the hermitian space is $n-1$. The n-dimensional case is treated analogously to what done above. If ϕ is a normal endomorphism of H^n, its characteristic polynomial $p_{\phi}(T)$ has at least a non zero complex root, λ say, with v a corresponding normalised eigenvector: $\phi(v) = \lambda v$, with $\|v\| = 1$. (Again, a normal endomorphism of H^n with only the zero eigenvalue is the null endomorphism.) We have $H^n = V_{\lambda} \oplus V_{\lambda}^{\perp}$ and v can be completed to an orthonormal basis $\mathcal{C} = (v, w_1, \ldots, w_n)$ for H^n. If $w \in V_{\lambda}^{\perp}$ we compute as above

$$\langle \phi(w)|v\rangle = \langle w|\phi^{\dagger}(v)\rangle = \langle w|v\rangle\bar{\lambda} = 0.$$

This shows that ϕ maps $V_{\lambda}^{\perp}$ to itself, while also $\phi^{\dagger}$ maps $V_{\lambda}^{\perp}$ to itself since,

$$\langle \phi^{\dagger}(w)|v\rangle = \langle w|\phi(v)\rangle = \langle w|v\rangle\lambda = 0.$$

The restriction of ϕ to $V_{\lambda}^{\perp}$ is then a normal operator on a $(n-1)$ dimensional hermitian space, and by assumption there exists an orthonormal basis $(u_1, \ldots, u_{n-1})$ for $V_{\lambda}^{\perp}$ made of eigenvectors for ϕ. The basis $\mathcal{E} = (v, u_1, \ldots, u_{n-1})$ is an orthonormal basis for H^n of eigenvectors for ϕ. □

Remark 12.2.3 Since the field of real numbers is not algebraically closed (and the fundamental theorem of algebra is valid on $\mathbb{C}$), it is worth stressing that an analogue of this theorem for normal endomorphisms on euclidean spaces does not hold. A matrix $A \in \mathbb{R}^{n,n}$ such that $({}^tA)\,A = A\,({}^tA)$, needs not be diagonalisable. An example is given by an antisymmetric (skew-adjoint, see Sect. 11.1) matrix A, which clearly commutes with ${}^tA$, being nonetheless not diagonalisable.

We showed in the Remark 12.1.8 that self-adjoint and unitary endomorphisms are normal. Within the set of normal endomorphisms, they can be characterised in terms of their spectrum.

If λ is an eigenvalue of a self-adjoint endomorphism ϕ, with $\phi(v) = \lambda v$, then

$$\lambda v = \phi(v) = \phi^{\dagger}(v) = \bar{\lambda} v$$

and thus one has $\lambda = \bar{\lambda}$. If λ is an eigenvalue for a unitary operator ϕ, with $\phi(v) = \lambda v$, then

$$\|v\|^2 = \|\phi(v)\|^2 = |\lambda|^2 \|v\|^2,$$

which gives $|\lambda| = 1$. It is easy to show also the converse of these claims, so to have the following.

Theorem 12.2.4 *A normal operator on H^n is self-adjoint if and only if its eigenvalues are real. A normal operator on H^n is unitary if and only if its eigenvalues have modulus 1.*

As a corollary, by merging the previous two theorems, we have a characterisation of self-adjoint and unitary operators in terms of their spectral properties, as follows.

Corollary 12.2.5 *An endomorphism ϕ on H^n is self-adjoint if and only if its spectrum is real and there exists an orthonormal basis for H^n of eigenvectors for ϕ. An endomorphism ϕ on H^n is unitary if and only if its spectrum is a subset of the unit circle in $\mathbb{C}$, and there exists an orthonormal basis for H^n of eigenvectors for ϕ.*

Exercise 12.2.6 Consider the hermitian space H^2, with $\mathcal{E} = (e_1, e_2)$ its canonical orthonormal basis, and the endomorphism ϕ represented with respect to $\mathcal{E}$ by

$$M_{\phi}^{\mathcal{E},\mathcal{E}} = \begin{pmatrix} 0 & a \\ -a & 0 \end{pmatrix} \qquad \text{with } a \in \mathbb{R}.$$

This endomorphism is not diagonalisable over $\mathbb{R}$, since it is antisymmetric (see Sect. 11.1) and the Remark 12.2.3. Being normal with respect to the hermitian structure in H^2, there exists an orthonormal basis for H^2 of eigenvectors for ϕ. The eigenvalue equation is $p_{\phi}(T) = T^2 + a^2 = 0$, so the eigenvalues are $\lambda_{\pm} = \pm ia$, with normalised eigenvectors $u_{\pm}$ given by

$$\lambda_{\pm} = \pm \mathrm{i}\, a \qquad u_{\pm} = \tfrac{1}{\sqrt{2}}(1, \pm \mathrm{i})_{\mathcal{E}},$$

while the unitary conjugation that diagonalises the matrix $M_{\phi}^{\mathcal{E},\mathcal{E}}$ is given by

$$\tfrac{1}{2}\begin{pmatrix} 1 & -\mathrm{i} \\ 1 & \mathrm{i} \end{pmatrix} \begin{pmatrix} 0 & a \\ -a & 0 \end{pmatrix} \begin{pmatrix} 1 & 1 \\ \mathrm{i} & -\mathrm{i} \end{pmatrix} = \begin{pmatrix} \mathrm{i}a & 0 \\ 0 & -\mathrm{i}a \end{pmatrix}.$$

The comparison of this with the content of the Example 12.1.16 follows by writing the matrix giving the change of basis from $\mathcal{E}$ to $\mathcal{B} = (u_+, u_-)$ as

$$M^{\mathcal{B},\mathcal{E}} = \frac{1}{2}\begin{pmatrix} 1 & -\mathrm{i} \\ 1 & \mathrm{i} \end{pmatrix} = \begin{pmatrix} \langle u_+|e_1\rangle & \langle u_+|e_2\rangle \\ \langle u_-|e_1\rangle & \langle u_-|e_2\rangle \end{pmatrix}.$$

We next study a family of normal endomorphisms, which will be useful when considering the properties of unitary matrices. The following definition comes naturally from the Definition 12.1.7.

Definition 12.2.7 An endomorphism ϕ in H^n is named *skew-adjoint* if $\langle u|\phi(v)\rangle + \langle\phi^\dagger(u)|v\rangle = 0$ for any $u, v \in H^n$. A matrix $A \in \mathbb{C}^{n,n}$ is named *skew-adjoint* if $A^\dagger = -A$.

We list some important results on skew-adjoint endomorphisms and matrices.

(a) It is clear that an endomorphism ϕ on H^n is skew-adjoint if and only if there exists an orthonormal basis $\mathcal{E}$ for H^n with respect to which the matrix $M_\phi^{\mathcal{E},\mathcal{E}}$ is skew-adjoint.
(b) Skew-adjoint endomorphisms are normal. We know from the Proposition 12.2.1 point (c), that if λ is an eigenvalue for the endomorphism ϕ, then $\bar{\lambda}$ is an eigenvalue for $\phi^\dagger$. This means that if λ is an eigenvalue For a skew-adjoint endomorphism ϕ, then $\bar{\lambda} = -\lambda$, so any eigenvalue for a skew-adjoint endomorphism is either purely imaginary or zero.
(c) There exists an orthonormal basis $\mathcal{E} = (e_1, \ldots, e_n)$ of eigenvectors for ϕ such that

$$\phi = \sum_{a=1}^{n} \mathrm{i}\lambda_a \, |e_a\rangle\langle e_a| \qquad \text{with} \quad \lambda_a \in \mathbb{R}.$$

(d) The real vector space of skew-adjoint matrices $A = -A^\dagger \in \mathbb{C}^{n,n}$ is a matrix Lie algebra (see the Definition 11.1.6), that is the commutator of skew-adjoint matrices is a skew-adjoint matrix; it is denoted $\mathfrak{u}(n)$ and it has dimension n.

Remark 12.2.8 In parallel with the Remark 11.1.9, self-adjoint matrices do not make up a Lie algebra since the commutator of two self-adjoint matrices is a skew-adjoint matrix.

Exercise 12.2.9 On the hermitian space H^3 we consider the endomorphism ϕ whose representing matrix is, with respect to the canonical basis $\mathcal{E}$, given by

$$M_\phi^{\mathcal{E},\mathcal{E}} = \begin{pmatrix} 0 & \mathrm{i} & a \\ \mathrm{i} & 0 & 0 \\ -a & 0 & 0 \end{pmatrix},$$

with a a real parameter. Since $(M_\phi^{\mathcal{E},\mathcal{E}})^\dagger = -M_\phi^{\mathcal{E},\mathcal{E}}$, then ϕ is skew-adjoint (and thus normal). Its characteristic equation

$$p_\phi(T) = -T(1 + a^2 + T^2) = 0$$

has solutions $\lambda = 0$ and $\lambda_\pm = \pm\mathrm{i}\sqrt{1 + a^2}$. Explicit calculations show that the eigenspaces are given by $\ker(\phi) = V_{\lambda=0} = \mathcal{L}(u_0)$ and $V_{\lambda_\pm} = \mathcal{L}(u_\pm)$ with

$$\begin{aligned} u_0 &= \tfrac{1}{\sqrt{1+a^2}}\,(0, \mathrm{i}a, 1), \\ u_\pm &= \tfrac{1}{\sqrt{2(1+a^2)}}\,(\sqrt{1+a^2}, \pm 1, \pm \mathrm{i}a). \end{aligned}$$

It is immediate to see that the set $\mathcal{B} = (u_0, u_\pm)$ gives an orthonormal basis for H^3.

Exercise 12.2.10 We close this section by studying an endomorphism which is *not* normal, and indeed diagonalisable with an eigenvector basis which is not orthonormal. In H^2 with respect to $\mathcal{E} = (e_1, e_2)$, consider the endomorphism whose representing matrix is

$$M = \begin{pmatrix} 0 & 1 \\ a & 0 \end{pmatrix}$$

with $a \in \mathbb{R}$. Than M is normal if and only if $a = 1$. The characteristic equation is

$$p_M(T) = T^2 - a = 0$$

so its spectral decomposition is given by

$$\lambda_\pm = \pm\sqrt{a}, \qquad V_{\lambda_\pm} = \mathcal{L}(u_\pm) \;\; \text{with} \;\; u_\pm = (1, \pm\sqrt{a})_{\mathcal{E}}.$$

Being $\langle u_+ | u_- \rangle = 1 - a$, the eigenvectors are orthogonal if and only if M is normal.

12.3 The Unitary Group

If $A, B \in \mathbb{C}^{n,n}$ are two unitary matrices, $A^\dagger A = I_n$ and $B^\dagger B = I_n$ (see the Definition 12.1.7), one has $(AB)^\dagger AB = B^\dagger A^\dagger AB = I_n$. Furthermore, $\det(A^\dagger) = \overline{\det(A)}$, so from $\det(AA^\dagger) = 1$ we have $|\det(A)| = 1$. Clearly, the identity matrix I_n is unitary and these leads to the following definition.

Definition 12.3.1 The collection of $n \times n$ unitary matrices is a group, called the *unitary group* of order n and denoted $\mathrm{U}(n)$. The subset $\mathrm{SU}(n) = \{A \in \mathrm{U}(n) : \det(A) = 1\}$ is a subgroup of $\mathrm{U}(n)$, called the *special unitary group* of order n.

Remark 12.3.2 With the the natural inclusion of real matrices as complex matrices whose entries are invariant under complex conjugation, it is clear that $\mathrm{O}(n)$ is a subgroup of $\mathrm{U}(n)$ and $\mathrm{SO}(n)$ is a subgroup of $\mathrm{SU}(n)$.

Now, the exponential of a matrix as in the Definition 11.2.1 can be extended to complex matrices. Thus, for a matrix $A \in \mathbb{C}^{n,n}$, its exponential is defined by by the expansion,

$$e^A = \sum_{k=0}^{\infty} \tfrac{1}{k!} A^k .$$

Then, all properties in the Proposition 11.2.2 have a counterpart for complex matrices, with point (e) there now reading $e^{A^\dagger} = (e^A)^\dagger$.

Theorem 12.3.3 *Let $M, U \in \mathbb{C}^{n,n}$. One has the following results.*

(a) If $M^\dagger = -M$, then $e^M \in \mathrm{U}(n)$. If $M^\dagger = -M$ and $\mathrm{tr}(M) = 0$, then $e^M \in \mathrm{SU}(n)$.
(b) Conversely, if $UU^\dagger = I_n$, there exists a skew-adjoint matrix $M = -M^\dagger$ such that $U = e^M$. If U is a special unitary matrix, there exists a skew-adjoint traceless matrix, $M = -M^\dagger$ with $\mathrm{tr}(M) = 0$, such that $U = e^M$.

Proof Let M be a skew-adjoint matrix. From the previous section we know that there exists a unitary matrix V such that $M = V\Delta_M V^\dagger$, with $\Delta_M = \mathrm{diag}(\mathrm{i}\rho_1, \dots, \mathrm{i}\rho_n)$ for $\rho_a \in \mathbb{R}$. We can then write

$$e^M = e^{V\Delta_M V^\dagger} = V\, e^{\Delta_M}\, V^\dagger$$

with $e^{\Delta_M} = \mathrm{diag}(e^{\mathrm{i}\rho_1}, \dots, e^{\mathrm{i}\rho_n})$. This means that e^{Δ_M} is a unitary matrix, and we can conclude that the starting matrix e^M is unitary. If $\mathrm{tr}(M) = 0$, then e^M is a special unitary matrix.

Alternatively, the result can be shown as follows. If $M = -M^\dagger$, then

$$(e^M)^\dagger = e^{M^\dagger} = e^{-M} = (e^M)^{-1}.$$

This concludes the proof of point (a).

Consider then a unitary matrix U. Since U is normal, there exists a unitary matrix V such that $U = V\Delta_U V^\dagger$ with $\Delta_U = \mathrm{diag}(e^{\mathrm{i}\varphi_1}, \dots, e^{\mathrm{i}\varphi_n})$, where $e^{\mathrm{i}\varphi_k}$ are the modulus 1 eigenvalues of U. Clearly, the matrix Δ_U can be written as

$$\Delta_U = e^{\delta_U}$$

with $\delta_U = \mathrm{diag}(\mathrm{i}\varphi_1, \dots, \mathrm{i}\varphi_n) = -(\delta_U)^\dagger$. This means that

$$U = V\, e^{\delta_U}\, V^\dagger = e^{V\delta_U V^\dagger}$$

with $(V\delta_U V^\dagger)^\dagger = -(V\delta_U V^\dagger)$. If $U \in \mathrm{SU}(n)$, then one has $\mathrm{tr}(V\delta_U V^\dagger) = 0$. This establishes point (b) and concludes the proof. □

Exercise 12.3.4 Consider the matrix, with $a, b \in \mathbb{R}$,

$$A = \begin{pmatrix} b & a \\ a & b \end{pmatrix} = A^{\dagger}.$$

Its eigenvalues λ are given by the solutions of the characteristic equation

$$p_A(T) = (b - T)^2 - a^2 = (b - T - a)(b - T + a) = 0.$$

Its spectral decomposition turns out to be

$$\lambda_{\pm} = b \pm a, \qquad V_{\lambda_{\pm}} = \mathcal{L}((1, \pm 1)).$$

To exponentiate the skew-adjoint matrix $\mathrm{i}A$ we can follow two ways.

- By normalising the eigenvectors, we have the conjugation with its diagonal form $A = V\Delta_A V^{\dagger}$,

$$\begin{pmatrix} b & a \\ a & b \end{pmatrix} = \frac{1}{\sqrt{2}}\begin{pmatrix} 1 & 1 \\ -1 & 1 \end{pmatrix}\begin{pmatrix} b-a & 0 \\ 0 & b+a \end{pmatrix}\frac{1}{\sqrt{2}}\begin{pmatrix} 1 & -1 \\ 1 & 1 \end{pmatrix}$$

so we have

$$\begin{aligned} e^{\mathrm{i}A} &= e^{\mathrm{i}V\Delta_A V^{\dagger}} = Ve^{\mathrm{i}\Delta_A}V^{\dagger} = \frac{1}{2}\begin{pmatrix} 1 & 1 \\ -1 & 1 \end{pmatrix}\begin{pmatrix} e^{\mathrm{i}(b-a)} & 0 \\ 0 & e^{\mathrm{i}(b+a)} \end{pmatrix}\begin{pmatrix} 1 & -1 \\ 1 & 1 \end{pmatrix} \\ &= \frac{1}{2}\begin{pmatrix} e^{\mathrm{i}(b-a)} + e^{\mathrm{i}(a+b)} & -e^{\mathrm{i}(b-a)} + e^{\mathrm{i}(a+b)} \\ -e^{\mathrm{i}(b-a)} + e^{\mathrm{i}(a+b)} & e^{\mathrm{i}(b-a)} + e^{\mathrm{i}(a+b)} \end{pmatrix} = \begin{pmatrix} e^{\mathrm{i}b}\cos a & \mathrm{i}e^{\mathrm{i}b}\sin a \\ \mathrm{i}e^{\mathrm{i}b}\sin a & e^{\mathrm{i}b}\cos a \end{pmatrix}. \end{aligned}$$

Notice that $\det(e^{\mathrm{i}A}) = e^{2\mathrm{i}b} = e^{\mathrm{i}\,\mathrm{tr}(A)}$.

- By setting

$$A = \tilde{A} + \tilde{B} = \begin{pmatrix} 0 & a \\ a & 0 \end{pmatrix} + \begin{pmatrix} b & 0 \\ 0 & b \end{pmatrix}$$

we see that A is the sum of two commuting matrices, since $\tilde{B} = bI_2$. So we can write

$$e^{\mathrm{i}A} = e^{\mathrm{i}(\tilde{A}+\tilde{B})} = e^{\mathrm{i}\tilde{A}}e^{\mathrm{i}\tilde{B}}.$$

Since $\tilde{B}$ is diagonal, $e^{\mathrm{i}\tilde{B}} = \mathrm{diag}(e^{\mathrm{i}b}, e^{\mathrm{i}ib})$. Computing as in the Exercise 11.2.4 we have

$$\tilde{A}^{2k} = \begin{pmatrix} (-1)^k a^{2k} & 0 \\ 0 & (-1)^k a^{2k} \end{pmatrix}, \qquad \tilde{A}^{2k+1} = \begin{pmatrix} 0 & (-1)^k \mathrm{i}a^{2k+1} \\ (-1)^k \mathrm{i}a^{2k+1} & 0 \end{pmatrix}$$

so that

$$e^{\mathrm{i}\tilde{A}} = \begin{pmatrix} \cos a & \mathrm{i}\sin a \\ \mathrm{i}\sin a & \cos a \end{pmatrix}$$

and

$$e^{iA} = \begin{pmatrix} \cos a & i \sin a \\ i \sin a & \cos a \end{pmatrix} \begin{pmatrix} e^{ib} & 0 \\ 0 & e^{ib} \end{pmatrix} = \begin{pmatrix} e^{ib} \cos a & ie^{ib} \sin a \\ ie^{ib} \sin a & e^{ib} \cos a \end{pmatrix}.$$

Exercise 12.3.5 In this exercise we describe how to reverse the construction of the previous one. That is, given the unitary matrix

$$U = \tfrac{1}{\sqrt{2}} \begin{pmatrix} 1 & 1 \\ -1 & 1 \end{pmatrix},$$

we determine the self-adjoint matrix $A = A^\dagger$ such that $U = e^{iA}$. Via the usual techniques it is easy to show that the spectral decomposition of U is given by

$$\lambda_\pm = \frac{a \pm i}{\sqrt{1 + a^2}}, \quad \text{with} \quad V_{\lambda_\pm} = \mathcal{L}((1, \pm i)).$$

Notice that $|\lambda_\pm| = 1$ so we can write $\lambda_\pm = e^{i\varphi_\pm}$ and, by normalising the eigenvectors for U,

$$U = V \Delta_U V^\dagger = \tfrac{1}{2} \begin{pmatrix} 1 & 1 \\ -i & i \end{pmatrix} \begin{pmatrix} e^{i\varphi_-} & 0 \\ 0 & e^{i\varphi_+} \end{pmatrix} \begin{pmatrix} 1 & i \\ 1 & -i \end{pmatrix},$$

with $V^\dagger V = I_2$. Since $\Delta_U = e^{i\delta_U}$ with $\delta_U = \delta_U^\dagger = \text{diag}(\varphi_-, \varphi_+)$, we write

$$U = V e^{i\delta_U} V^\dagger = e^{iV\delta_U V^\dagger} = e^{iA}$$

where $A = A^\dagger$ with

$$A = V \delta_U V^\dagger = \tfrac{1}{2} \begin{pmatrix} 1 & 1 \\ -i & i \end{pmatrix} \begin{pmatrix} e^{i\varphi_-} & 0 \\ 0 & e^{i\varphi_+} \end{pmatrix} \begin{pmatrix} 1 & i \\ 1 & -i \end{pmatrix} = \tfrac{1}{2} \begin{pmatrix} \varphi_- + \varphi_+ & i(\varphi_- - \varphi_+) \\ i(\varphi_- - \varphi_+) & \varphi_- + \varphi_+ \end{pmatrix}.$$

Notice that the matrix A is not uniquely determined by U, since the angular variables $\varphi_\pm$ are defined up to 2π periodicity by

$$\cos \varphi_\pm = \frac{a}{\sqrt{1 + a^2}}, \qquad \sin \varphi_\pm = \pm \frac{1}{\sqrt{1 + a^2}}.$$

We close this section by considering one parameter groups of unitary matrices. We start with a self-adjoint matrix $A = A^\dagger \in \mathbb{C}^{n,n}$, and define the matrix

$$U_s = e^{isA}, \quad \text{for} \quad s \in \mathbb{R}.$$

From the properties of the exponential of a matrix, it is easy to show that, for any real s, s', the following identities hold.

(i) $(U_s)^\dagger U_s = I_n$,
that is U_s is unitary,
(ii) $U_0 = I_n$,
(iii) $(U_s)^\dagger = U_{-s}$,
(iv) $U_{s+s'} = U_s U_{s'} = U_{s'} U_s$,
thus in particular, these unitary matrices commute for different values of the parameter.

The map $\mathbb{R} \to \mathrm{U}(n)$ given by $s \mapsto U_s$ is, according to the definition in the Appendix A.4, a group homomorphism between $(\mathbb{R}, +)$ and $\mathrm{U}(n)$ (with group multiplication), that is between the abelian group $\mathbb{R}$ with respect to the sum and the non abelian group $\mathrm{U}(n)$ with respect to the matrix product. This leads to the following definition.

Definition 12.3.6 If U_s is a family (labelled by a real parameter s) of elements in $\mathrm{U}(n)$ such that, for any value of $s \in \mathbb{R}$, the above identities ii) – iv) are fulfilled, then U_s is called a *one parameter group of unitary matrices* of order n.

For any self-adjoint matrix A, we have a one parameter group of unitary matrices given by $U_s = e^{\mathrm{i}sA}$. The matrix A is usually called the *infinitesimal generator* of the one parameter group.

Proposition 12.3.7 *For any $A = A^\dagger \in \mathbb{C}^{n,n}$, the elements $U_s = e^{\mathrm{i}sA}$ give a one parameter group of unitary matrices in H^n. Conversely, if U_s is a one parameter group of unitary matrices in H^n, there exists a self-adjoint matrix $A = A^\dagger$ such that $U_s = e^{\mathrm{i}sA}$.*

Proof Let $U_s \in \mathrm{U}(n)$ be a one parameter group of unitary matrices. For each value $s \in \mathbb{R}$ the matrix U_s can be diagonalised, and since U_s commutes with any $U_{s'}$, it follows that there exists an orthonormal basis $\mathcal{B}$ for H^n of common eigenvectors for any U_s. So there is a unitary matrix V (providing the change of basis from $\mathcal{B}$ to the canonical base $\mathcal{E}$) such that

$$U_s = V\{\mathrm{diag}(e^{\mathrm{i}\varphi_1(s),\dots,\mathrm{i}\varphi_n(s)})\}V^\dagger$$

where $e^{\mathrm{i}\varphi_k(s)}$ are the eigenvalues of U_s. From the condition $U_s U_{s'} = U_{s+s'}$ it follows that the dependence of the eigenvalues on the parameter s is linear, and from $U_0 = I_n$ we know that $\varphi_k(s=0) = 0$. We can eventually write

$$U_s = V\{\mathrm{diag}(e^{\mathrm{i}s\varphi_1,\dots,\mathrm{i}s\varphi_n})\}V^\dagger = Ve^{\mathrm{i}s\,\delta}V^\dagger = e^{\mathrm{i}s\,V\delta V^\dagger}$$

where $\delta = \mathrm{diag}(\varphi_1, \dots, \varphi_n)$ is a self-adjoint matrix. We then set $A = V\delta V^\dagger = A^\dagger$ to be the infinitesimal generator of the given one parameter group of unitary matrices. □

Chapter 13
Quadratic Forms

13.1 Quadratic Forms on Real Vector Spaces

In Sect. 3.1 the notion of scalar product on a finite dimensional real vector space has been introducedas a bilinear symmetric map $\cdot\,:\,V\times V\,\rightarrow\,\mathbb{R}$ with additional properties. Such additional properties are that $v\cdot v\geq 0$ for $v\in V$, with $v\cdot v=0\ \Leftrightarrow\ v=0_V$. This is referred to as positive definiteness.

We start by introducing the more general notion of *quadratic form*.

Definition 13.1.1 Let V be a finite dimensional real vector space. A *quadratic form* on V is a map

$$\mathcal{Q}\,:\,V\times V\longrightarrow\mathbb{R}\qquad(v,w)\mapsto\mathcal{Q}(v,w)$$

that fulfils the following properties. For any $v,w,v_1,v_2\in V$ and $a_1,a_2\in\mathbb{R}$ it holds that:

(Q1) $\mathcal{Q}(v,w)=\mathcal{Q}(w,v)$,
(Q2) $\mathcal{Q}((a_1v_1+a_2v_2),w)=a_1\mathcal{Q}(v_1,w)+a_2\mathcal{Q}(v_2,w)$.

When a quadratic form is positive definite, that is for any $v\in V$ the additional conditions

(E1) $\mathcal{Q}(v,v)\geq 0$;
(E2) $\mathcal{Q}(v,v)=0\quad\Leftrightarrow\quad v=0_V$.

are satisfied, then $\mathcal{Q}$ is a scalar product, and we say that V is an euclidean space.

With respect to a basis $\mathcal{B}=(u_1,\ldots,u_n)$ for V, the conditions Q1 and Q2 are clearly satisfied if and only if there exists a symmetric matrix $F=(F_{ab})\in\mathbb{R}^{n,n}$ such that

$$\mathcal{Q}(v,w)\;=\;\mathcal{Q}\big((v_1,\ldots,v_n)_{\mathcal{B}},(w_1,\ldots,w_n)_{\mathcal{B}}\big)\;=\;\sum_{a,b=1}^{n}F_{ab}\,v_aw_b.$$

G. Landi and A. Zampini, *Linear Algebra and Analytic Geometry for Physical Sciences*, Undergraduate Lecture Notes in Physics,
https://doi.org/10.1007/978-3-319-78361-1_13

This expression can be also written as

$$\mathcal{Q}(v, w) = \begin{pmatrix} v_1 & \cdots & v_n \end{pmatrix} \begin{pmatrix} F_{11} & \cdots & F_{1n} \\ \vdots & & \vdots \\ F_{n1} & \cdots & F_{nn} \end{pmatrix} \begin{pmatrix} w_1 \\ \vdots \\ w_n \end{pmatrix}$$

Not surprisingly, the matrix representing the action of the quadratic form $\mathcal{Q}$ depends on the basis considered in V. Under a change of basis $\mathcal{B} \to \mathcal{B}'$ with $\mathcal{B}' = (u'_1, \ldots, u'_n)$ and corresponding matrix $M^{\mathcal{B}',\mathcal{B}}$, as we know, the components of the vectors v, w are transformed as

$$\begin{pmatrix} v'_1 \\ \vdots \\ v'_n \end{pmatrix} = M^{\mathcal{B}',\mathcal{B}} \begin{pmatrix} v_1 \\ \vdots \\ v_n \end{pmatrix}$$

and analogously for w. So we write the action of the quadratic form $\mathcal{Q}$ as

$$\mathcal{Q}(v, w) = \begin{pmatrix} v'_1 & \cdots & v'_n \end{pmatrix} \left({}^t\!M^{\mathcal{B}',\mathcal{B}}\, F\, M^{\mathcal{B}',\mathcal{B}} \right) \begin{pmatrix} w'_1 \\ \vdots \\ w'_n \end{pmatrix}.$$

If we write the dependence on the basis as $\mathcal{Q} \to F^{\mathcal{B}}$, we have then shown the following result.

Proposition 13.1.2 *Given a quadratic form $\mathcal{Q}$ on the finite dimensional real vector space V, with $F^{\mathcal{B}}$ and $F^{\mathcal{B}'}$ the matrices representing $\mathcal{Q}$ on V with respect to the bases $\mathcal{B}$ and $\mathcal{B}'$, it holds that*

$$F^{\mathcal{B}'} = {}^t\!M^{\mathcal{B}',\mathcal{B}}\, F^{\mathcal{B}}\, M^{\mathcal{B}',\mathcal{B}}.$$

Corollary 13.1.3 *Since the matrix $F^{\mathcal{B}}$ associated with the quadratic form $\mathcal{Q}$ on V for the basis $\mathcal{B}$ is symmetric, it is evident from the Proposition 4.1.20 that the matrix $F^{\mathcal{B}'}$ associated with $\mathcal{Q}$ with respect to any other basis $\mathcal{B}'$ is symmetric as well.*

The Proposition 13.1.2 is the counterpart of the Proposition 7.9.9 which related the matrices of a linear maps in different bases. This transformation is not the same as the one for the matrix of an endomorphism as described at the beginning of Chap. 9. To parallel the definition there, one is led to the following definition.

Definition 13.1.4 The symmetric matrices $A, B \in \mathbb{R}^{n,n}$ are called *quadratically equivalent* (or simply *equivalent*) if there exists a matrix $P \in \mathrm{GL}(n)$, such that $B = {}^t\!PAP$. Analogously, the quadratic forms $\mathcal{Q}$ and $\mathcal{Q}'$ defined on a real finite dimensional vector space V are called *equivalent* if their representing matrices are (quadratically) equivalent.

Exercise 13.1.5 Let us consider the symmetric matrices

$$A = \begin{pmatrix} 1 & 0 \\ 0 & 2 \end{pmatrix}, \qquad B = \begin{pmatrix} 1 & 0 \\ 0 & 3 \end{pmatrix}.$$

They are *not* similar, since for example $\det(A) = 2 \neq \det(B) = 3$ (recall that if two matrices are similar, then their determinants must coincide, from the Binet Theorem 5.1.16). They are indeed *quadratically equivalent*: the matrix

$$P = \begin{pmatrix} 1 & 0 \\ 0 & \sqrt{\frac{3}{2}} \end{pmatrix}$$

gives ${}^tPAP = B$.

In parallel with the Remark 9.1.4 concerning similarity of matrices, it is easy to show that the quadratic equivalence is an equivalence relation within the collection of symmetric matrices in $\mathbb{R}^{n,n}$. It is then natural to look for a canonical representative in any equivalence class.

Proposition 13.1.6 *Any quadratic form $\mathcal{Q}$ is equivalent to a diagonal quadratic form, that is one whose representing matrix is diagonal.*

Proof This is just a consequence of the fact that symmetric matrices are orthogonally diagonalisable. From the Proposition 10.5.1 we know that for any symmetric matrix $A \in \mathbb{R}^{n,n}$ there exists a matrix P which is orthogonal, that is $P^{-1} = {}^tP$, such that

$${}^tPAP = \Delta_A$$

where Δ_A is a diagonal matrix whose entries are the eigenvalues of A. □

Without any further requirements on the quadratic form, the matrix Δ_A may have a number μ of positive eigenvalues, a number ν of negative eigenvalues, and also the zero eigenvalue with multiplicity $m_0 = m_{\lambda=0}$. We can order the eigenvalues as follows

$$\Delta_A = \operatorname{diag}(\lambda_{p_1}, \cdots, \lambda_{p_\mu}, \lambda_{n_1}, \cdots, \lambda_{n_\nu}, 0, \cdots, 0)$$

As in the Exercise 13.1.5, we know that the diagonal matrix

$$Q = \operatorname{diag}\left(\frac{1}{\sqrt{\lambda_{p_1}}}, \ldots, \frac{1}{\sqrt{\lambda_{p_\mu}}}, \frac{1}{\sqrt{|\lambda_{n_1}|}}, \ldots, \frac{1}{\sqrt{|\lambda_{n_\nu}|}}, 1, \ldots, 1\right)$$

is such that

$${}^tQ\Delta_A Q = \operatorname{diag}(1, \ldots, 1, -1, \ldots, -1, 0, \ldots, 0) = \mathcal{D}_A$$

with the expected multiplicities μ for $+1$, ν for -1 and m_0 for 0. Since we are considering only transformations between real basis, these multiplicities are constant in each equivalence class of symmetric matrices.

For quadratic forms, this means that any quadratic form $\mathcal{Q}$ on V is equivalent to a diagonal one whose diagonal matrix has a number of μ times $+1$, a number of ν times -1 and a number of $m_0 = \dim(V) - \mu - \nu$ times 0. The multiplicities μ and ν depend only on the equivalence class. Equivalently, for a quadratic form $\mathcal{Q}$ on V, there is a basis for V with respect to which the matrix representing $\mathcal{Q}$ is diagonal, with diagonal entries given $+1$ repeated μ times, -1 repeated ν times and m_0 multiplicity of 0.

Definition 13.1.7 Given a symmetric matrix A on $\mathbb{R}^{n,n}$, we call $\mathcal{D}_A$ its *canonical form* (or *reduced form*). If $\mathcal{Q}$ is a quadratic form on $\mathbb{R}^n$ whose matrix $F^{\mathcal{B}}$ is canonical, then one has

$$\mathcal{Q}(v, w) = v_{p_1} w_{p_1} + \cdots + v_{p_\mu} w_{p_\mu} - (v_{n_1} w_{n_1} + \cdots + v_{n_\nu} w_{n_\nu})$$

with $v = (v_{p_1}, \ldots, v_{p_\mu}, v_{n_1}, \ldots, v_{n_\nu}, \widetilde{v}_1, \ldots, \widetilde{v}_{m_0})$ and analogously for w. This is the *canonical form* for the quadratic form $\mathcal{Q}$. The triple $\mathrm{sign}(\mathcal{Q}) = (\mu, \nu, m_0)$ is called the *signature* of the quadratic form $\mathcal{Q}$. In particular, the quadratic form $\mathcal{Q}$ is called *positive definite* if $\mathrm{sign}(\mathcal{Q}) = (\mu = n, 0, 0)$, and *negative definite* if $\mathrm{sign}(\mathcal{Q}) = (0, \nu = n, 0)$.

Exercise 13.1.8 On $V = \mathbb{R}^3$ consider the quadratic form

$$\mathcal{Q}(v, w) = v_1 w_2 + v_2 w_1 + v_1 w_3 + v_3 w_1 + v_2 w_3 + v_3 w_2$$

where $v = (v_1, v_2, v_3)_{\mathcal{B}}$ and $w = (w_1, w_2, w_3)_{\mathcal{B}}$ with respect to a given basis (u_1, u_2, u_3). Its action is represented by the matrix

$$F^{\mathcal{B}} = \begin{pmatrix} 0 & 1 & 1 \\ 1 & 0 & 1 \\ 1 & 1 & 0 \end{pmatrix}$$

To diagonalise it, we compute its eigenvalues from the characteristic polynomial,

$$p_{F^{\mathcal{B}}}(T) = -T^3 + 3T + 2 = (2 - T)(1 + T)^2.$$

The eigenvalue $\lambda = 2$ is simple, with eigenspace $V_{\lambda=2} = \mathcal{L}((1, 1, 1))$, while the eigenvalue $\lambda = -1$ has multiplicity $m_{\lambda=-1} = 2$, with corresponding eigenspace $V_{\lambda=-1} = \mathcal{L}((1, -1, 0), (1, 1, -2))$. If we define

$$P = M^{\mathcal{B}',\mathcal{B}} = \tfrac{1}{\sqrt{6}} \begin{pmatrix} \sqrt{2} & \sqrt{3} & 1 \\ \sqrt{2} & -\sqrt{3} & 1 \\ \sqrt{2} & 0 & -2 \end{pmatrix}$$

we see that

$$
{}^{t}PF^{\mathcal{B}}P = \begin{pmatrix} 2 & 0 & 0 \\ 0 & -1 & 0 \\ 0 & 0 & -1 \end{pmatrix} = \Delta_A = F^{\mathcal{B}'}
$$

with respect to the basis $\mathcal{B}' = (u_1', u_2', u_3')$ of eigenvectors given explicitly by

$$
\begin{aligned}
u_1' &= \tfrac{1}{\sqrt{3}}(u_1 + u_2 + u_3), \\
u_2' &= \tfrac{1}{\sqrt{2}}(u_1 - u_2), \\
u_3' &= \tfrac{1}{\sqrt{6}}(u_1 + u_2 - 2u_3).
\end{aligned}
$$

With respect to the basis $\mathcal{B}'$ the quadratic form is written as

$$
\mathcal{Q}(v, w) = 2v_1'w_1' - (v_2'w_2' + v_3'w_3').
$$

Motivated by the Exercise 13.1.5, with the matrix

$$
Q = \begin{pmatrix} \frac{1}{\sqrt{2}} & 0 & 0 \\ 0 & 1 & 0 \\ 0 & 0 & 1 \end{pmatrix}
$$

we have that

$$
{}^{t}QF^{\mathcal{B}'}Q = \begin{pmatrix} 1 & 0 & 0 \\ 0 & -1 & 0 \\ 0 & 0 & -1 \end{pmatrix} = F^{\mathcal{B}''}
$$

on the basis $\mathcal{B}'' = (u_1'' = \frac{1}{\sqrt{2}}u_1', u_2'' = u_2', u_3'' = u_3')$. With respect to $\mathcal{B}''$ the quadratic form is

$$
\mathcal{Q}(v, w) = v_1''w_1'' - v_2''w_2'' - v_3''w_3'',
$$

in terms of the components of v, w in the basis $\mathcal{B}''$. Its signature is $\mathrm{sign}(\mathcal{Q}) = (1, 2, 0)$.

Exercise 13.1.9 On the vector space $\mathbb{R}^4$ with canonical basis $\mathcal{E}$, consider the quadratic form

$$
\mathcal{Q}(v, w) = u_1w_1 + u_2w_2 + u_1w_2 + u_2w_1 + u_3w_4 + u_4w_3 - u_3w_3 - u_4w_4,
$$

for any two vectors v, w in $\mathbb{R}^4$. Its representing matrix is

$$
F^{\mathcal{E}} = \begin{pmatrix} 1 & 1 & 0 & 0 \\ 1 & 1 & 0 & 0 \\ 0 & 0 & -1 & 1 \\ 0 & 0 & 1 & -1 \end{pmatrix},
$$

which has been already studied in the Exercise 10.5.3. We can then immediately write

$$P = \frac{1}{\sqrt{2}} \begin{pmatrix} 1 & 0 & 1 & 0 \\ -1 & 0 & 1 & 0 \\ 0 & 1 & 0 & 1 \\ 0 & 1 & 0 & -1 \end{pmatrix}, \qquad {}^tPF^{\mathcal{E}}P = \begin{pmatrix} 0 & 0 & 0 & 0 \\ 0 & 0 & 0 & 0 \\ 0 & 0 & -2 & 0 \\ 0 & 0 & 0 & 2 \end{pmatrix} = F^{\mathcal{E}'},$$

with the basis $\mathcal{E}' = (e'_1, e'_2, e'_3, e'_4)$ given by

$$\begin{aligned} e'_1 &= \tfrac{1}{\sqrt{2}}(e_1 - e_2), \\ e'_2 &= \tfrac{1}{\sqrt{2}}(e_3 + e_4), \\ e'_3 &= \tfrac{1}{\sqrt{2}}(e_1 + e_2), \\ e'_4 &= \tfrac{1}{\sqrt{2}}(e_3 - e_4). \end{aligned}$$

With respect to the basis $\mathcal{E}'' = (e''_1 = e'_1, e''_2 = e'_2, e''_3 = \frac{1}{\sqrt{2}}e'_3, e''_4 = \frac{1}{\sqrt{2}}e'_4)$ it is clear that the matrix representing the action of $\mathcal{Q}$ is $F^{\mathcal{E}''} = \mathrm{diag}(0, 0, -1, 1)$, so that the canonical form of the quadratic form $\mathcal{Q}$ reads

$$\mathcal{Q}(v, w) = -v''_3 w''_3 + v''_4 w''_4$$

with $v = (v''_1, v''_2, v''_3, v''_4)_{\mathcal{E}''}$ and analogously for w. It signature is $\mathrm{sign}(\mathcal{Q}) = (1, 1, 2)$

Remark 13.1.10 Once the dimension n of the real vector space V is fixed, the collection of inequivalent quadratic forms, that is the quotient of the symmetric matrices by the quadratic equivalence relation of the Definition 13.1.7, is labelled by the possible signatures of the quadratic forms, or equivalently by the signatures of the symmetric matrices, written as $\mathrm{sign}(\mathcal{Q}) = (\mu, \nu, n - \mu - \nu)$.

Finally, we state the conditions for a quadratic form to provides a scalar product for a finite dimensional real vector space V. Since we have discussed the topics at length, we omit the proof of the following proposition.

Proposition 13.1.11 *A quadratic form $\mathcal{Q}$ on a finite dimensional real vector space V provides a scalar product if and only if it is positive definite. In such a case we denote the scalar product by*

$$v \cdot w = \mathcal{Q}(v, w).$$

Exercise 13.1.12 With respect to the canonical basis $\mathcal{E}$ on $\mathbb{R}^2$ we consider the quadratic form

$$\mathcal{Q}(v, w) = a v_1 w_1 + v_1 w_2 + v_2 w_1, \qquad \text{with} \quad a \in \mathbb{R},$$

for $v = (v_1, v_2)$ and $w = (w_1, w_2)$. The matrix representing $\mathcal{Q}$ is given by

$$F^{\mathcal{E}} = \begin{pmatrix} a & 1 \\ 1 & 0 \end{pmatrix}$$

and its characteristic polynomial, $p_{F^{\mathcal{E}}}(T) = T^2 - aT - 1$, gives eigenvalues

$$\lambda_{\pm} = \tfrac{1}{2}(a \pm \sqrt{a^2 + 4}).$$

Since for any real value of a there is one positive eigenvalue and one negative eigenvalue, we conclude that the signature of the quadratic form is $\text{sign}(\mathcal{Q}) = (1, 1, 0)$.

Exercise 13.1.13 Consider, from the Exercise 11.1.11, the three dimensional vector space V of antisymmetric matrices in $\mathbb{R}^{3,3}$. If we set

$$\mathcal{Q}(L, L') = -\tfrac{1}{2}\,\text{tr}\,(LL')$$

with $L, L' \in V$, it is immediate to verify that $\mathcal{Q}$ is a quadratic form. Also, the basis elements L_a given in the Exercise 11.1.11 are orthonormal,

$$\mathcal{Q}(L_a, L_b) = \delta_{ab}.$$

Then, the space of real antisymmetric 3×3 matrices is an euclidean space for this scalar product.

Exercise 13.1.14 On $\mathbb{R}^2$ again with the canonical basis, we consider the quadratic form

$$\mathcal{Q}(v, w) = v_1 w_1 + v_2 w_2 + a(v_1 w_2 + v_2 w_1), \qquad \text{with} \quad a \in \mathbb{R},$$

whose representing matrix is

$$F^{\mathcal{E}} = \begin{pmatrix} 1 & a \\ a & 1 \end{pmatrix}.$$

Its characteristic polynomial is $p_{F^{\mathcal{E}}} = (1 - T)^2 - a^2 = (1 - T - a)(1 - T + a)$, with eigenvalues

$$\lambda_{\pm} = 1 \pm a\,.$$

We have the following cases:

- for $a > 1$, it is $\text{sign}(\mathcal{Q}) = (1, 1, 0)$;
- for $a = \pm 1$, it is $\text{sign}(\mathcal{Q}) = (1, 0, 1)$;
- for $a < -1$, it is $\text{sign}(\mathcal{Q}) = (1, 1, 0)$;
- for $-1 < a < 1$, it is $\text{sign}(\mathcal{Q}) = (2, 0, 0)$.

In this last case, the quadratic form endows $\mathbb{R}^2$ with a scalar product. The eigenspaces are

$$\lambda_- = (1 - a), \qquad V_{\lambda_-} = \mathcal{L}((1, -1)),$$
$$\lambda_+ = (1 + a), \qquad V_{\lambda_+} = \mathcal{L}((1, 1)),$$

so we can define the matrix

$$M^{\mathcal{E}',\mathcal{E}} = \tfrac{1}{\sqrt{2}}\begin{pmatrix} 1 & 1 \\ -1 & 1 \end{pmatrix}$$

which gives

$${}^tM^{\mathcal{E}',\mathcal{E}}F^{\mathcal{E}}M^{\mathcal{E}',\mathcal{E}} = \begin{pmatrix} 1-a & 0 \\ 0 & 1+a \end{pmatrix}.$$

With respect to the basis $\mathcal{E}' = \frac{1}{\sqrt{2}}(e_1' = (e_1 - e_2),\, e_2' = \frac{1}{\sqrt{2}}(e_1 + e_2))$ the quadratic form is

$$\mathcal{Q}(v, w) = (1-a)v_1'w_1' + (1+a)v_2'w_2'.$$

We obtain the canonical form for $\mathcal{Q}$ if we consider the basis $\mathcal{E}''$ given by

$$e_1'' = \frac{1}{\sqrt{1-a}}e_1', \qquad e_2'' = \frac{1}{\sqrt{1+a}}e_2'.$$

The basis $\mathcal{E}''$ is orthonormal with respect to the scalar product defined by $\mathcal{Q}$.

Exercise 13.1.15 This exercise puts the results of the previous one in a more general context.

(a) From Exercise 13.1.14 we know that the symmetric matrix

$$S = \begin{pmatrix} 1 & a \\ a & 1 \end{pmatrix},$$

with $a \in \mathbb{R}$, is quadratically equivalent to the diagonal matrix

$$S' = \begin{pmatrix} 1-a & 0 \\ 0 & 1+a \end{pmatrix}.$$

Let us consider S and S' as matrices in $\mathbb{C}^{2,2}$ with real entries (recall that $\mathbb{R}$ is a subfield of $\mathbb{C}$). We can then write

$$\begin{pmatrix} (1-a)^{-1/2} & 0 \\ 0 & (1+a)^{-1/2} \end{pmatrix}\begin{pmatrix} 1-a & 0 \\ 0 & 1+a \end{pmatrix}\begin{pmatrix} (1-a)^{-1/2} & 0 \\ 0 & (1+a)^{-1/2} \end{pmatrix} = \begin{pmatrix} 1 & 0 \\ 0 & 1 \end{pmatrix}$$

for any $a \in \mathbb{R}$. This means that, by complexifying the entries of the real symmetric matrix S, there exists a transformation

$$S \quad \mapsto {}^tPSP = I_2$$

with $P \in \mathrm{GL}(n, \mathbb{C})$ (the group of invertible $n \times n$ complex matrices), which transforms S to I_2.

(b) From the Exercise 13.1.8 we know that the symmetric matrix

$$S = \begin{pmatrix} 0 & 1 & 1 \\ 1 & 0 & 1 \\ 1 & 1 & 0 \end{pmatrix}$$

is quadratically equivalent to

$$S' = \begin{pmatrix} 1 & 0 & 0 \\ 0 & -1 & 0 \\ 0 & 0 & -1 \end{pmatrix}.$$

By again considering them as complex matrices, we can write

$$I_3 = \begin{pmatrix} 1 & 0 & 0 \\ 0 & \mathrm{i} & 0 \\ 0 & 0 & \mathrm{i} \end{pmatrix} \begin{pmatrix} 1 & 0 & 0 \\ 0 & -1 & 0 \\ 0 & 0 & -1 \end{pmatrix} \begin{pmatrix} 1 & 0 & 0 \\ 0 & \mathrm{i} & 0 \\ 0 & 0 & \mathrm{i} \end{pmatrix}.$$

Thus, S is quadratically equivalent to I_3 via an invertible matrix $P \in \mathbb{C}^{n,n}$.

If A is a symmetric matrix with real entries, from the Proposition 13.1.6 we know that it is quadratically equivalent to

$$\Delta_A = \mathrm{diag}\,(\lambda_{p_1}, \cdots, \lambda_{p_\mu}, \lambda_{n_1}, \cdots, \lambda_{n_\nu}, 0, \cdots, 0),$$

with $\lambda_{p_j} > 0$ and $\lambda_{n_j} < 0$. Given the invertible matrix

$$P = \mathrm{diag}\,(\tfrac{1}{\sqrt{\lambda_{p_1}}}, \ldots, \tfrac{1}{\sqrt{\lambda_{p_\mu}}}, \tfrac{\mathrm{i}}{\sqrt{|\lambda_{n_1}|}}, \ldots, \tfrac{\mathrm{i}}{\sqrt{|\lambda_{n_\nu}|}}, 1, \ldots, 1)$$

in $\mathbb{C}^{n,n}$, one finds that

$${}^tP\Delta_A P = \mathrm{diag}\,(1, \ldots, 1, 1, \ldots, 1, 0, \ldots, 0) = \widetilde{\mathcal{D}}_A,$$

where the number of non zero terms $+1$ is given by the rank of A.

If we now define that two symmetric matrices $A, B \in \mathbb{C}^{n,n}$ are quadratically equivalent if there exists a matrix $P \in \mathrm{GL}(n, \mathbb{C})$ such that $B = {}^tPAP$, we can conclude that any real symmetric matrix A is quadratically equivalent to a diagonal matrix $\widetilde{\mathcal{D}}_A$ as above.

The diagonal matrix $\widetilde{\mathcal{D}}_A$ above gives a canonical form for A with respect to quadratic equivalence *after complexification*. Notice that, since $(\mathrm{i}I_n)A(\mathrm{i}I_n) = -A$, we have that A is quadratically equivalent to $-A$. This means that a notion of *complex* signature does not carry much information since it cannot measure the signs of the eingenvalues of A, but only its rank. If $A = {}^tA = \bar{A}$, then we set $\mathrm{sign}(A) = (\mathrm{rk}(A), \dim\ker(A))$.

We conclude by observing that what we have sketched above gives the main properties of a *real quadratic form on a complex* finite dimensional vector space, whose definition is as follows.

Definition 13.1.16 A *real* quadratic forms on a *complex* finite dimensional vector spaces is a map

$$\mathcal{S} : \mathbb{C}^n \times \mathbb{C}^n \longrightarrow \mathbb{C}, \qquad (v, w) \mapsto \mathcal{S}(v, w)$$

such that, for any $v, w, v_1, v_2 \in \mathbb{C}^n$ and $a_1, a_2 \in \mathbb{C}$ it holds that:

(S1) $\mathcal{S}(v, w) = \mathcal{S}(w, v)$,
(S2) $\mathcal{S}(v, w) \in \mathbb{R}$ if and only if $v = \bar{v}$ and $w = \bar{w}$,
(S3) $\mathcal{S}((a_1 v_1 + a_2 v_2), w) = a_1 \mathcal{S}(v_1, w) + a_2 \mathcal{S}(v_2, w)$.

It is clear that $\mathcal{S}$ is a real quadratic form on $\mathbb{C}^n$ if and only if there exists a real basis $\mathcal{B}$ for $\mathbb{C}^n$, that is a basis which is invariant under complex conjugation, with respect to which the matrix $S^{\mathcal{B}} \in \mathbb{C}^{n,n}$ representing $\mathcal{S}$ is symmetric with real entries.

In order to have a more elaborate notion of signature for a bilinear form on complex vector spaces, one needs the notion of hermitian form as explained in the next section.

13.2 Quadratic Forms on Complex Vector Spaces

It is straightforward to generalise to $\mathbb{C}^n$ the main results of the theory of quadratic forms on $\mathbb{R}^n$. The following definition comes naturally after Sects. 3.4 and 8.2.

Definition 13.2.1 Let V be a finite dimensional complex vector space. A *hermitian form* on V is a map

$$\mathcal{H} : V \times V \longrightarrow \mathbb{C}, \qquad (v, w) \mapsto \mathcal{H}(v, w)$$

that fulfils the following properties. For any $v, w, v_1, v_2 \in V$ and $a_1, a_2 \in \mathbb{C}$ it holds that:

(H1) $\mathcal{H}(v, w) = \overline{\mathcal{H}(w, v)}$,
(H2) $\mathcal{H}((a_1 v_1 + a_2 v_2), w) = \overline{a}_1 \mathcal{H}(v_1, w) + \overline{a}_2 \mathcal{Q}(v_2, w)$.

When a hermitian form is positive definite, that is for any $v \in V$ the additional conditions

(E1) $\mathcal{H}(v, v) \geq 0$;
(E2) $\mathcal{H}(v, v) = 0 \quad \Leftrightarrow \quad v = 0_V$.

are satisfied, then $\mathcal{H}$ is a hermitian product, and we say that V is a hermitian space.

We list the properties of hermitian forms in parallel with those of the real case.

(a) With respect to any given basis $\mathcal{B} = (u_1, \ldots, u_n)$ of V, the conditions H1 and H2 are satisfied if and only if there exists a selfadjoint matrix $H = (H_{ab}) \in \mathbb{C}^{n,n}$, $H = H^\dagger$, such that

$$\mathcal{H}(v, w) = \sum_{a,b=1}^{n} H_{ab}\, \overline{v}_a w_b.$$

If we denote by $H^{\mathcal{B}}$ the dependence on the basis of V for the matrix giving the action of $\mathcal{H}$, under a change of bases $\mathcal{B} \to \mathcal{B}'$ we have

$$H^{\mathcal{B}'} = (M^{\mathcal{B}',\mathcal{B}})^\dagger\, H^{\mathcal{B}}\, M^{\mathcal{B}',\mathcal{B}} = (H^{\mathcal{B}'})^\dagger. \tag{13.1}$$

(b) Two selfadjoint matrices $A, B \in \mathbb{C}^{n,n}$ are defined to be *equivalent* if there exists an invertible matrix P such that $B = P^\dagger A P$. This is an equivalence relation within the set of selfadjoint matrices. Analogously, two hermitian forms $\mathcal{H}$ and $\mathcal{H}'$ on $\mathbb{C}^n$ are defined to be equivalent if their representing matrices are equivalent.
(c) From the spectral theory for selfadjoint matrices it is clear that any hermitian form $\mathcal{H}$ is equivalent to a hermitian form whose representing matrix is diagonal. Referring to the relation (13.1), there exists a unitary matrix $U = M^{\mathcal{B}',\mathcal{B}}$ of the change of basis from $\mathcal{B}$ to $\mathcal{B}'$ such that $H^{\mathcal{B}'} = \mathrm{diag}(\lambda_1, \ldots, \lambda_n)$, with $\lambda_j \in \mathbb{R}$ giving the spectrum of $H^{\mathcal{B}}$.
(d) The matrix $H^{\mathcal{B}'}$ is further reduced to its *canonical form* via the same conjugation operation described for the real case after the Proposition 13.1.6.

Since, as in real case, no conjugation as in (13.1) can alter the signs of the eigenvalues of a given selfadjoint matrix, the notion of signature is meaningful for hermitian forms. Such a signature characterises equivalence classes of selfadjoint matrices (and then of hermitian forms) via the equivalence relation we are considering.
(e) A hermitian form $\mathcal{H}$ equips $\mathbb{C}^n$ with a hermitian product if and only if it is positive definite.

Exercise 13.2.2 On $\mathbb{C}^2$ we consider the basis $\mathcal{B} = (u_1, u_2)$ and the hermitian form

$$\mathcal{H}(v, w) = a(v_1 w_1 + v_2 w_2) + \mathrm{i}\, b(v_1 w_2 - v_2 w_1), \qquad \text{with} \quad a, b \in \mathbb{R}$$

for $v = (v_1, v_2)_{\mathcal{B}}$ and $w = (w_1, w_2)_{\mathcal{B}}$. The hermitian form is represented by the matrix

$$H^{\mathcal{B}} = \begin{pmatrix} a & \mathrm{i}\, b \\ -\mathrm{i}\, b & a \end{pmatrix} = (H^{\mathcal{B}})^\dagger.$$

The spectral resolution of this matrix gives

$$\lambda_\pm = a \pm b, \qquad V_{\lambda_\pm} = \mathcal{L}(u_\pm)$$

with normalised eigenvectors

$$u_{\pm} = \tfrac{1}{\sqrt{2}}(\pm \mathrm{i}, 1)_{\mathcal{B}},$$

and with respect to the basis $\mathcal{B}' = (b'_1 = u_+,\ b'_2 = u_-)$ one finds

$$H^{\mathcal{B}'} = \begin{pmatrix} a+b & 0 \\ a-b & 0 \end{pmatrix}.$$

We reduce the hermitian form $\mathcal{H}$ to its canonical form by defining a basis

$$\mathcal{B}'' = \big(\tfrac{1}{\sqrt{|a+b|}}\, b'_1,\ \tfrac{1}{\sqrt{|a-b|}}\, b'_2\big)$$

so to have

$$M^{\mathcal{B}''} = \begin{pmatrix} \frac{a+b}{|a+b|} & 0 \\ 0 & \frac{a-b}{|a-b|} \end{pmatrix}.$$

We see that the signature of $\mathcal{H}$ depends on the relative moduli of a and b. It endows $\mathbb{C}^2$ with a hermitian product if and only if $|a| > |b|$, with $\mathcal{B}''$ giving an orthonormal basis for it.

13.3 The Minkowski Spacetime

We now describe the quadratic form used for a geometrical description of the electromagnetism and for the special theory of relativity.

Let V be a four dimensional real vector space equipped with a quadratic form $\mathcal{Q}$ with signature $\mathrm{sign}(\mathcal{Q}) = (3, 1, 0)$. From the theory we have developed in Sect. 13.1 there exists a (canonical) basis $\mathcal{E} = (e_0, e_1, e_2, e_3)$ with respect to which the action of $\mathcal{Q}$ is given by[1]

$$\mathcal{Q}(v, w) = -v_0 w_0 + v_1 w_1 + v_2 w_2 + v_3 w_3$$

with $v = (v_0, v_1, v_2, v_3)$ and $w = (w_0, w_1, w_2, w_3)$.

Definition 13.3.1 The equivalence class of quadratic forms on $\mathbb{R}^4$ characterised by the signature $(3, 1, 0)$ is said to provide $\mathbb{R}^4$ a *Minkowski* quadratic form, that we denote by η. The datum $(\mathbb{R}^4, \eta)$ is called the Minkowski spacetime, using the name from physics. We shall denote it by M^4 and with a slight abuse of terminology, we shall also denote the action of η as a scalar product

$$v \cdot w = \eta(v, w)$$

and refer to it as the *(Minkowski) scalar product* in M^4.

[1] The reason why we denote the first element by e_0 and the corresponding component of a vector v by v_0 comes from physics, since such components is identified with the *time* coordinate of an event.

Definition 13.3.2 We list the natural generalisations to M^4 of well known definitions in E^n.

(a) For any $v \in (\mathbb{R}^4, \eta)$, the quantity $\|v\|^2 = v \cdot v$ is the square of the (Minkowski) norm of $v \in \mathbb{R}^4$;

the vector v is called *space-like* if $\|v\|^2 > 0$,

the vector v is called *light-like* if $\|v\| = 0$,

the vector v is called *time-like* if $\|v\|^2 < 0$.

(b) Two vectors $v, w \in M^4$ are orthogonal if $v \cdot w = 0$; thus a light-like vector is orthogonal to itself.

(c) A basis $\mathcal{B}$ for $\mathbb{R}^4$ is *orthonormal* if the action of η with respect to $\mathcal{B}$ is diagonal, that is if and only if the matrix $\eta^{\mathcal{B}}$ has the form

$$\eta^{\mathcal{B}} = \begin{pmatrix} -1 & 0 & 0 & 0 \\ 0 & 1 & 0 & 0 \\ 0 & 0 & 1 & 0 \\ 0 & 0 & 0 & 1 \end{pmatrix}.$$

We simply denote $\eta_{\mu\nu} = (\eta^{\mathcal{B}})_{\mu\nu}$ with $\mathcal{B}$ orthonormal.

(d) A matrix $A \in \mathbb{R}^{4,4}$ is a *Lorentz* matrix if its columns yield an orthonormal basis for M^4.

We omit the proof of the following results, which generalise to M^4 analogous results valid in E^n.

Proposition 13.3.3 *Let $\mathcal{B} = (e_0, e_1, e_2, e_3)$ be an orthonormal basis for M^4, with $A \in \mathbb{R}^{4,4}$ and $\phi \in \mathrm{End}(M^4)$.*

(a) The matrix A is a Lorentz matrix if and only if ${}^t\!A\, \eta\, A = \eta$.

(b) It holds that $\phi(v) \cdot \phi(w) = v \cdot w$ for any $v, w \in M^4$ if and only if $M_\phi^{\mathcal{B},\mathcal{B}}$ is a Lorentz matrix.

(c) The system $\mathcal{B}' = (\phi(e_0), \ldots, \phi(e_3))$ is an orthonormal basis for M^4 if and only if for any $v, w \in M^4$ one has $\phi(v) \cdot \phi(w) = v \cdot w$, that is if and only if

$$\phi(e_\mu) \cdot \phi(e_\nu) = e_\mu \cdot e_\nu = \eta_{\mu\nu}.$$

As an immediate consequence of such proposition, one proves that, if $u \in M^4$ is a space-like vector, there exists an orthonormal basis $\mathcal{B}'$ for M^4 with respect to which $u = (0, u'_1, u'_2, u'_3)_{\mathcal{B}'}$. Analogously, if u is a time-like vector, there exists a basis $\mathcal{B}''$ with respect to which $u = (u''_0, 0, 0, 0)_{\mathcal{B}''}$.

Indeed it is straightforward to prove that the set of Lorentz matrices form a group, for matrix multiplication, denoted by $\mathrm{O}(3, 1)$ and called the *Lorentz group*. If the endomorphism ϕ is represented, with respect to an orthonormal basis for M^4 by a

Lorentz matrix, then ϕ is said to be a Lorentz transformation. This means that the set of Lorentz transformations is a group isomorphic to the Lorentz group.

Example 13.3.4 In the special theory of relativity the position of a point mass at a given time t is represented with a vector $x = (x_0 = ct, x_1, x_2, x_3)_{\mathcal{B}}$ in M^4 with respect to an orthonormal basis $\mathcal{B}$, with (x_1, x_2, x_3) giving the so called *spatial* components of x and c denoting the speed of light. Such a vector x is also called an *event*. The linear map

$$\begin{pmatrix} x'_0 \\ x'_1 \\ x'_2 \\ x'_3 \end{pmatrix} = \begin{pmatrix} \gamma & -\beta\gamma & 0 & 0 \\ -\beta\gamma & \gamma & 0 & 0 \\ 0 & 0 & 1 & 0 \\ 0 & 0 & 0 & 1 \end{pmatrix} \begin{pmatrix} x_0 \\ x_1 \\ x_2 \\ x_3 \end{pmatrix}$$

with

$$\beta = v/c \qquad \text{and} \qquad \gamma = (1-\beta^2)^{-1/2},$$

yields the components of the vector x with respect to an orthonormal basis $\mathcal{B}'$ corresponding to an *inertial reference system*, (an inertial observer) which is moving with constant spatial velocity v along the direction e_1. Notice that, being c a limit value for the velocity, we have $|\beta| < 1$ and then $\gamma \geq 1$. It is easy to see that this map is a Lorentz transformation, and that the matrix gives the change of basis $M^{\mathcal{B}',\mathcal{B}}$ in M^4.

From the identity ${}^t\!A\,\eta\,A = \eta$ one gets $\det A = \pm 1$ for a Lorentz matrix A. The set of Lorentz matrices whose determinant is positive is the (sub)group SO(3, 1) of *proper* Lorentz matrices.

If $A_{\mu\nu}$ denotes the entries of a Lorentz matrix A, then from the same identity we can write that

$$-A_{00}^2 + \sum_{k=1}^{3} A_{k0}^2 = -1 \qquad \text{and} \qquad -A_{00}^2 + \sum_{k=1}^{3} A_{0k}^2 = -1,$$

thus proving that $A_{00}^2 \geq 1$. Lorentz matrices with $A_{00} > 1$ are called *ortochronous*. We omit the proof that the set of ortochronous Lorentz matrices form a group as well. Proper *and* ortochronous Lorentz matrices form therefore a group, that we denote by

$$\mathrm{SO}(3,1)^{\uparrow} = \{A \in \mathrm{O}(3,1) \,:\, \det A = 1,\ A_{00} > 1\}.$$

Notice that the Lorentz matrix given in Example 13.3.4 is proper and ortochronous. Given the physical interpretation of the components of a vector in M^4 mentioned before, it is natural to call the endomorphisms represented by the Lorentz matrices

$$P = \begin{pmatrix} 1 & 0 & 0 & 0 \\ 0 & -1 & 0 & 0 \\ 0 & 0 & -1 & 0 \\ 0 & 0 & 0 & -1 \end{pmatrix}, \qquad T = \begin{pmatrix} -1 & 0 & 0 & 0 \\ 0 & 1 & 0 & 0 \\ 0 & 0 & 1 & 0 \\ 0 & 0 & 0 & 1 \end{pmatrix}$$

the *(spatial) parity* and the *time reversal*. The matrix P is improper and ortochronous, while T is improper and antichronous.

We can generalise the final remark from Example 11.3.1 to the Lorentz group case. If A is an improper ortochronous Lorentz matrix, then it is given by the product PA' with $A' \in \mathrm{SO}(3,1)^{\uparrow}$. If A is an improper antichronous Lorentz matrix, then it is given by the product TA' with $A' \in \mathrm{SO}(3,1)^{\uparrow}$. If A is the product PTA' with $A' \in \mathrm{SO}(3,1)^{\uparrow}$, it is called a proper antichronous Lorentz matrix.

Let us describe the structure of the group $\mathrm{SO}(3,1)^{\uparrow}$ in more details. Firstly, notice that if $R \in \mathrm{SO}(3)$ then all matrices of the form

$$A_R = \begin{pmatrix} 1 & 0 & 0 & 0 \\ 0 & & & \\ 0 & & R & \\ 0 & & & \end{pmatrix}$$

are elements in $\mathrm{SO}(3,1)^{\uparrow}$. The set of such matrices A is clearly isomorphic to the group SO(3), so we can refer to SO(3) as the subgroup of *spatial rotations* within the Lorentz group.

The Lorentz matrix in the Example 13.3.4 is not such a rotation. From the Exercise 11.2.3 we write

$$e^{uS_1} = \begin{pmatrix} \gamma & \beta\gamma & 0 & 0 \\ \beta\gamma & \gamma & 0 & 0 \\ 0 & 0 & 1 & 0 \\ 0 & 0 & 0 & 1 \end{pmatrix} \qquad \text{with} \qquad S_1 = \begin{pmatrix} 0 & 1 & 0 & 0 \\ 1 & 0 & 0 & 0 \\ 0 & 0 & 0 & 0 \\ 0 & 0 & 0 & 0 \end{pmatrix}$$

with $\sinh u = \beta\gamma$ and $\cosh u = \gamma$ so that $\operatorname{tgh} u = v/c$.

We therefore have a closer look at the exponential of symmetric matrices of the form

$$S(u) = \begin{pmatrix} 0 & u_1 & u_2 & u_3 \\ u_1 & 0 & 0 & 0 \\ u_2 & 0 & 0 & 0 \\ u_3 & 0 & 0 & 0 \end{pmatrix} = u_1 S_1 + u_2 S_2 + u_3 S_3, \tag{13.2}$$

with $u = (u_1, u_2, u_3)$ a triple of real parameters. If the matrix $R = (R_{ij})$ represents a spatial rotation, a direct computation shows that

$$A_{R^{-1}}\, S\, A_R = \begin{pmatrix} 0 & \sum_{k=1}^3 R_{k1}u_k & \sum_{k=1}^3 R_{k2}u_k & \sum_{k=1}^3 R_{k3}u_k \\ \sum_{k=1}^3 R_{k1}u_k & 0 & 0 & 0 \\ \sum_{k=1}^3 R_{k2}u_k & 0 & 0 & 0 \\ \sum_{k=1}^3 R_{k3}u_k & 0 & 0 & 0 \end{pmatrix}:$$

We see that $u = (u_1, u_2, u_3)$ transforms like a vector in a three dimensional euclidean space, and therefore we write the identity above as

$$S(R^{-1}u) \;=\; A_{R^{-1}} S(u) A_R .$$

This identity allows us to write (see the Proposition 11.2.2)

$$e^{S(u)} \;=\; A_{R^{-1}} e^{S(Ru)} A_R .$$

If R is a proper rotation mapping $u \mapsto (\|u\|_E, 0, 0)$, with $\|u\|_E^2 = u_1^2 + u_2^2 + u_3^2$ the square of the euclidean three-norm, we get

$$e^{S(u)} \;=\; A_{R^{-1}} e^{(\|u\|_E S_1)} A_R .$$

Alternatively, one shows by direct computations that

$$S^2(u) \;=\; \begin{pmatrix} \|u\|_E^2 & 0 & 0 & 0 \\ 0 & & & \\ 0 & & Q & \\ 0 & & & \end{pmatrix} \qquad \text{and} \quad S^3(u) \;=\; \|u\|_E^2 \, S(u)$$

$$\Rightarrow \qquad S^{2k}(u) = \|u\|_E^{2(k-1)} S(u), \quad S^{2k+1}(u) = \|u\|_E^{2k} S(u),$$

where $Q \in \mathbb{R}^{3,3}$ has entries $Q_{ij} = u_i u_j$, so that $Q^2 = \|u\|_E^2 \, Q$. These identities give then

$$e^{S(u)} \;=\; 1 + \tfrac{1}{\|u\|_E^2} \, (\cosh \|u\|_E^2 \; - \; 1) \, S^2(u) \; + \; \tfrac{1}{\|u\|_E} \, \sinh \|u\|_E \, S(u).$$

It is easy to show that $e^{S(u)} \in \mathrm{SO}(3,1)^{\uparrow}$. Such transformations are called *Lorentz boosts*, or hyperbolic rotations. They give the matrices of change of bases $M^{\mathcal{B},\mathcal{B}'}$ where $\mathcal{B}'$ is the orthonormal basis corresponding to an inertial reference system moving with constant velocity $v = (v_1, v_2, v_3)$ in the physical euclidean three dimensional space with respect to the reference system represented by $\mathcal{B}$, by identifying for the velocity,

$$c \, (\mathrm{tgh} \, \|u\|_E) = \|v\|_E .$$

From the properties of the group SO(3) we know that each proper spatial rotation is the exponential of a suitable antisymmetric matrix, that is $A_R = e^{\widetilde{L}}$ where $\widetilde{L}$ is an element in the three dimensional vector space spanned by the matrices $\widetilde{L}_j \subset \mathbb{R}^{4,4}$ of the form

$$\widetilde{L}_j \;=\; \begin{pmatrix} 0 & 0 & 0 & 0 \\ 0 & & & \\ 0 & & L_j & \\ 0 & & & \end{pmatrix}$$

with the antisymmetric matrices L_j, $j = 1, 2, 3$, those of the Exercise 11.1.10, the generators of the Lie algebra $\mathfrak{so}(3)$. With the symmetric matrices S_j in (13.2), we compute the commutators to be

$$[\widetilde{L}_i, \widetilde{L}_j] = \sum_{i,j=1}^{3} \varepsilon_{ijk}\widetilde{L}_k$$

$$[S_i, S_j] = -\sum_{i,j=1}^{3} \varepsilon_{ijk}\widetilde{L}_k$$

$$[S_i, \widetilde{L}_j] = \sum_{i,j=1}^{3} \varepsilon_{ijk}S_k,$$

thus proving that the six dimensional vector space $\mathcal{L}(\widetilde{L}_1, \widetilde{L}_2, \widetilde{L}_3, S_1, S_2, S_3)$ is a matrix Lie algebra (see the Definition 11.1.7) which is denoted $\mathfrak{so}(3, 1)$. What we have discussed gives the proof of the first part of the following proposition, which is analogous of the Proposition 11.2.6.

Proposition 13.3.5 *If M is a matrix in $\mathfrak{so}(3, 1)$, then $e^M \in \mathrm{SO}(3, 1)^{\uparrow}$. When restricted to $\mathfrak{so}(3, 1)$, the exponential map is surjective onto $\mathrm{SO}(3, 1)^{\uparrow}$.*

This means that the group of proper and ortochronous Lorentz matrices is given by spatial rotations, hyperbolic rotations (that is boosts) and their products.

13.4 Electro-Magnetism

By recalling the framework of Sect. 1.4, in the standard euclidean formulation on the space E^3 representing the physical space $\mathcal{S}$ (and with an orthonormal basis) one describes the three dimensional electric $\mathbf{E}(t, \mathbf{x})$ field and the magnetic field $\mathbf{B}(t, \mathbf{x})$ as depending on both the three dimensional position vector $\mathbf{x} = (x_1, x_2, x_3)$ and the time coordinate t. In this section we show that the Maxwell equations for electro-magnetism can be naturally formulated in terms of the geometry of the Minkowski space M^4.

Example 13.4.1 The Maxwell equations in vacuum for the pair $(\mathbf{E}(t, \mathbf{x}), \mathbf{B}(\mathbf{t}, \mathbf{x}))$ are written as

$$\begin{aligned} \operatorname{div}\mathbf{B} &= 0, & \operatorname{rot}\mathbf{E} + \tfrac{\partial \mathbf{B}}{\partial t} &= 0 \\ \operatorname{div}\mathbf{E} &= \tfrac{\rho}{\varepsilon_0}, & \operatorname{rot}\mathbf{B} &= \mu_0\mathbf{J} + \mu_0\varepsilon_0\tfrac{\partial \mathbf{E}}{\partial t} \end{aligned}$$

where ε_0 and μ_0 are the vacuum permittivity and permeability, with $c^2\varepsilon_0\mu_0 = 1$. The *sources* of the fields are the electric charge density ρ (a scalar field) and the current density $\mathbf{J}$ (a vector field).

For the homogeneous Maxwell equations (the first two) the vector fields $\mathbf{E}$ and $\mathbf{B}$ can be written in terms of a vector potential $\mathbf{A}(t, \mathbf{x}) = (A_1(t, \mathbf{x}), A_2(t, \mathbf{x}), A_3(t, \mathbf{x}))$ and a scalar potential $\phi(t, \mathbf{x})$, as

$$\mathbf{B} = \text{rot}\,\mathbf{A}, \qquad \mathbf{E} = -\text{grad}\,\phi - \tfrac{\partial\,\mathbf{A}}{\partial\,t},$$

that makes the homogeneous equations automatically satisfied from the identity div (rot) $= 0$ and rot (grad) $= 0$, in Exercise 1.4.1. If the potentials satisfy the so called *Lorentz gauge condition*

$$\text{div}\,\mathbf{A} + \tfrac{1}{c^2}\tfrac{\partial\,\phi}{\partial\,t} = 0,$$

the two Maxwell equations depending on the sources can be written as

$$\begin{aligned} \nabla^2 A_j - \tfrac{1}{c^2}\tfrac{\partial^2 A_j}{\partial t^2} &= -\mu_0 J_j, \quad \text{for} \quad j = 1, 2, 3, \\ \nabla^2\phi - \tfrac{1}{c^2}\tfrac{\partial^2\phi}{\partial t^2} &= -\tfrac{\rho}{\varepsilon_0} \end{aligned}$$

where $\nabla^2 = \sum_{k=1}^{3}\partial_k^2$ is the spatial Laplacian operator with $\partial_k = \partial/\partial x_k$.

If we define the four-potential as $A = (A_0 = -\frac{\phi}{c}, \mathbf{A})$, then the Lorentz gauge condition is written as (recall the Definition 13.3.2 for the metric $\eta_{\mu\nu}$)

$$\sum_{\mu,\nu=0}^{3} \eta_{\mu\nu}\,\partial_\mu A_\nu = 0,$$

where we also define $\partial_0 = \partial/\partial x_0 = \partial/c\partial t$. In terms of the four-current $J = (J_0 = -\rho/c\varepsilon_0, \mu_0\mathbf{J})$, the inhomogeneous Maxwell equations are written as

$$\sum_{\mu,\nu}^{3} \eta_{\mu\nu}\partial_\mu\partial_\nu A_\rho = -J_\rho.$$

Using the four-dimensional 'nabla' operator $\boldsymbol{\nabla} = (\partial_0, \partial_1, \partial_2, \partial_3)$ we can then write the Lorentz gauge condition as

$$\boldsymbol{\nabla}\cdot A = \sum_{\mu,\nu=0}^{3} \eta_{\mu\nu}\partial_\mu A_\nu = 0,$$

and the inhomogeneous Maxwell equations as

$$\boldsymbol{\nabla}^2 A_\rho = \sum_{\mu,\nu=0}^{3} \eta_{\mu\nu}\partial_\mu\partial_\nu A_\rho = -J_\rho, \qquad \text{for} \quad \rho = 0, 1, 2, 3,$$

thus generalising to the Minkowski spacetime the analogous operations written for the euclidean space E^3 in Sect. 1.4.

Example 13.4.2 From the relations defining the vector fields **E** and **B** in terms of the four-potential vector A, we can write for their components in the physical space

$$B_a = \sum_{b,c=1}^{3} \varepsilon_{abc} \partial_b A_c$$
$$E_a = c(\partial_a A_0 - \partial_0 A_a)$$

for $a = 1, 2, 3$. This shows that the quantity

$$F_{\mu\nu} = \partial_\mu A_\nu - \partial_\nu A_\mu$$

with $\mu, \nu \in \{0, \ldots, 3\}$, defines the entries of the antisymmetric *field strength matrix* (or more precisely field strength 'tensor') F given by

$$F = (F_{\mu\nu}) = \begin{pmatrix} 0 & -E_1/c & -E_2/c & -E_3/c \\ E_1/c & 0 & B_3 & -B_2 \\ E_2/c & -B_3 & 0 & B_1 \\ E_3/c & B_2 & -B_1 & 0 \end{pmatrix}.$$

Merging the definition of F with the Lorentz gauge condition we have

$$\sum_{\mu,\nu}^{3} \eta_{\mu\nu} \partial_\mu \partial_\nu A_\rho = \sum_{\mu,\nu=0}^{3} \eta_{\mu\nu} \partial_\mu (F_{\nu\rho} + \partial_\rho A_\nu)$$
$$= \sum_{\mu,\nu=0}^{3} \eta_{\mu\nu} \partial_\mu F_{\nu\rho} + \partial_\rho \Big(\sum_{\mu,\nu=0}^{3} \eta_{\mu\nu} \partial_\mu A_\nu \Big) = \sum_{\mu,\nu=0}^{3} \eta_{\mu\nu} \partial_\mu F_{\nu\rho}$$

so we can write the inhomogeneous Maxwell equations as

$$\sum_{\mu,\nu=0}^{3} \eta_{\mu\nu} \partial_\mu F_{\nu\rho} = -J_\rho \qquad \text{for} \quad \rho = 0, 1, 2, 3.$$

The homogeneous Maxwell equation can be written in a similar way by means of another useful quantity, the *dual field strength matrix* (or tensor) $\widetilde{F}_{\mu\nu}$. For this one needs the (four dimensional) *totally antisymmetric* symbol $\varepsilon_{a_1 a_2 a_3 a_4}$ with indices $a_j = 0, 1, 2, 3$ and defined by

$$\varepsilon_{a_1 a_2 a_3 a_4} = \begin{cases} +1 & \text{if } (a_1, a_2, a_3, a_4) \text{ is an even permutation of } (0,1,2,3) \\ -1 & \text{if } (a_1, a_2, a_3, a_4) \text{ is an odd permutation of } (0,1,2,3) \\ 0 & \text{if any two indices are equal} \end{cases}.$$

Also, let $\eta^{-1} = (\eta^{\mu\nu})$ be the inverse of the matrix $\eta = (\eta_{\mu\nu})$. The dual field strength matrix is the antisymmetric matrix defined by

$$\widetilde{F} = (\widetilde{F}_{\mu\nu}), \qquad \widetilde{F}_{\mu\nu} = \tfrac{1}{2} \sum_{\alpha,\beta,\gamma,\delta=0}^{3} \varepsilon_{\mu\nu\gamma\delta}\, \eta^{\gamma\alpha}\eta^{\delta\beta} F_{\alpha\beta} = \begin{pmatrix} 0 & B_1 & B_2 & B_3 \\ -B_1 & 0 & -E_3/c & E_2/c \\ -B_2 & E_3/c & 0 & -E_1/c \\ -B_3 & -E_2/c & E_1/c & 0 \end{pmatrix}.$$

Notice that the elements of $\widetilde{F}$ are obtained from those of F by the exchange $\mathbf{E} \leftrightarrow -c\mathbf{B}$.

A straightforward computation shows that the homogeneous Maxwell equations can be written as

$$\sum_{\mu,\nu=0}^{3} \eta_{\mu\nu} \partial_\mu \widetilde{F}_{\nu\rho} = 0, \qquad \text{for} \quad \rho = 0, 1, 2, 3.$$

In terms of $F_{\mu\nu}$ rather then $\widetilde{F}_{\mu\nu}$, these homogeneous equation are the four equations

$$\partial_\rho F_{\mu\nu} + \partial_\mu F_{\nu\rho} + \partial_\nu F_{\rho\mu} = 0$$

for μ, ν, ρ any three of the integers 0, 1, 2, 3.

We now have a glimpse at the geometric nature of the four-potential A and of the antisymmetric matrix F, that is we study how they transform under a change of basis from $\mathcal{B}$ to $\mathcal{B}'$ for M^4. If two inertial observers (for the orthonormal bases $\mathcal{B}$ and $\mathcal{B}'$ for M^4) relate their spacetime components as in the Example (13.3.4), we know from physics that for the transformed electric and magnetic fields $\mathbf{E}'$ and $\mathbf{B}'$ one has

$$\begin{aligned} E_1' &= E_1, & B_1' &= B_1 \\ E_2' &= \gamma(E_2 - vB_3), & B_2' &= \gamma(B_2 + (v/c^2)E_3) \\ E_3' &= \gamma(E_3 + vB_2), & B_2' &= \gamma(B_3 - (v/c^2)E_2) \end{aligned}$$

For the transformed potential $A' = (A_\rho')$ and matrix $F' = (F_{\mu\nu}')$ with $F_{\mu\nu}' = \partial_k' A_s' - \partial_s' A_k'$ (where $\partial_a' = \partial/\partial x_a'$), one then finds

$$A' = M^{\mathcal{B}',\mathcal{B}} A$$

and

$$F' = {}^t(M^{\mathcal{B},\mathcal{B}'})\, F\, M^{\mathcal{B},\mathcal{B}'}.$$

It is indeed possible to check that such identities are valid for any proper and ortochronous Lorentz matrix giving the change of orthonormal basis $\mathcal{B} \to \mathcal{B}'$.

If we denote by M^{4*} the space dual to $(\mathbb{R}^4, \eta)$ with $\{\epsilon_0, \epsilon_1, \epsilon_2, \epsilon_3\}$ the basis dual to $\mathcal{B} = (e_0, \ldots, e_3)$, the definition

$$\eta(\epsilon_a, \epsilon_b) = \eta(e_a, e_b)$$

clearly defines a Minkowski quadratic form on $\mathbb{R}^{4*}$, making then the space M^{4*}. Also, if $\mathcal{B}$ is orthonormal, then $\mathcal{B}^*$ is orthonormal as well.

Recall now the results described in Sect. 8.1 on the dual of a vector space. The previous relations, when compared with the Example 13.3.4, show that the vectors $A = (A_0, \mathbf{A})$ is indeed an element in the dual space M^{4*} to M^4 with respect to the dual basis $\mathcal{B}^*$ to $\mathcal{B}$. From the Proposition 13.1.2 we see also that the matrix elements F transform as the entries of a quadratic form in M^{4*} (although F is antisymmetric). All this means that the components of the electro-magnetic fields **E**, **B** are the entries of an antisymmetric matrix F which transform as a 'contravariant' quadratic form under (proper and orthochronous) Lorentz transformations.

Chapter 14
Affine Linear Geometry

14.1 Affine Spaces

Intuitively, an affine space is a vector space without a 'preferred origin', that is as a set of points such that at each of these there is associated a model (a reference) vector space.

Definition 14.1.1 The *real affine space of dimension* n, denoted by $\mathbb{A}^n(\mathbb{R})$ or simply $\mathbb{A}^n$, is the set $\mathbb{R}^n$ equipped with the map

$$\alpha : \mathbb{A}^n \times \mathbb{A}^n \to \mathbb{R}^n$$

given by

$$\alpha((a_1, \ldots, a_n), (b_1, \ldots, b_n)) = (b_1 - a_1, \ldots, b_n - a_n).$$

Notice that the domain of α is the cartesian product of $\mathbb{R}^n \times \mathbb{R}^n$, while the range of α is the vector space $\mathbb{R}^n$. The notation $\mathbb{A}^n$ stresses the differences between an affine space structure and a vector space structure on the same set $\mathbb{R}^n$. The n-tuples of $\mathbb{A}^n$ are called *points*.

By $\mathbb{A}^1$ we have the *affine real line*, by $\mathbb{A}^2$ the *affine real plane*, by $\mathbb{A}^3$ the *affine real space*. There is an analogous notion of complex affine space $\mathbb{A}^n(\mathbb{C})$, modelled on the vector space $\mathbb{C}^n$.

Remark 14.1.2 The following properties for $\mathbb{A}^n$ easily follows from the above definition:

(p1) for any point $P \in \mathbb{A}^n$ and for any vector $v \in \mathbb{R}^n$, there exists a unique point Q in $\mathbb{A}^n$ such that $\alpha(P, Q) = v$,

(p2) for any triple P, Q, R of points in $\mathbb{A}^n$, it holds that $\alpha(P, Q) + \alpha(Q, R) = \alpha(P, R)$.

G. Landi and A. Zampini, *Linear Algebra and Analytic Geometry for Physical Sciences*, Undergraduate Lecture Notes in Physics,
https://doi.org/10.1007/978-3-319-78361-1_14

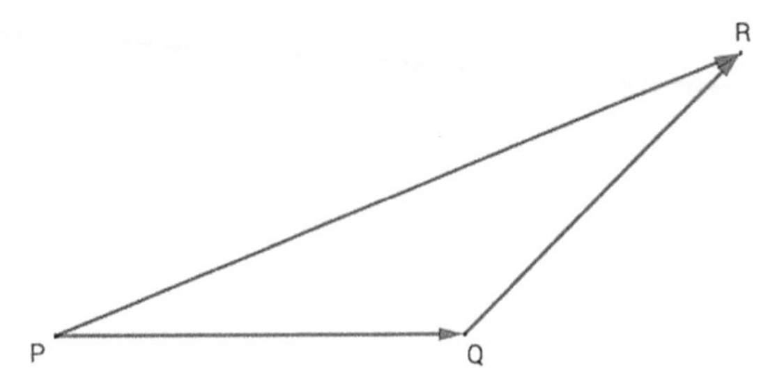

Fig. 14.1 The sum rule $(Q - P) + (R - Q) = R - P$

The property (p2) amounts to the sum rule of vectors (see Fig. 14.1).

Remark 14.1.3 Given the points $P, Q \in \mathbb{A}^n$ and the definition of the map α, the vector $\alpha(P, Q)$ will be also denoted by

$$v = \alpha(P, Q) = Q - P.$$

Then, from the property (p1), we shall write

$$Q = P + v.$$

And property (p2), the sum rule for vectors in $\mathbb{R}^n$, is written as

$$(Q - P) + (R - Q) = R - P.$$

Remark 14.1.4 Given an affine space $\mathbb{A}^n$, from (p2) we have that

(a) for any $P \in \mathbb{A}^n$ it is $\alpha(P, P) = 0_{\mathbb{R}^n}$ (setting $P = Q = R$),
(b) for any pair of points $P, Q \in \mathbb{A}^n$ it is (setting $R = P$), $\alpha(P, Q) = -\alpha(Q, P)$.

A reference system in an affine space is given by selecting a point $O \in \mathbb{A}^n$ so that from (p1) we have a bijection

$$\alpha_O : \mathbb{A}^n \to \mathbb{R}^n, \qquad \alpha_O(P) = \alpha(O, P) = P - O, \tag{14.1}$$

and then a basis $\mathcal{B} = (v_1, \ldots, v_n)$ for $\mathbb{R}^n$.

Definition 14.1.5 The datum $(O, \mathcal{B})$ is called an *affine coordinate system* or an *affine reference system* for $\mathbb{A}^n$ with origin O and basis $\mathcal{B}$. With respect to a reference system $(O, \mathcal{B})$ for $\mathbb{A}^n$, if

$$P - O = (x_1, \ldots, x_n)_{\mathcal{B}} = x_1 v_1 + \cdots + x_n v_n$$

we call $(x_1, \ldots, x_n)$ the *coordinates* of the point $P \in \mathbb{A}^n$ and often write $P = (x_1, \ldots, x_n)$. If $\mathcal{E}$ is the canonical basis for $\mathbb{R}^n$, then $(O, \mathcal{E})$ is the *canonical reference system* for $\mathbb{A}^n$.

Remark 14.1.6 Once an origin has been selected, the affine space $\mathbb{A}^n$ has the structures of $\mathbb{R}^n$ as a vector space. Given a reference system $(O, \mathcal{B})$ for $\mathbb{A}^n$, with $\mathcal{B} = (b_1, \ldots, b_n)$, the points A_i in $\mathbb{A}^n$ given by

$$A_i = O + b_i$$

for $i = 1, \ldots, n$, are called the *coordinate points* of $\mathbb{A}^n$ with respect to $\mathcal{B}$. They have coordinates

$$A_1 = (1, 0, \ldots, 0)_{\mathcal{B}}, \quad A_2 = (0, 1, \ldots, 0)_{\mathcal{B}}, \quad \ldots \quad A_n = (0, 0, \ldots, 1)_{\mathcal{B}}.$$

With the canonical basis $\mathcal{E} = (e_1, \ldots, e_n)$, for $\mathbb{R}^n$ the coordinates points $A_i = O + e_i$ will have coordinates

$$A_1 = (1, 0, \ldots, 0), \quad A_2 = (0, 1, \ldots, 0), \quad \ldots \quad A_n = (0, 0, \ldots, 1).$$

Definition 14.1.7 With $w \in \mathbb{R}^n$, the map

$$T_w : \mathbb{A}^n \to \mathbb{A}^n, \qquad T_w(P) = P + w.$$

is called the *translation of* $\mathbb{A}^n$ *along* ws.

It is clear that T_w is a bijection between $\mathbb{A}^n$ and itself, since T_{-w} is the inverse map to T_w. Once a reference system has been introduced in $\mathbb{A}^n$, a translation can be described by a set of equations, as the following exercise shows.

Exercise 14.1.8 Let us fix the canonical cartesian coordinate system $(O, \mathcal{E})$ for $\mathbb{A}^3$, and consider the vector $w = (1, -2, 1)$. If $P = (x, y, z) \in \mathbb{A}^3$, then $P - O = xe_1 + ye_2 + ze_3$ and we write

$$\begin{aligned} T_w(P) - O &= (P + w) - O \\ &= (P - O) + w \\ &= (xe_1 + ye_2 + ze_3) + (e_1 - 2e_2 + e_3) \\ &= (x + 1)e_1 + (y - 2)e_2 + (z + 1)e_3, \end{aligned}$$

so $T_w\big((x, y, z)) = (x + 1, y - 2, z + 1\big)$.

Following this exercise, it is easy to obtain the equations for a generic translation.

Proposition 14.1.9 *Let $\mathbb{A}^n$ be an affine space with the reference system $(O, \mathcal{B})$. With a vector $w = (w_1, \ldots, w_n)_{\mathcal{B}}$ in $\mathbb{R}^n$, the translation T_w has the following equations*

$$T_w\big((x_1, \ldots, x_n)_{\mathcal{B}}\big) = (x_1 + w_1, \ldots, x_n + w_n)_{\mathcal{B}}.$$

Remark 14.1.10 The translation T_w induces an isomorphism of vector spaces $\phi : \mathbb{R}^n \to \mathbb{R}^n$ given by

$$P - O \quad \mapsto \quad T_w(P) - T_w(O).$$

It is easy to see that ϕ is the *identity* isomorphism. By fixing the orthogonal cartesian reference system $(O, \mathcal{E})$ for $\mathbb{A}^n$, with corresponding coordinates $(x_1, \ldots, x_n)$ for a point P, and a vector $w = w_1 e_1 + \cdots + w_n e_n$, we can write

$$\mathbb{R}^n \ni P - O = x_1 e_1 + \cdots + x_n e_n$$

and

$$T_w(P) = (x_1 + w_1, \ldots, x_n + w_n), \qquad T_w(O) = (w_1, \ldots, w_n),$$

so that we compute

$$\begin{aligned} T_w(P) - T_w(O) &= (T_w(P) - O) - (T_w(O) - O) \\ &= \big((x_1 + w_1)e_1 + \cdots (x_n + w_n)e_n\big) - (w_1 e_1 + \cdots + w_n e_n) \\ &= x_1 e_1 + \cdots + x_n e_n = P - O. \end{aligned}$$

More precisely, such an isomorphism is defined between two distinct copies of the vector space $\mathbb{R}^n$, those associated to the points O and $O' = T_w(O)$ in $\mathbb{A}^n$ thought of as the origins of two different reference systems for $\mathbb{A}^n$. This is depicted in Fig. 14.2.

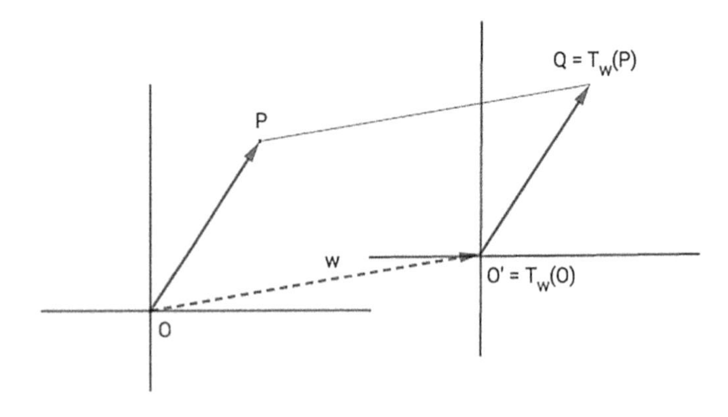

Fig. 14.2 The translation T_w

14.2 Lines and Planes

From the notion of vector line in $\mathbb{R}^2$, using the bijection $\alpha_O : \mathbb{A}^2 \mapsto \mathbb{R}^2$, given in (14.1), it is natural to define a *(straight) line by the origin* the subset in $\mathbb{A}^2$ that corresponds to $\mathcal{L}(v)$ in $\mathbb{R}^2$.

Exercise 14.2.1 Consider $v = (1, 2) \in \mathbb{R}^2$. The corresponding line by the origin in $\mathbb{A}^2$ is the set

$$\{P \in \mathbb{A}^2 \;:\; (P - O) \in \mathcal{L}(v)\} \;=\; \{(x, y) = \lambda(1, 2),\;\; \lambda \in \mathbb{R}\}.$$

Based on this, we have the following definition.

Definition 14.2.2 A (straight) line by the origin in $\mathbb{A}^n$ is the subset

$$r_O \;=\; \{P \in \mathbb{A}^n \;:\; (P - O) \in \mathcal{L}(v)\}$$

for a vector $v \in \mathbb{R}^n \setminus \{0\}$. The vector v is called the *direction vector* of r_O.

Using the identification between $\mathbb{A}^n$ and $\mathbb{R}^n$ given in (14.1) we write

$$r_O = \{P \in \mathbb{A}^n \;:\; P = \lambda v,\;\; \lambda \in \mathbb{R}\},$$

or even

$$r_O : \;\; P = \lambda v\,, \qquad \lambda \in \mathbb{R}.$$

We call such an expression the *vector equation* for the line r_O. Once a reference system $(O, \mathcal{B})$ for $\mathbb{A}^n$ is chosen, via the identification of the components of $P - O$ with respect to $\mathcal{B}$ with the coordinates of a point P, we write the vector equation above as

$$r_O : \;\; (x_1, \ldots, x_n) = \lambda(v_1, \ldots, v_n)\,, \qquad \text{with} \quad \lambda \in \mathbb{R}$$

with $v = (v_1, \ldots, v_n)$ providing the direction of the line.

Remark 14.2.3 It is clear that the subset r_O coincides with $\mathcal{L}(v)$, although they belong to different spaces, that is $r_O \subset \mathbb{A}^n$ while $\mathcal{L}(v) \subset \mathbb{R}^n$. With such a caveat, these sets will be often identified.

Exercise 14.2.4 The line r_O in $\mathbb{A}^3$ with direction vector $v = (1, 2, 3)$ has the vector equation,

$$r_O : \;\; (x, y, z) = \lambda(1, 2, 3), \quad \lambda \in \mathbb{R}.$$

Exercise 14.2.5 Consider the affine space $\mathbb{A}^2$ with the orthogonal reference system $(O, \mathcal{E})$. The subset given by

$$r = \{(x, y) = (1, 2) + \lambda(0, 1),\;\; \lambda \in \mathbb{R}\}$$

clearly represents a line that runs parallel to the second reference axis. Under the translation T_u with $u = (-1, -2)$ we get the set

$$\begin{aligned} T_u(r) &= \{P + u,\ P \in r\} \\ &= \{(x, y) = \lambda(0, 1),\ \lambda \in \mathbb{R}\}, \end{aligned}$$

which is a line by the origin (indeed the second axis of the reference system). If $r_O = T_u(r)$, a line by the origin, it is clear that $r = T_w(r_O)$, with $w = -u$.

This exercise suggests the following definition.

Definition 14.2.6 A set $r \subset \mathbb{A}^n$ is called a *line* if there exist a translation T_w in $\mathbb{A}^n$ and a line r_O by the origin such that $r = T_w(r_O)$.

Being the sets r_O and $\mathcal{L}(v)$ in $\mathbb{R}^n$ coincident, we shall refer to $\mathcal{L}(v)$ as the *direction* of r, and we shall denote it by S_r (with the letter S referring to the fact that $\mathcal{L}(v)$ is a vector subspace in $\mathbb{R}^n$). Notice that, for a line, it is $\dim(S_r) = 1$.

The equation for an arbitrary line follows easily from that of a line by the origin. Let us consider a line by the origin,

$$r_O : \ P = \lambda v\ , \quad \lambda \in \mathbb{R},$$

and the translation T_w with $w \in \mathbb{R}^n$. If $w = Q - O$, the line $r = T_w(r_O)$ is given by

$$\begin{aligned} r &= \{P \in \mathbb{A}^n \ : \ P = T_w(P_O),\ P_O \in r_O\} \\ &= \{P \in \mathbb{A}^n \ : \ P = Q + \lambda v,\ \lambda \in \mathbb{R}\}, \end{aligned}$$

so we write

$$r : \ P = Q + \lambda v. \tag{14.2}$$

With respect to a reference system $(O, \mathcal{B})$, where $Q = (q_1, \ldots, q_n)_{\mathcal{B}}$ and $v = (v_1, \ldots, v_n)_{\mathcal{B}}$, the previous equation can be written as

$$r : \quad (x_1, \ldots, x_n) = (q_1, \ldots, q_n) + \lambda(v_1, \ldots, v_n), \tag{14.3}$$

or equivalently

$$r : \ \begin{cases} x_1 = q_1 + \lambda v_1 \\ \vdots \\ x_n = q_n + \lambda v_n \end{cases} . \tag{14.4}$$

Definition 14.2.7 The expression (14.2) (or equivalently 14.3) is called the *vector equation* of the line r, while the expression (14.4) is called the *parametric equation* of r (stressing that λ is a real parameter).

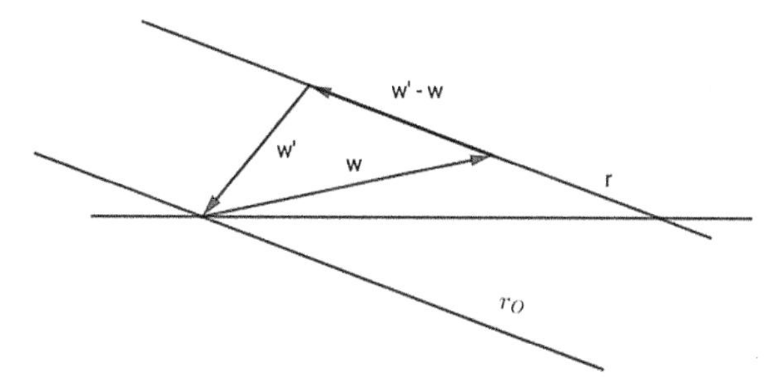

Fig. 14.3 The translation $T_{w'}$ with $w' - w \in \mathcal{L}(v)$ maps r into r_O

Remark 14.2.8 Consider the line whose vector equation is $r \; : \; P = Q + \lambda v$.

(a) We have a unique point in r for each value of λ, and selecting a point of r gives a unique value for λ. The point in r is Q if and only if $\lambda = 0$;
(b) The direction of r is clearly the vector line $\mathcal{L}(v)$. This means that the direction vector v is not uniquely determined by the equation, since each element $v' \in \mathcal{L}(v)$ is a direction vector for r. This arbitrariness can be re-absorbed by a suitable rescaling of the parameter λ: with a rescaling the equation for r can always be written in the given form with v its direction vector.
(c) The point Q is not unique. As the Fig. 14.3 shows, if $Q = O + w$ is a point in r, then any translation $T_{w'}$ with $w' - w \in \mathcal{L}(v)$ maps r into the same line by the origin.

Exercise 14.2.9 We check whether the following lines coincide:

$$\begin{aligned} r : \quad & (x, y) = (1, 2) + \lambda(1, -1), \\ r' : \quad & (x, y) = (2, 1) + \mu(1, -1). \end{aligned}$$

Clearly r and r' have the same direction, which is $S_r = S_{r'} = \mathcal{L}((1, -1)) = r_O$. If we consider $Q = (1, 2) \in r$ and $Q' = (2, 1) \in r'$ with $w = Q - O = (1, 2)$ and $w' = Q' - O = (2, 1)$ we compute,

$$r = T_w(r_O), \qquad r' = T_{w'}(r_O).$$

We have that r coincides with r': as described in the remark above, $w - w' = (-1, 1) \in \mathcal{L}((1, -1))$.

In analogy with the definition of affine lines, one defines planes in $\mathbb{A}^n$.

Definition 14.2.10 A *plane through the origin* in $\mathbb{A}^n$ is any subset of the form

$$\pi_O = \{P \in \mathbb{A}^n \; : \; (P - O) \in \mathcal{L}(u, v)\},$$

with two linearly independent vectors $u, v \in \mathbb{R}^n$.

With the usual identification of a point $P \in \mathbb{A}^n$ with its image $\alpha(P) \in \mathbb{R}^n$ (see 14.1), we write

$$\pi_O = \{P \in \mathbb{A}^n \ : \ P = \lambda u + \mu v, \quad \lambda, \mu \in \mathbb{R}\},$$

or also

$$\pi_O : \ P = \lambda u + \mu v$$

with λ, μ real parameters.

Definition 14.2.11 A subset $\pi \subset \mathbb{A}^n$ is called a *plane* if there exist a translation map T_w in $\mathbb{A}^n$ and a plane π_O through the origin such that $\pi = T_w(\pi_O)$. Since we can identify the elements in π_O with the vectors in $\mathcal{L}(u, v) \subset \mathbb{R}^n$, generalising the analogue Definition 14.2.6 for a line, we define the space $S_\pi = \mathcal{L}(u, v)$ to be the *direction* of π. Notice that $\dim(S_\pi) = 2$.

If $Q = T_w(O)$, that is $w = Q - O$, the points $P \in \pi$ are characterised by

$$P = Q + \lambda u + \mu v. \tag{14.5}$$

Let $(O, \mathcal{B})$ be a reference system for $\mathbb{A}^n$. If $Q = (q_1, \ldots, q_n)_\mathcal{B} \in \mathbb{A}^n$, with $u = (u_1, \ldots, u_n)_\mathcal{B}$ and $v = (v_1, \ldots, v_n)_\mathcal{B} \in \mathbb{R}^n$, the above equation can be written as

$$\pi : \begin{cases} x_1 = q_1 + \lambda u_1 + \mu v_1 \\ \vdots \\ x_n = q_n + \lambda u_v + \mu v_n \end{cases} . \tag{14.6}$$

The relation (14.5) is the *vector equation* of the plane π, while (14.6) is a parametric equation of π.

Exercise 14.2.12 Given the linearly independent vectors $v_1 = (1, 0, 1)$ and $v_2 = (1, -1, 0)$ with respect to the basis $\mathcal{B}$ in $\mathbb{R}^3$, the plane π_O through the origin associated to them is the set of points $P \in \mathbb{A}^3$ given by the vector equation

$$P = \lambda_1 v_1 + \lambda_2 v_2, \quad \lambda_1, \lambda_2 \in \mathbb{R}.$$

With the reference system $(O, \mathcal{B})$, with $P = (x, y, z)$ its parametric equations is

$$(x, y, z) = \lambda_1(1, 0, 1) + \lambda_2(1, -1, 0) \quad \Leftrightarrow \quad \pi : \begin{cases} x = \lambda_1 + \lambda_2 \\ y = -\lambda_2 \\ z = \lambda_1 \end{cases} .$$

Exercise 14.2.13 Given the translation T_w in $\mathbb{A}^3$ with $w = (1, -1, 2)$ in a basis $\mathcal{B}$, the plane π_O of the previous exercise is mapped into the plane $\pi = T_w(\pi_O)$ whose vector equation is

$$\pi : \quad P = Q + \lambda_1 v_1 + \lambda_2 v_2,$$

with $Q = T_w(O) = (1, -1, 2)$. We can equivalently represent the points in π as

$$\pi : \quad (x, y, z) = (1, -1, 2) + \lambda_1(1, 0, 1) + \lambda_2(1, -1, 0).$$

Exercise 14.2.14 Let us consider the vectors v_1, v_2 in $\mathbb{R}^4$ with the following components

$$v_1 = (1, 0, 1, 0), \quad v_2 = (2, 1, 0, -1)$$

in a basis $\mathcal{B}$, and the point Q in $\mathbb{A}^4$ with coordinates

$$Q = (2, 1, 1, 2).$$

in the corresponding reference system $(O, \mathcal{B})$. The plane $\pi \subset \mathbb{A}^4$ through Q whose direction is $S_\pi = \mathcal{L}(v_1, v_2)$ has the vector equation

$$\pi : \quad (x_1, x_2, x_3, x_4) = (2, 1, 1, 2) + \lambda_1(1, 0, 1, 0) + \lambda_2(2, 1, 0, -1)$$

and parametric equation

$$\pi : \quad \begin{cases} x_1 = 2 + \lambda_1 + 2\lambda_2 \\ x_2 = 1 + \lambda_2 \\ x_3 = 1 + \lambda_1 \\ x_4 = 2 - \lambda_2 \end{cases} .$$

Remark 14.2.15 The natural generalisation of the Remark 14.2.8 holds for planes as well. A vector equation for a given plane π is not unique. If

$$\pi : \; P = Q + \lambda u + \mu v$$
$$\pi' : \; P = Q' + \lambda u' + \mu v'$$

are two planes in $\mathbb{A}^n$, then

$$\pi = \pi' \quad \Leftrightarrow \quad \begin{cases} S_\pi = S_{\pi'} \quad (\text{that is } \mathcal{L}(u, v) = \mathcal{L}(u', v')) \\ Q - Q' \in S_\pi \end{cases}$$

Proposition 14.2.16 *Given two distinct points A, B in $\mathbb{A}^n$ (with $n \geq 2$), there is only one line through A and B. A vector equation is*

$$r_{AB} : \; P = A + \lambda(B - A).$$

Proof Being $A \neq B$, this vector equation gives a line since $B - A$ is a non zero vector and the set of points $P - A$ is a one dimensional vector space (that is the direction is one dimensional). The equation r_{AB} contains A (for $\lambda = 0$) and B (for $\lambda = 1$). This shows there exists a line through A and B.

Let us consider another line r_A through A. Its vector equation will be $P = A + \mu v$, with $v \in \mathbb{R}^n$ and μ a real parameter. The point B is contained in r_A if and only if there exists a value μ_0 of the parameter such that $B = A + \mu_0 v$, that is $B - A = \mu_0 v$. Thus the direction of r_A would be $S_{r_A} = \mathcal{L}(v) = \mathcal{L}(B - A) = S_{r_{AB}}$. The line r_A then coincides with r_{AB}. □

Exercise 14.2.17 The line in $\mathbb{A}^2$ through the points $A = (1, 2)$ and $B = (1, -2)$ has equation

$$P = (x, y) = (1, 2) + \lambda(0, -4).$$

Exercise 14.2.18 Let the points A and B in $\mathbb{A}^3$ have coordinates $A = (1, 1, 1)$ and $B = (1, 2, -2)$. The line r_{AB} through them has the vector

$$(x, y, z) = (1, 1, 1) + \lambda(0, 1, -3).$$

Does the point $P = (1, 0, 4)$ belong to r_{AB}? In order to answer this question we need to check whether there is a $\lambda \in \mathbb{R}$ that solves the linear system

$$\begin{cases} 1 = 1 \\ 0 = 1 + \lambda \\ 4 = 1 - 3\lambda. \end{cases}$$

It is evident that $\lambda = -1$ is a solution, so P is a point in r_{AB}.

An analogue of the Proposition 14.2.16 holds for three points in an affine space.

Proposition 14.2.19 *Let A, B, C be three points in an affine space $\mathbb{A}^n$ (with $n \geq 3$). If they are not contained in the same line, there exists a unique plane π_{ABC} through them, with a vector equation given by*

$$\pi_{ABC} : \quad P = A + \lambda(B - A) + \mu(C - A).$$

Proof The vectors $B - A$ and $C - A$ are linearly independent, since they are not contained in the same line. The direction of π_{ABC} is then two dimensional, with $S_{\pi_{ABC}} = \mathcal{L}(B - A, C - A)$. The point A is in π_{ABC}, corresponding to $P(\lambda = \mu = 0)$; the point B is in π_{ABC}, corresponding to $P(\lambda = 1, \mu = 0)$; the point C is in π_{ABC} corresponding to $P(\lambda = 0, \mu = 1)$.

We have then proven that a plane through A, B, C exists. Let us suppose that

$$\pi' : \; P = A + \lambda u + \mu v.$$

gives a plane through the points A, B, C (which are not on the same line) with u and v linearly independent (so its direction is given by $S_{\pi'} = \mathcal{L}(u, v)$). This means that

$B - A \in \mathcal{L}(u, v)$ and $C - A \in \mathcal{L}(u, v)$. Since the spaces are both two dimensional, this reads $\mathcal{L}(B - A, C - A) = \mathcal{L}(u, v)$, proving that π' coincides with π_{ABC}. □

Exercise 14.2.20 Let $A = (1, 2, 0)$, $B = (1, 1, 1)$ and $C = (0, 1, -1)$ be three points in $\mathbb{A}^3$. They are not on the same line, since the vectors $B - A = (0, -1, 1)$ and $C - A = (-1, -1, -1)$ are linearly independent. A vector equation of the plane π_{ABC} is

$$\pi : \quad (x, y, z) = (1, 2, 0) + \lambda(0, -1, 1) + \mu(-1, -1, -1).$$

14.3 General Linear Affine Varieties and Parallelism

The natural generalisation of (straight) lines and planes leads to the definition of a linear affine variety L in $\mathbb{A}^n$, where the direction of L is a subspace in $\mathbb{R}^n$ of dimension greater than 2.

Definition 14.3.1 A *linear affine variety of dimension* k in $\mathbb{A}^n$ is a set

$$L = \{P \in \mathbb{A}^n \ : \ (P - Q) \in V\},$$

where Q is a point in the affine space $\mathbb{A}^n$ and $V \subset \mathbb{R}^n$ is a vector subspace of dimension k in $\mathbb{R}^n$. The vector subspace V is called the *direction* of the variety L, and denoted by $S_L = V$. If $V = \mathcal{L}(v_1, \ldots, v_k)$, a *vector equation* for L is

$$L : \ P = Q + \lambda_1 v_1 + \cdots + \lambda_k v_k$$

for scalars $\lambda_1, \ldots, \lambda_k$ in $\mathbb{R}$.

Remark 14.3.2 It is evident that a line is a one dimensional linear affine variety, while a plane is a two dimensional linear affine variety.

Definition 14.3.3 An linear affine variety of dimension $n - 1$ in $\mathbb{A}^n$ is called a *hyperplane* in $\mathbb{A}^n$.

It is clear that a line is a hyperplane in $\mathbb{A}^2$, while a plane is a hyperplane in $\mathbb{A}^3$.

Exercise 14.3.4 We consider the affine space $\mathbb{A}^4$, the point Q with coordinates $Q = (2, 1, 1, 2)$ with respect to a given reference system $(O, \mathcal{B})$, and the vector subspace $S = \mathcal{L}(v_1, v_2, v_3)$ in $\mathbb{R}^4$ with generators $v_1 = (1, 0, 1, 0)$, $v_2 = (2, 1, 0, -1)$, $v_3 = (0, 0, -1, 1)$ with respect to $\mathcal{B}$. The vector equation of the linear affine variety L with direction $S_L = \mathcal{L}(v_1, v_2, v_3)$ and containing Q is

$$L : (x_1, x_2, x_3, x_4) = (2, 1, 1, 2) + \lambda_1(1, 0, 1, 0) + \lambda_2(2, 1, 0, -1) + \lambda_3(0, 0, -1, 1),$$

while its parametric equation reads

$$L : \begin{cases} x_1 = 2 + \lambda_1 + 2\lambda_2 \\ x_2 = 1 + \lambda_2 \\ x_3 = 1 + \lambda_1 - \lambda_3 \\ x_4 = 2 - \lambda_2 + \lambda_3 \end{cases}.$$

Definition 14.3.5 Let L, L' be two linear affine varieties of the same dimension k in $\mathbb{A}^n$. We say that L is *parallel* to L' if they have the same directions, that is if $S_L = S_{L'}$.

Exercise 14.3.6 Let $L_O \subset \mathbb{A}^n$ be a line through the origin. A line L in $\mathbb{A}^n$ is parallel to L_O if and only if $L = T_w(L_O)$, for $w \in \mathbb{R}^n$. From the Remark 14.2.15 we know that $L = L_O$ if and only if $w \in S_L$.

Let us consider the line through the origin in $\mathbb{A}^2$ given by $L_O : (x, y) = \lambda(3, -2)$. A line will be parallel to L_O if and only if its vector equation is given by

$$L : \quad (x, y) = (\alpha, \beta) + \lambda(3, -2),$$

with $(\alpha, \beta) \in \mathbb{R}^2$. The line L is moreover distinct from L' if and only if $(\alpha, \beta) \notin S_L$.

Definition 14.3.7 Let us consider in $\mathbb{A}^n$ a linear affine variety L of dimension k and a second linear affine variety L' of dimension k', with $k > k'$. The variety L is said to be *parallel* to L' if $S_{L'} \subset S_L$, that is if the direction of L' is a subspace of the direction of L.

Exercise 14.3.8 Let us consider in $\mathbb{A}^3$ the plane given by

$$\pi : \quad (x, y, z) = (0, 2, -1) + \lambda_1(1, 0, 1) + \lambda_2(0, 1, 1).$$

We check whether the following lines,

$$\begin{aligned} r_1 &: \quad (x, y, z) = \lambda(1, -1, 0) \\ r_2 &: \quad (x, y, z) = (0, 3, 0) + \lambda(1, 1, 2) \\ r_3 &: \quad (x, y, z) = (1, -1, 1) + \lambda(1, 1, 1), \end{aligned}$$

are parallel to π.

If S_π denotes the direction of π, we clearly have that $S_\pi = \mathcal{L}(w_1, w_2) = \mathcal{L}((1, 0, 1), (0, 1, 1))$, while we denote by v_i a vector spanning the direction S_{r_i} of the line r_i, $i = 1, 2, 3$. To verify whether $S_{r_i} \subset S_\pi$ it is sufficient to compute the rank of the matrix whose rows are given by (w_1, w_2, v_i).

- For $i = 1$, after a reduction procedure we have,

$$\begin{pmatrix} w_1 \\ w_2 \\ v_1 \end{pmatrix} = \begin{pmatrix} 1 & 0 & 1 \\ 0 & 1 & 1 \\ 1 & -1 & 0 \end{pmatrix} \mapsto \begin{pmatrix} 1 & 0 & 1 \\ 0 & 1 & 1 \\ 0 & 1 & 1 \end{pmatrix}.$$

Since this matrix has rank 2, we have that $v_1 \in \mathcal{L}(w_1, w_2)$, that is $S_{r_1} \subset S_\pi$. We conclude that r_1 is parallel to π. One also checks that $r_1 \not\subset \pi$, since $(0,0,0) \in r_1$ but $(0,0,0) \notin \pi$. To show this, one notices that the origin $(0,0,0)$ is contained in π if and only if the linear system

$$(0,0,0) = (0,2,-1) + \lambda_1(1,0,1) + \lambda_2(0,1,1) \quad \Rightarrow \quad \begin{cases} 0 = \lambda_1 \\ 0 = 2 + \lambda_2 \\ 0 = -1 + \lambda_1 + \lambda_2 \end{cases}$$

has a solution. It is evident that such a solution does not exist.

- For $i = 2$ we proceed as above. The following reduction by rows

$$\begin{pmatrix} w_1 \\ w_2 \\ v_2 \end{pmatrix} = \begin{pmatrix} 1 & 0 & 1 \\ 0 & 1 & 1 \\ 1 & 1 & 2 \end{pmatrix} \quad \mapsto \quad \begin{pmatrix} 1 & 0 & 1 \\ 0 & 1 & 1 \\ 0 & 1 & 1 \end{pmatrix}$$

shows that $v_2 \in \mathcal{L}(w_1, w_2)$, thus r_2 is parallel to π. Now $r_2 \subset \pi$: a point P is in r_2 if and only there exists a $\lambda \in \mathbb{R}$ such that $P = (\lambda, \lambda + 3, 2\lambda)$. For any value of λ, the linear system

$$(\lambda, \lambda+3, 2\lambda) = (0,2,-1) + \lambda_1(1,0,1) + \lambda_2(0,1,1) \quad \Rightarrow \quad \begin{cases} \lambda = \lambda_1 \\ \lambda + 3 = 2 + \lambda_2 \\ 2\lambda \end{cases}$$

has the unique solution $\lambda_1 = \lambda,\ \lambda_2 = \lambda + 1$.

- For $i = 3$ the following reduction by rows

$$\begin{pmatrix} w_1 \\ w_2 \\ v_3 \end{pmatrix} = \begin{pmatrix} 1 & 0 & 1 \\ 0 & 1 & 1 \\ 1 & 1 & 1 \end{pmatrix} \quad \mapsto \quad \begin{pmatrix} 1 & 0 & 1 \\ 0 & 1 & 1 \\ 0 & 1 & 0 \end{pmatrix}$$

shows that the matrix ${}^t(w_1, w_2, v_3)$ has rank 3, so r_3 is not parallel to π.

Definition 14.3.9 Let $L, L' \subseteq \mathbb{A}^n$ two distinct linear affine varieties. We say that L and L' are *incident* if their intersection is non empty, while they are said to be *skew* if they are neither parallel nor incident.

Remark 14.3.10 It is easy to see that two lines (or a line and a plane) are incident if they have a common point. Two distinct planes in $\mathbb{A}^n$ (with $n \geq 3$) are incident if they have a common line.

Exercise 14.3.11 In the affine space $\mathbb{A}^3$ we consider the line r_3 and the plane π as in the Exercise 14.3.8. We know already that they are not parallel, and a point $P = (x, y, z)$ belongs to the intersection $r_3 \cap \pi$ if and only if there exists a λ such that

$P = (1+\lambda, -1+\lambda, 1+\lambda) \in r_3$ and there exist scalars λ_1, λ_2 such that $P = (\lambda_1, 2+\lambda_2, -1+\lambda_1+\lambda_2) \in \pi$. These conditions are equivalent to the linear system

$$\begin{cases} 1+\lambda = \lambda_1 \\ -1+\lambda = 2+\lambda_2 \\ 1+\lambda = -1+\lambda_1+\lambda_2 \end{cases}$$

that has the unique solution $(\lambda = 4, \lambda_1 = 5, \lambda_2 = 1)$. This corresponds to $P = (5, 3, 5) \in r_3 \cap \pi$.

Exercise 14.3.12 Consider again the lines r_1 and r_2 in the Exercise 14.3.8. We know they are not parallel, since $v_1 \notin \mathcal{L}(v_2)$. They are not incident: there are indeed no values of λ and μ such that a point $P = \lambda(1, -1, 0)$ in r_1 coincides with a point $P = (0, 3, 0) + \mu(1, 1, 2)$ in r_2, since the following linear system

$$\begin{cases} \lambda = \mu \\ -\lambda = 3+\mu \\ 0 = 2\mu \end{cases}$$

has no solution. Thus r_1 and r_2 are skew.

Exercise 14.3.13 Given the planes

$$\begin{aligned} \pi :\ & (x, y, z) = (0, 2, -1) + \lambda_1(1, 0, 1) + \lambda_2(0, 1, 1) \\ \pi' :\ & (x, y, z) = (1, -1, 1) + \lambda_1(0, 0, 1) + \lambda_2(2, 1, -1) \end{aligned}$$

in $\mathbb{A}^3$, we determine all the lines which are parallel to *both* π *and* π'.

We denote by r a generic line satisfying such a condition. From the Definition 14.3.5, we require that $S_r \subseteq S_\pi \cap S_{\pi'}$ for the direction S_r of r. Since $S_\pi = \mathcal{L}((1, 0, 1), (0, 1, 1))$ while $S_{\pi'} = \mathcal{L}((0, 0, 1), (2, 1, -1))$, in order to compute $S_\pi \cap S_{\pi'}$ we write the condition

$$\alpha(1, 0, 1) + \beta(0, 1, 1) = \alpha'(0, 0, 1) + \beta'(2, 1, -1)$$

as the linear homogeneous system for $(\alpha, \beta, \alpha', \beta')$ given by $\Sigma :\ AX = 0$ with

$$A = \begin{pmatrix} 1 & 0 & 0 & 2 \\ 0 & 1 & 0 & 1 \\ 1 & 1 & 1 & -1 \end{pmatrix}, \qquad X = \begin{pmatrix} \alpha \\ \beta \\ -\alpha' \\ -\beta' \end{pmatrix}.$$

The space of solution for such a linear system is easily found to be

$$S_\Sigma = \{(\alpha, \beta, -\alpha', -\beta') = t(2, 1, -4, -1)\ :\ t \in \mathbb{R}\},$$

so we have that the intersection $S_\pi \cap S_{\pi'}$ is one dimensional and spanned by the vector

$$2(1, 0, 1) + (0, 1, 1) = 4(0, 0, 1) + (2, 1, -1) = (2, 1, 3).$$

This gives that $S_r = \mathcal{L}((2, 1, 3))$, so we finally write

$$r : \quad (x, y, z) = (a, b, c) + \lambda(2, 1, 3).$$

for an arbitrary $(a, b, c) \in \mathbb{A}^3$.

14.4 The Cartesian Form of Linear Affine Varieties

In the previous sections we have seen that a linear affine variety can be described either with a vector equation or a parametric equation. In this section we relate linear affine varieties to systems of linear equations.

Proposition 14.4.1 *A linear affine variety $L \subseteq \mathbb{A}^n$ corresponds to the space of the solutions of an associated linear system with m equations in n unknowns, that is*

$$\Sigma_L : \quad AX = B, \qquad \textit{for} \;\; A \in \mathbb{R}^{m,n}. \tag{14.7}$$

Moreover, the space of solutions of the corresponding homogeneous linear system describes the direction space $S_L = L_O$ of L, that is

$$\Sigma_{L_O} : \quad AX = 0.$$

We say that the linear system Σ_L given in (14.7) is the *cartesian equation* for the linear affine variety L of dimension $n - \mathrm{rk}(A)$. By computing the space of the solutions of Σ_L in terms of $n - \mathrm{rk}(A)$ parameters, one gets the parametric equation for L. Conversely, given the parametric equation of L, its corresponding cartesian equation is given by consistently 'eliminating' all the parameters in the parametric equation. This linear affine variety can be represented both by a cartesian equation and by a parametric equation, which are related as

$$\begin{array}{ccc} \text{linear system } \Sigma : AX = B & \Longleftrightarrow & \text{space of the solutions for } \Sigma : AX = B \\ (\textit{cartesian equation}) & & (\textit{parametric equation}) \end{array}$$

Notice that for a linear affine variety L a cartesian equation is not uniquely determined: any linear system Σ' which is equivalent to Σ_L (that is for which the spaces of the solutions for Σ_L and Σ' coincide) describe the same linear affine variety. An analogue result holds for the direction space of L, which is equivalently described by any homogenous linear system Σ'_O equivalent to Σ_{L_O}.

We avoid an explicit proof of the Proposition 14.4.1 in general, and analyse the equivalence between the two descriptions via the following examples.

Exercise 14.4.2 Let us consider the line $r \subset \mathbb{A}^2$ with parametric equation

$$r : \quad \begin{cases} x = 1 + \lambda \\ y = 2 - \lambda \end{cases}.$$

We can express the parameter λ in terms of x from the first relation, that is $\lambda = x - 1$, and replace this in the second relation, having

$$x + y - 3 = 0.$$

We set

$$s = \{(x, y) \in \mathbb{A}^2 \, : \, x + y - 3 = 0\}$$

and show that s coincides with r. Clearly $r \subseteq s$, since a point with coordinates $(1 + \lambda, 2 - \lambda) \in r$ solves the linear equation for s:

$$(1 + \lambda) + (2 - \lambda) - 3 = 0.$$

In order to prove that $s \subseteq r$, consider a point $P = (x, y) \in s$, so that $P = (x, y = 3 - x)$ for any value of x: this means considering x as a real parameter. By writing $\lambda = x - 1$, we have $P = (x = \lambda + 1, y = 2 - \lambda)$ for any $\lambda \in \mathbb{R}$, so $P \in r$. We have then $s = r$ as linear affine varieties.

Proposition 14.4.3 *Given a, b, c in $\mathbb{R}$ with $(a, b) \neq (0, 0)$, the solutions of the equation*

$$\Sigma_r : \quad ax + by + c = 0 \tag{14.8}$$

provide the coordinates of all the points $P = (x, y)$ of a line r in $\mathbb{A}^2$ whose direction $S_r = \mathcal{L}((-b, a))$ is given by the solutions of the associated linear homogenous equation

$$\Sigma_{r_O} : \quad ax + by = 0.$$

Moreover, if $r \subset \mathbb{A}^2$ is a line with direction $S_r = \mathcal{L}((-b, a))$, then there exists $c \in \mathbb{R}$ such that the cartesian form for the equation of r is given by (14.8).

Proof We start by showing that the solutions of (14.8) give the coordinates of the points representing the line with direction $\mathcal{L}((-b, a))$ in parametric form.

Let us assume $a \neq 0$. We can then write the space of the solutions for (14.8) as

$$(x, y) = (-\frac{b}{a}\mu - \frac{c}{a}, \mu)$$

where $\mu \in \mathbb{R}$ is a parameter. By rescaling the parameter, that is defining $\lambda = \mu/a$, we write the space of solutions as the points having coordinates,

$$\begin{aligned}(x, y) &= (-b\lambda - \frac{c}{a}, a\lambda) \\ &= (-\frac{c}{a}, 0) + \lambda(-b, a).\end{aligned}$$

This expression gives the vector (and the parametric) equation of a line through $(-c/a, 0)$ with direction $S_r = \mathcal{L}((-b, a))$.

If $a = 0$, we can write the space of the solutions for (14.8) as

$$(x, y) = (\mu, -\frac{c}{b})$$

where $\mu \in \mathbb{R}$ is a parameter. By rescaling the parameter, we can write

$$(x, y) = (-\lambda b, -\frac{c}{b}) = (0, -\frac{c}{b}) + \lambda(-b, 0),$$

giving the vector equation of a line through the point $(0, -c/b)$ with direction $S_r = \mathcal{L}((-b, 0))$.

Now let r be a line in $\mathbb{A}^2$ with direction $S_r = \mathcal{L}((-b, a))$. Its parametric equation is of the form

$$r: \quad \begin{cases} x = x_0 - b\lambda \\ y = y_0 + a\lambda \end{cases}$$

where (x_0, y_0) is an arbitrary point in $\mathbb{A}^2$. If $a \neq 0$, we can eliminate λ by setting

$$\lambda = \frac{y - y_0}{a}$$

from the second relation and then

$$x = x_0 - \frac{b}{a}(y - y_0),$$

resulting into the linear equation

$$ax + by + c = 0$$

with $c = -(ax_0 + by_0)$.

If $a = 0$ then $b \neq 0$, so by rescaling the parameter as $\mu = x_0 - \lambda b$, the points of the line r are $(x = \mu, y = y_0)$. This is indeed the set of the solutions of the linear equation

$$ax + by + c = 0$$

with $a = 0$ and $c = -by_0$. We have then shown that any line with a given direction has the cartesian form given by a suitable linear equation (14.8). □

The equation $ax + by + c = 0$ is called the *cartesian equation of a line* in $\mathbb{A}^2$.

Remark 14.4.4 As already mentioned, a line does not uniquely determine its cartesian equation. With $ax + by + c = 0$ the cartesian equation for r, any other linear equation

$$\rho ax + \rho by + \rho c = 0, \qquad \text{with} \quad \rho \neq 0$$

yields a cartesian equation for the same line, since

$$\rho ax + \rho by + \rho c = 0 \quad \Leftrightarrow \quad \rho(ax + by + c) = 0 \quad \Leftrightarrow \quad ax + by + c = 0.$$

Exercise 14.4.5 The line $\Sigma_r : 2x - y + 3 = 0$ in $\mathbb{A}^2$ has direction $\Sigma_{r_O} : 2x - y = 0$, or $S_r = \mathcal{L}((1, 2))$.

Exercise 14.4.6 We turn now to the description of a plane in the three dimensional affine space in terms of a cartesian equation. Let us consider the plane $\pi \subset \mathbb{A}^3$ with parametric equation

$$\pi : \quad \begin{cases} x = 1 + 2\lambda + \mu \\ y = 2 - \lambda - \mu \\ z = \mu \end{cases} .$$

We eliminate the parameter μ by setting $\mu = z$ from the third relation, and write

$$\pi : \quad \begin{cases} x = 1 + 2\lambda + z \\ y = 2 - \lambda - z \\ \mu = z \end{cases} .$$

We can then eliminate the parameter λ by using the second (for example) relation, so to have $\lambda = 2 - y - z$ and write

$$\pi : \quad \begin{cases} x = 1 + 2(2 - y - z) + z \\ \lambda = 2 - y - z \\ \mu = z \end{cases} .$$

Since these relations are valid for any choice of the parameters λ and μ, we have a resulting linear equation with three unknowns:

$$\Sigma_\pi : \quad x + 2y + z - 5 = 0.$$

Such an equation still represents π, since every point $P \in \pi$ solves the equation (as easily seen by taking $P = (1 + 2\lambda + \mu, 2 - \lambda - \mu, \mu)$) and the space of solutions of Σ_π coincides with the set π.

This example has a general validity for representing in cartesian form a plane in $\mathbb{A}^3$. A natural generalisation of the proof of the previous Proposition 14.4.3 allows one to show the following result.

Proposition 14.4.7 *Given a, b, c, d in $\mathbb{R}$ with $(a, b, c) \neq (0, 0, 0)$, the solutions of the equation*

$$\Sigma_\pi : \quad ax + by + cz + d = 0 \tag{14.9}$$

provide the coordinates of all the points $P = (x, y, z)$ of a plane π in $\mathbb{A}^3$ whose direction S_π is given by the solutions of the associated linear homogenous equation

$$\Sigma_{\pi_O} : \quad ax + by + cz = 0. \tag{14.10}$$

If $\pi \subset \mathbb{A}^3$ is a plane with direction $S_\pi = \mathbb{R}^2$ given by the space of the solutions of (14.10)*, then there exists $d \in \mathbb{R}$ such that the cartesian form for the equation of π is given by* (14.9).

The equation

$$ax + by + cz + d = 0$$

is called the *cartesian equation of a plane* in $\mathbb{A}^3$.

Remark 14.4.8 Analogously to what we noticed in the Remark 14.4.4, the cartesian equation of a plane π in $\mathbb{A}^3$ is not uniquely determined, since it can be again multiplied by a non zero scalar.

Exercise 14.4.9 We next look for a cartesian equation for a line in $\mathbb{A}^3$. As usual, by way of an example, we start by considering the parametric equation of the line $r \subset \mathbb{A}^3$ given by

$$r : \quad \begin{cases} x = 1 + 2\lambda \\ y = 2 - \lambda \\ z = \lambda \end{cases} .$$

By eliminating the parameter λ via (for example) the third relation $\lambda = z$ we have

$$r : \quad \begin{cases} x = 1 + 2z \\ y = 2 - z \\ \lambda = z \end{cases} .$$

Since the third relations formally amounts to redefine a parameter, we write

$$\Sigma_r : \quad \begin{cases} x - 2z - 1 = 0 \\ y + z - 2 = 0 \end{cases} ,$$

which is a linear system with three unknowns and rank 2, thus having ∞^1 solutions. In analogy with the procedure used above for the other examples, it is easy to show that the space of solutions of Σ_r coincides with the line r in $\mathbb{A}^3$.

The following result is the natural generalisation of the Propositions 14.4.3 and 14.4.7.

Proposition 14.4.10 *Given the (complete, see the Definition 6.1.5) matrix*

$$(A, B) = \begin{pmatrix} a_1 & b_1 & c_1 & -d_1 \\ a_2 & b_2 & c_2 & -d_2 \end{pmatrix} \in \mathbb{R}^{2,4}$$

with

$$\mathrm{rk}(A) = \mathrm{rk}\begin{pmatrix} a_1 & b_1 & c_1 \\ a_2 & b_2 & c_2 \end{pmatrix} = 2,$$

the solutions of the linear system

$$\Sigma_\pi : AX = B \quad \Leftrightarrow \quad \begin{cases} a_1x + b_1y + c_1z + d_1 = 0 \\ a_2x + b_2y + c_2z + d_2 = 0 \end{cases} \tag{14.11}$$

provide the coordinates of all the points $P = (x, y, z)$ of a line r in $\mathbb{A}^3$ whose direction S_r is given by the solutions of the associated linear homogenous system

$$\Sigma_{r_O} : \quad AX = 0. \tag{14.12}$$

If $r \subset \mathbb{A}^3$ is a line whose direction $S_r = \mathbb{R}$ is given by the space of the solutions of the linear homogenous system (14.12) *with $A \in \mathbb{R}^{3,2}$ and* $\mathrm{rk}(A) = 2$, *then there exists a vector $B = {}^t(-d_1, -d_2)$ such that the cartesian form for the equation of r is given by* (14.11).

The linear system

$$\Sigma_r : \quad \begin{cases} a_1x + b_1y + c_1z + d_1 = 0 \\ a_2x + b_2y + c_2z + d_2 = 0 \end{cases}$$

with $\mathrm{rk}\begin{pmatrix} a_1 & b_1 & c_1 \\ a_2 & b_2 & c_2 \end{pmatrix} = 2$ is called the *cartesian equation of the line* r in $\mathbb{A}^3$.

Remark 14.4.11 We notice again that the cartesian form (14.11) is not uniquely determined by the line r, since any linear system Σ' which is equivalent to Σ_r describes the same line.

We now a few examples of linear affine varieties described by cartesian equations obtained via removing parameters in their parametric equations.

Exercise 14.4.12 We consider the hyperplane in $\mathbb{A}^4$ with parametric equation

$$H : \quad \begin{cases} x = 1 + \lambda + \mu + \nu \\ y = \lambda - \mu \\ z = \mu + \nu \\ t = \nu \end{cases} .$$

Let us eliminate the parameters: we start by eliminating μ via the fourth relations, then ν by the third relation and eventually λ via the second relation. We have then

$$H: \quad \begin{cases} x = 1 + \lambda + \mu + t \\ y = \lambda - \mu \\ z = \mu + t \\ \nu = t \end{cases} \quad \Leftrightarrow \quad \begin{cases} x = 1 + \lambda + (z - t) + t \\ y = \lambda - (z - t) \\ \mu = z - t \\ \nu = t \end{cases}$$

$$\Leftrightarrow \quad \begin{cases} x = 1 + (y + z - t) + (z - t) + t \\ \lambda = y + z - t \\ \mu = z - t \\ \nu = t \end{cases}.$$

As we have noticed previously, since these relations are valid for each value of the parameters λ, μ, ν, the computations amount to a redefinition of the parameters to y, z, t, so we consider only the first relation, and write

$$\Sigma_H: \quad x - y - 2z + t - 1 = 0$$

as the cartesian equation of the hyperplane H in $\mathbb{A}^4$ with the starting parametric equation. The direction $S_H = \mathbb{R}^3$ of such a hyperplane is given by the vector space corresponding to the space of the solutions of the homogeneous linear equation

$$x - y - 2z + t = 0.$$

Exercise 14.4.13 We consider the plane π in $\mathbb{A}^3$ whose vector equation is given by

$$\pi: \quad P = Q + \lambda v_1 + \mu v_2,$$

with $Q = (2, 3, 0)$ and $v_1 = (1, 0, 1)$, $v_2 = (1, -1, 0)$. By denoting the coordinates $P = (x, y, z)$ we write

$$\begin{pmatrix} x \\ y \\ z \end{pmatrix} = \begin{pmatrix} 2 \\ 3 \\ 0 \end{pmatrix} + \lambda \begin{pmatrix} 1 \\ 0 \\ 1 \end{pmatrix} + \mu \begin{pmatrix} 1 \\ -1 \\ 0 \end{pmatrix},$$

which reads as the parametric equation

$$\pi: \quad \begin{cases} x = 2 + \lambda + \mu \\ y = 3 - \mu \\ z = \lambda \end{cases}.$$

If we eliminate the parameters we write

$$H: \quad \begin{cases} \lambda = z \\ \mu = 3 - y \\ x = 2 + z + 3 - y \end{cases}$$

so to have the following cartesian equation for π:

$$\Sigma_{\pi}: \quad x + y - z - 5 = 0.$$

The direction $S_{\pi} = \mathbb{R}^2$ of the plane π is the space of the solutions of the homogeneous equation

$$x + y - z = 0,$$

and it is easy to see that $S_{\pi} = \mathcal{L}(v_1, v_2)$.

Exercise 14.4.14 We consider the line $r : P = Q + \lambda v$ in $\mathbb{A}^4$, with $Q = (1, -1, 2, 1)$ and direction vector $v = (1, 2, 2, 1)$. Its parametric equation is given by

$$r: \quad \begin{cases} x_1 = 1 + \lambda \\ x_2 = 2 - \lambda \\ x_3 = 2 + 2\lambda \\ x_4 = 1 + \lambda \end{cases}.$$

If we use the first relation to eliminate the parameter λ, we write

$$r: \quad \begin{cases} \lambda = x_1 - 1 \\ x_2 = 2 - (x_1 - 1) \\ x_3 = 2 + 2(x_1 - 1) \\ x_4 = 1 + (x_1 - 1) \end{cases}$$

which amounts to the following cartesian equation

$$\Sigma_r: \quad \begin{cases} x_1 + x_2 - 3 = 0 \\ 2x_1 + x_3 = 0 \\ x_1 + x_4 = 0 \end{cases}.$$

Again, the direction $S_r = \mathbb{R}$ of the line r is given by the space of the solutions for the homogeneous linear system

$$\begin{cases} x_1 + x_2 = 0 \\ 2x_1 + x_3 = 0. \\ x_1 + x_4 = 0 \end{cases}$$

It is easy to see that $S_{r_O} = \mathcal{L}(v)$.

Exercise 14.4.15 We consider the plane $\pi \subseteq \mathbb{A}^3$ whose cartesian equation is

$$\Sigma_\pi : \quad 2x - y + z - 1 = 0.$$

By choosing as free unknowns x, y, we have $z = -2x + y + 1$, that is $P = (x, y, z) \in \pi$ if and only if

$$(x, y, z) = (a, b, -2a + b + 1) = (0, 0, 1) + a(1, 0, -2) + b(0, 1, 1)$$

for any choice of the real parameters a, b. The former relation is then the vector equation of π.

Exercise 14.4.16 We consider the line $r \subseteq \mathbb{A}^3$ with cartesian equation

$$\Sigma_r : \quad \begin{cases} x - y + z - 1 = 0 \\ 2x + y + 2 = 0 \end{cases}.$$

In order to have a vector equation for r we solve such a linear system, getting

$$\Sigma_r : \quad \begin{cases} y = -2x - 2 \\ z = -3x - 3 \end{cases}.$$

Then the space of the solutions for Σ_r is given by the elements

$$(x, y, z) = (a, -2a - 2, -3a - 3) = (0, -2, -3) + a(1, -2, -3).$$

This relation yields a vector equation for r.

We conclude this section by rewriting the Proposition 14.4.1, whose formulation should appear now clearer.

Proposition 14.4.17 *Given the matrix*

$$(A, B) = \begin{pmatrix} a_{11} & a_{12} & \dots & a_{1n} & -b_1 \\ a_{21} & a_{22} & \dots & a_{2n} & -b_2 \\ \vdots & & & \vdots & \vdots \\ a_{m1} & a_{m2} & \dots & a_{mn} & -b_m \end{pmatrix} \in \mathbb{R}^{m,n}$$

with

$$\mathrm{rk}(A) = \mathrm{rk} \begin{pmatrix} a_{11} & a_{12} & \dots & a_{1n} \\ a_{21} & a_{22} & \dots & a_{2n} \\ \vdots & & & \vdots \\ a_{m1} & a_{m2} & \dots & a_{mn} \end{pmatrix} = m < n,$$

the solutions of the linear system

$$\Sigma_L : AX = B \quad \Leftrightarrow \quad \begin{cases} a_{11}x_1 + a_{12}x_2 + \cdots + a_{1n}x_n + b_1 = 0 \\ \cdots \\ a_{m1}x_1 + a_{m2}x_2 + \cdots + a_{mn}x_n + b_m = 0 \end{cases} \tag{14.13}$$

give the coordinates of all points $P = (x_1, x_2, \dots, x_n)$ *of a linear affine variety* L *in* $\mathbb{A}^n$ *of dimension* $k = n - m$ *and whose direction* S_L *is given by the solutions of the associated linear homogenous system*

$$\Sigma_{L_O} : \quad AX = 0. \tag{14.14}$$

If $L \subset \mathbb{A}^n$ *is a linear affine variety of dimension* k*, whose direction* $S_L \cong \mathbb{R}^k$ *is the space of solutions of the linear homogenous system* $AX = 0$ *with* $A \in \mathbb{R}^{m,n}$ *and* $\mathrm{rk}(A) = m < n$*, then there is a vector* $B = {}^t(-b_1, \dots, -b_m)$ *such that the cartesian form for the equation of* L *is given by* (14.13).

14.5 Intersection of Linear Affine Varieties

In this section, by studying particular examples, we introduce some aspects of the general problem of the intersection (that is of the mutual position) of different linear affine varieties.

14.5.1 *Intersection of two lines in* $\mathbb{A}^2$
Let r and r' be the lines in $\mathbb{A}^2$ given by the cartesian equations

$$\Sigma_r : \quad ax + by + c = 0; \qquad \Sigma_{r'} : \quad a'x + b'y + c' = 0.$$

Their intersection is given by the solutions of the linear system

$$\Sigma_{r \cap r'} : \quad \begin{cases} ax + by = -c \\ a'x + b'y = -c' \end{cases}.$$

By defining

$$A = \begin{pmatrix} a & b \\ a' & b' \end{pmatrix}, \qquad (A, B) = \begin{pmatrix} a & b & -c \\ a' & b' & -c' \end{pmatrix}$$

the matrices associated to such a linear system, we have three different possibilities:

- if $\mathrm{rk}(A) = \mathrm{rk}((A, B)) = 1$, the system $\Sigma_{r \cap r'}$ is solvable, with the space of solutions $S_{\Sigma_{r \cap r'}}$ containing ∞^1 solutions. This means that $r = r'$, the two lines coincide;
- if $\mathrm{rk}(A) = \mathrm{rk}((A, B)) = 2$, the system $\Sigma_{r \cap r'}$ is solvable, with the space of solutions $S_{\Sigma_{r \cap r'}}$ made of only one solution, the point $P = (x_0, y_0)$ of intersection between the lines r and r';

- if $\text{rk}(A) = 1$ and $\text{rk}((A, B)) = 2$, the system $\Sigma_{r\cap r'}$ is not solvable, which means that $r \cap r' = \emptyset$; the lines r and r' are therefore parallel, with common direction given by $\mathcal{L}((-b, a))$.

We can summarise such cases as in the following table

$\text{rk}(A)$	$\text{rk}((A, B))$	$S_{\Sigma_{r\cap r'}}$	$r \cap r'$
1	1	∞^1	$r = r'$
2	2	1	$P = (x_0, y_0)$
1	2	$\emptyset$	$\emptyset$

The following result comes easily from the analysis above.

Corollary 14.5.2 *Given the lines r and r' in $\mathbb{A}^2$ with cartesian equations $\Sigma_r : ax + by + c = 0$ and $\Sigma_{r'} : a'x + b'y + c' = 0$, we have that*

$$r = r' \quad \Longleftrightarrow \quad \text{rk}\begin{pmatrix} a & b & -c \\ a' & b' & -c' \end{pmatrix} = 1.$$

Exercise 14.5.3 Given the lines r and s on $\mathbb{A}^2$ whose cartesian equations are

$$\Sigma_r : \quad x + y - 1 = 0, \qquad \Sigma_s : \quad x + 2y + 2 = 0,$$

we study their mutual position. We consider therefore the linear system

$$\Sigma_{r\cap s} : \quad \begin{cases} x + y = 1 \\ x + 2y = -2 \end{cases}.$$

The reduction

$$(A, B) = \begin{pmatrix} 1 & 1 & 1 \\ 1 & 2 & -2 \end{pmatrix} \quad \mapsto \quad \begin{pmatrix} 1 & 1 & 1 \\ 0 & 1 & -3 \end{pmatrix} = (A', B')$$

proves that $\text{rk}(A, B) = \text{rk}(A', B') = 2$ and $\text{rk}(A) = \text{rk}(A') = 2$. The lines r and s have a unique point of intersection, which is computed to be $r \cap s = \{(4, -3)\}$.

Exercise 14.5.4 Consider the lines r and s_α given by their cartesian equations

$$\Sigma_r : \quad x + y - 1 = 0, \qquad \Sigma_{s_\alpha} : \quad x + \alpha y + 2 = 0$$

with $\alpha \in \mathbb{R}$ a parameter. We study the mutual position of r and s_α as depending on the value of α. We therefore study the linear system

$$\Sigma_{r\cap s_\alpha} : \quad \begin{cases} x + y = 1 \\ x + \alpha y = -2 \end{cases}.$$

We use the reduction

$$(A, B) = \begin{pmatrix} 1 & 1 & 1 \\ 1 & \alpha & -2 \end{pmatrix} \mapsto \begin{pmatrix} 1 & 1 & 1 \\ 0 & \alpha - 1 & -3 \end{pmatrix} = (A', B'),$$

proving that $\text{rk}(A, B) = \text{rk}(A', B') = 2$ for any value of α, while $\text{rk}(A) = \text{rk}(A') = 2$ if and only if $\alpha \neq 1$. This means that r is parallel to s_α if and only if $\alpha = 1$ (being in such a case $\Sigma_{s_1} : x + y + 2 = 0$), while for any $\alpha \neq 1$ the two lines intersects in one point, whose coordinates are computed to be

$$r \cap s_\alpha = (\frac{\alpha + 2}{\alpha - 1}, \frac{3}{1 - \alpha}).$$

The following examples show how to study the mutual position of two lines which are not given in the cartesian form. They present different methods without the need to explicitly transforming a parametric or a vector equation into its cartesian form.

Exercise 14.5.5 We consider the line r in $\mathbb{A}^2$ with vector equation

$$r : \quad (x, y) = (1, 2) + \lambda(1, -1),$$

and the line s whose cartesian equation is

$$\Sigma_s : \quad 2x - y - 6 = 0.$$

These line intersect for each value of the parameter λ giving a point in r whose coordinates solve the equation Σ_s. From

$$r : \quad \begin{cases} x = 1 + \lambda \\ y = 2 - \lambda \end{cases}$$

we have

$$2(1 + \lambda) - (2 - \lambda) - 6 = 0 \quad \Leftrightarrow \quad \lambda = 2.$$

This means that r and s intersects in one point, the one with coordinates $(x = 3, y = 0)$.

Exercise 14.5.6 As in the exercise above we consider the line r given by the vector equation

$$r : \quad (x, y) = (1, -1) + \lambda(2, -1)$$

and the line s given by the cartesian equation

$$\Sigma_s : \quad x + 2y - 3 = 0.$$

Their intersections correspond to the value of the parameter λ which solve the equation

$$(1+2\lambda)+2(-1-\lambda)-3=0 \quad \Leftrightarrow \quad -4=0.$$

This means that $r \cap s = \emptyset$; these two lines are parallel.

Exercise 14.5.7 Consider the lines r and s in $\mathbb{A}^2$ both given by a vector equation, for example

$$r: \quad (x, y) = (1,0)+\lambda(1,-2), \qquad s: \quad (x, y) = (1,-1)+\mu(-1,1).$$

The intersection $r \cap s$ corresponds to values of the parameters λ and μ for which the coordinates of a point in r coincide with those of a point in s. We have then to solve the linear system

$$\begin{cases} 1+\lambda = 1-\mu \\ -2\lambda = -1+\mu \end{cases} \quad \Leftrightarrow \quad \begin{cases} \lambda = -\mu \\ 2\mu = -1+\mu \end{cases} \quad \Leftrightarrow \quad \begin{cases} \lambda = 1 \\ \mu = -1 \end{cases}.$$

Having such a linear system one solution, the intersection $s \cap r = P$ where the point P corresponds to the value $\lambda = 1$ in r or equivalently to the value $\mu = -1$ in s. Then $r \cap s = (2, -2)$.

Exercise 14.5.8 As in the previous exercise, we study the intersection of the lines

$$r: \quad (x, y) = (1,1)+\lambda(-1,2), \qquad s: \quad (x, y) = (1,2)+\mu(1,-2).$$

We proceed as above, and consider the linear system

$$\begin{cases} 1-\lambda = 1+\mu \\ 1+2\lambda = 2-2\mu \end{cases} \quad \Leftrightarrow \quad \begin{cases} -\lambda = \mu \\ 1-2\mu = 2-2\mu \end{cases} \quad \Leftrightarrow \quad \begin{cases} -\lambda = \mu \\ 1 = 2 \end{cases}.$$

Since this linear system is not solvable, we conclude that r does not intersect s, and since the direction of r and s coincide, we have that r is parallel to s.

14.5.9 *Intersection of two planes in* $\mathbb{A}^3$
Consider the planes π and π' in $\mathbb{A}^3$ with cartesian equations given by

$$\Sigma_\pi: \quad ax+by+cz+d=0, \qquad \Sigma_{\pi'}: \quad a'x+b'y+c'z+d'=0.$$

Their intersection is given by the solutions of the linear system

$$\Sigma_{\pi\cap\pi'}: \quad \begin{cases} ax+by+cz+d=0 \\ a'x+b'y+c'z+d'=0 \end{cases}$$

which is characterized by the matrices

$$A = \begin{pmatrix} a & b & c \\ a' & b' & c' \end{pmatrix}, \qquad (A, B) = \begin{pmatrix} a & b & c & -d \\ a' & b' & c' & -d' \end{pmatrix}.$$

We have the following possible cases.

$\mathrm{rk}(A)$	$\mathrm{rk}((A, B))$	$S_{\Sigma_{\pi\cap\pi'}}$	$\pi \cap \pi'$
1	1	∞^2	$\pi = \pi'$
2	2	∞^1	line
1	2	$\emptyset$	$\emptyset$

Notice that the case $\pi \cap \pi' = \emptyset$ corresponds to π parallel to π'.

The following corollary parallels the one in Corollary 14.5.2.

Corollary 14.5.10 *Consider two planes π and π' in $\mathbb{A}^3$ having cartesian equations $\Sigma_\pi : ax + by + cz + d = 0$ and $\Sigma_{\pi'} : a'x + b'y + c'z + d' = 0$. One has*

$$\pi = \pi' \iff \mathrm{rk} \begin{pmatrix} a & b & c & -d \\ a' & b' & c' & -d' \end{pmatrix} = 1.$$

Exercise 14.5.11 We consider the planes π and π' in $\mathbb{A}^3$ whose cartesian equations are

$$\Sigma_\pi : \quad x - y + 3z + 2 = 0 \qquad \Sigma_{\pi'} : \quad x - y + z + 1 = 0.$$

The intersection is given by the solutions of the system

$$\Sigma_{\pi\cap\pi'} : \quad \begin{cases} x - y + 3z = -2 \\ x - y + z = -1 \end{cases}.$$

By reducing the complete matrix of such a linear system,

$$(A, B) = \begin{pmatrix} 1 & -1 & 3 & -2 \\ 1 & -1 & 1 & -1 \end{pmatrix} \mapsto \begin{pmatrix} 1 & -1 & 3 & -2 \\ 0 & 0 & 2 & -1 \end{pmatrix},$$

we see that $\mathrm{rk}(A, B) = \mathrm{rk}(A) = 2$, so the linear system has ∞^1 solutions. The intersection $\pi \cap \pi'$ is therefore a line with cartesian equation given by $\Sigma_{\pi\cap\pi'}$.

Exercise 14.5.12 We consider the planes π and π' in $\mathbb{A}^3$ given by

$$\Sigma_\pi : \quad x - y + z + 2 = 0 \qquad \Sigma_{\pi'} : \quad 2x - 2y + 2z + 1 = 0.$$

As in the previous exercise, we reduce the complete matrix of the linear system $\Sigma_{\pi\cap\pi'}$,

$$(A, B) = \begin{pmatrix} 1 & -1 & 1 & -2 \\ 2 & -2 & 2 & -1 \end{pmatrix} \mapsto \begin{pmatrix} 1 & -1 & 1 & -2 \\ 0 & 0 & 0 & 3 \end{pmatrix},$$

to get $\mathrm{rk}(A) = 1$ while $\mathrm{rk}(A, B) = 2$, so $\pi \cap \pi' = \emptyset$. Since these planes are in $\mathbb{A}^3$, they are parallel.

Exercise 14.5.13 We consider the planes π, π', π'' in $\mathbb{A}^3$ whose cartesian equations are given by

$$\begin{aligned} \Sigma_{\pi} : &\quad x - 2y - z + 1 = 0 \\ \Sigma_{\pi'} : &\quad x + y - 2 = 0 \\ \Sigma_{\pi''} : &\quad 2x - 4y - 2z - 5 = 0\,. \end{aligned}$$

For the mutual positions of the pairs π, π' and π, π'', we start by considering the linear system

$$\Sigma_{\pi\cap\pi'} : \quad \begin{cases} x - 2y - z = -1 \\ x + y = 2 \end{cases}.$$

For the complete matrix

$$(A, B) = \begin{pmatrix} 1 & -2 & -1 & -1 \\ 1 & 1 & 0 & 2 \end{pmatrix}$$

we easily see that $\mathrm{rk}(A) = \mathrm{rk}(A, B) = 2$, so the intersection $\pi \cap \pi'$ is the line whose cartesian equation is the linear system $\Sigma_{\pi\cap\pi'}$.

For the intersections of π with π'' we consider the linear system

$$\Sigma_{\pi\cap\pi''} : \quad \begin{cases} x - 2y - z = -1 \\ 2x - 4y - 2z = 5 \end{cases}.$$

The complete matrix

$$(A, B) = \begin{pmatrix} 1 & -2 & 1 & -1 \\ 2 & -4 & -2 & 5 \end{pmatrix},$$

has $\mathrm{rk}(A) = 1$ and $\mathrm{rk}(A, B) = 2$. This means that $\Sigma_{\pi\cap\pi''}$ has no solutions, that is the planes π and π' are parallel, having the same direction given by the vector space solutions of $S_{\pi_O} : x - 2y - z = 0$.

14.5.14 *Intersection of a line with a plane in* $\mathbb{A}^3$
We consider the line r and the plane π in $\mathbb{A}^3$ given by the cartesian equations

$$\Sigma_r : \begin{cases} a_1x + b_1y + c_1z + d_1 = 0 \\ a_2x + b_2y + c_2z + d_2 = 0 \end{cases}, \qquad \Sigma_\pi : \quad ax + by + cz + d = 0.$$

Again, their intersection is given by the solutions of the linear system

$$\Sigma_{\pi\cap r} : \quad \begin{cases} a_1x + b_1y + c_1z = -d_1 \\ a_2x + b_2y + c_2z = -d_2 \\ ax + by + cz = -d \end{cases},$$

with its associated matrices

$$A = \begin{pmatrix} a_1 & b_1 & c_1 \\ a_2 & b_2 & c_2 \\ a & b & c \end{pmatrix}, \qquad (A, B) = \begin{pmatrix} a_1 & b_1 & c_1 & -d_1 \\ a_2 & b_2 & c_2 & -d_2 \\ a & b & c & -d \end{pmatrix}.$$

Since the upper two row vectors of both A and (A, B) matrices are linearly independent, because the corresponding equations represent a line in $\mathbb{A}^3$, only the following cases are possible.

$\mathrm{rk}(A)$	$\mathrm{rk}((A, B))$	$S_{\Sigma_{\pi\cap r}}$	$\pi \cap r$
2	2	∞^1	r
3	3	∞^0	point
2	3	$\emptyset$	$\emptyset$

Notice that, when $\mathrm{rk}(A) = \mathrm{rk}(A, B) = 2$, it is $r \subset \pi$, while, if $\mathrm{rk}(A) = 2$ and $\mathrm{rk}(A, B) = 3$, then r is parallel to π. Indeed, when $\mathrm{rk}(A) = 2$, then $S_r \subset S_\pi$, the direction of r is a subspace in the direction of π. In order to show this, we consider the linear systems for the directions S_r and S_π,

$$\Sigma_{r_O} : \begin{cases} a_1x + b_1y + c_1z = 0 \\ a_2x + b_2y + c_2z = 0 \end{cases}, \qquad \Sigma_{\pi_O} : \quad ax + by + cz = 0.$$

Since $\mathrm{rk}(A) = 2$ and the upper two row vectors are linearly independent, we can write

$$(a, b, c) = \lambda_1(a_1, b_1, c_1) + \lambda_2(a_2, b_2, c_2).$$

If $P = (x_0, y_0.z_0)$ is a point in S_r, then $a_i x_0 + b_i y_0 + c_i z_0 = 0$ for $i = 1, 2$. We can then write

$$\begin{aligned} ax_0 + by_0 + cz_0 &= (\lambda_1 a_1 + \lambda_2 a_2)x_0 + (\lambda_1 b_1 + \lambda_2 b_2)y_0 + (\lambda_1 c_1 + \lambda_2 c_2)z_0 \\ &= \lambda_1(a_1x_0 + b_1y_0 + c_1z_0) + \lambda_2(a_2x_0 + b_2y_0 + c_2z_0) \\ &= 0 \end{aligned}$$

and this proves that $P \in S_\pi$, that is the inclusion $S_r \subset S_\pi$.

Exercise 14.5.15 Given in $\mathbb{A}^3$ the line r and the plane π with cartesian equations

$$\Sigma_r : \quad \begin{cases} x - 2y - z + 1 = 0 \\ x + y - 2 = 0 \end{cases}, \qquad \Sigma_\pi : \quad 2x + y - 2z - 5 = 0,$$

their intersection is given by the solutions of the linear system $\Sigma_{\pi \cap r} : AX = B$ whose associated complete matrix, suitably reduced, reads

$$(A, B) = \begin{pmatrix} 1 & -2 & -1 & -1 \\ 1 & 1 & 0 & 2 \\ 2 & 1 & -2 & 5 \end{pmatrix} \mapsto \begin{pmatrix} 1 & -2 & -1 & -1 \\ 1 & 1 & 0 & 2 \\ 0 & 5 & 0 & 7 \end{pmatrix} = (A', B').$$

Then $\mathrm{rk}(A) = 3$ and $\mathrm{rk}(A, B) = 3$, so the linear system $\Sigma_{\pi \cap r}$ has a unique solution, which corresponds to the unique point P of intersection between r and π. The coordinates of P are easily computed to be $P = (\frac{3}{5}, \frac{7}{5}, -\frac{6}{5})$.

Exercise 14.5.16 We consider in $\mathbb{A}^3$ the line r and the plane π_h with equations

$$\Sigma_r : \quad \begin{cases} x - 2y - z + 1 = 0 \\ x + y - 2 = 0 \end{cases}, \qquad \Sigma_\pi : \quad 2x + hy - 2z - 5 = 0,$$

where h is a real parameter. The complete matrix of to the linear system $\Sigma_{\pi_h \cap r} : AX = B$ giving the intersection of π_h and r is

$$(A_h, B) = \begin{pmatrix} 1 & -2 & -1 & -1 \\ 1 & 1 & 0 & 2 \\ 2 & h & -2 & 5 \end{pmatrix}.$$

We notice that the rank of A_h is at least 2, with $\mathrm{rk}(A_h) = 3$ if and only if $\det(A_h) \neq 0$. It is $\det(A_h) = -h - 4$, so $\mathrm{rk}(A_h) = 3$ if and only if $h \neq -4$. In such a case $\mathrm{rk}(A_h) = 3 = \mathrm{rk}(A_h, B)$, and this means that r and $\pi_{h \neq -4}$ have a unique point of intersection.

If $h = -4$, then $\mathrm{rk}(A_{-4}) = 2$: the reduction

$$(A_{-4}, B) = \begin{pmatrix} 1 & -2 & -1 & -1 \\ 1 & 1 & 0 & 2 \\ 2 & -4 & -2 & 5 \end{pmatrix} \mapsto \begin{pmatrix} 1 & -2 & -1 & -1 \\ 1 & 1 & 0 & 2 \\ 0 & 0 & 0 & 7 \end{pmatrix}$$

shows that $\mathrm{rk}(A_{-4}, B) = 3$, so the linear system $A_{-4}X = B$ has no solutions, and r is parallel to π.

Exercise 14.5.17 As in the Exercise 14.5.15 we study the intersection of a plane π (represented by a cartesian equation) and a line r in $\mathbb{A}^3$ (represented by a parametric equation). Consider for instance,

$$r: \quad (x, y, z) = (3, -1, 5) + \lambda(1, -1, 2), \qquad \Sigma_\pi : \quad x + y - z + 1 = 0.$$

As before, the intersection $\pi \cap r$ corresponds to the values of the parameter λ for which the coordinates $P = (3 + \lambda, -1 - \lambda, 5 + 2\lambda)$ of a point in r solve the cartesian equation for π, that is

$$(3 + \lambda) + (-1 - \lambda) - (5 + 2\lambda) + 1 = 0 \quad \Rightarrow \quad -2\lambda - 2 = 0 \quad \Rightarrow \quad \lambda = -1.$$

We have then $r \cap \pi = (2, 0, 3)$.

14.5.18 *Intersection of two lines in* $\mathbb{A}^3$
We consider a line r and a line r' in $\mathbb{A}^3$ with cartesian equations

$$\Sigma_r : \quad \begin{cases} a_1 x + b_1 y + c_1 z + d_1 = 0 \\ a_2 x + b_2 y + c_2 z + d_2 = 0 \end{cases}, \qquad \Sigma'_r : \quad \begin{cases} a'_1 x + b'_1 y + c'_1 z + d'_1 = 0 \\ a'_2 x + b'_2 y + c'_2 z + d'_2 = 0 \end{cases}.$$

The intersection is given by the linear system $\Sigma_{r\cap r'}$ whose associated matrices are

$$A = \begin{pmatrix} a_1 & b_1 & c_1 \\ a_2 & b_2 & c_2 \\ a'_1 & b'_1 & c'_1 \\ a'_2 & b'_2 & c'_2 \end{pmatrix}, \qquad (A, B) = \begin{pmatrix} a_1 & b_1 & c_1 & -d_1 \\ a_2 & b_2 & c_2 & -d_2 \\ a'_1 & b'_1 & c'_1 & -d'_1 \\ a'_2 & b'_2 & c'_2 & -d'_2 \end{pmatrix}.$$

Once again, different possibilities depending on the mutual ranks of these. As we stressed in the previous case 14.5.14, since r and r' are lines, the upper two row vectors R_1 and R_2 of both A and (A, B) are linearly independent, as are the last two row vectors, R_3 and R_4. Then,

$\mathrm{rk}(A)$	$\mathrm{rk}((A, B))$	$S_{\Sigma_{r\cap r'}}$	$r \cap r'$
2	2	∞^1	r
3	3	∞^0	point
2	3	$\emptyset$	$\emptyset$
3	4	$\emptyset$	$\emptyset$

In the first case, with $\mathrm{rk}(A) = \mathrm{rk}(A, B) = 2$, the lines r, r' coincide, while in the second case, with $\mathrm{rk}(A) = \mathrm{rk}(A, B) = 3$, they have a unique point of intersection, whose coordinates are given by the solution of the system $AX = B$.

In the third and the fourth case, the condition $\mathrm{rk}(A) \neq \mathrm{rk}(A, B)$ means that the two lines do not intersect. If $\mathrm{rk}(A) = 2$, then the row vectors R_3 and R_4 of A are both linearly dependent of R_1 and R_2, and therefore the homogeneous linear systems

$$\Sigma_{r_O} : \begin{cases} a_1 x + b_1 y + c_1 z = 0 \\ a_2 x + b_2 y + c_2 z = 0 \end{cases}, \qquad \Sigma_{r'_O} : \begin{cases} a'_1 x + b'_1 y + c'_1 z = 0 \\ a'_2 x + b'_2 y + c'_2 z = 0 \end{cases},$$

are equivalent. We have then that $S_r = S_{r'}$, the direction of r coincide with that of r', that is r is parallel to r'. If $\mathrm{rk}(A) = 3$ (the fourth case in the table above) the lines are not parallel and do not intersect, so they are skew.

Exercise 14.5.19 We consider the line r and r' in $\mathbb{A}^3$ whose cartesian equations are

$$\Sigma_r : \begin{cases} x - y + 2z + 1 = 0 \\ x + z - 1 = 0 \end{cases}, \qquad \Sigma_{r'} : \begin{cases} y - z + 2 = 0 \\ x + y + z = 0 \end{cases}.$$

We reduce the complete matrix associated to the linear system $\Sigma_{r \cap r'}$, that is

$$(A, B) = \begin{pmatrix} 1 & -1 & 2 & -1 \\ 1 & 0 & 1 & 1 \\ 0 & 1 & -1 & -2 \\ 1 & 1 & 1 & 0 \end{pmatrix} \mapsto \begin{pmatrix} 1 & -1 & 2 & -1 \\ 0 & 1 & -1 & 2 \\ 0 & 1 & -1 & -2 \\ 0 & 2 & -1 & 1 \end{pmatrix}$$

$$\mapsto \begin{pmatrix} 1 & -1 & 2 & -1 \\ 0 & 1 & -1 & 2 \\ 0 & 0 & 0 & -4 \\ 0 & 0 & 1 & -3 \end{pmatrix} = (A', B').$$

Since $\mathrm{rk}(A') = 3$ and $\mathrm{rk}(A', B') = 4$, the two lines are skew.

Chapter 15
Euclidean Affine Linear Geometry

15.1 Euclidean Affine Spaces

In the previous chapter we have dealt with the (real and linear) affine space $\mathbb{A}^n$ as modelled on the vector space $\mathbb{R}^n$. In this chapter we study the additional structures on $\mathbb{A}^n$ that come when passing from $\mathbb{R}^n$ to the euclidean space E^n (see the Chap. 3). Taking into account the scalar product allows one to introduce metric notions (such as distances and angles) into an affine space.

Definition 15.1.1 The affine space $\mathbb{A}^n$ associated to the Euclidean vector space $E^n = (\mathbb{R}^n, \cdot)$ is called the *Euclidean affine space* and denoted $\mathbb{E}^n$. A reference system $(O, \mathcal{B})$ for $\mathbb{E}^n$ is called *cartesian orthogonal* if the basis $\mathcal{B}$ for E^n is orthonormal.

Recall that, if $\mathcal{B}$ is an orthonormal basis for E^n, the matrix of change of basis $M^{\mathcal{E},\mathcal{B}}$ (the matrix whose column vectors are the components of the vectors in $\mathcal{B}$ with respect to the canonical basis $\mathcal{E}$) is orthogonal by definition (see the Chap. 10, Definition 10.1.1), and thus $\det(M^{\mathcal{E},\mathcal{B}}) = \pm 1$.

In our analysis in this chapter we shall always consider cartesian orthogonal reference systems.

Exercise 15.1.2 Let r be the (straight) line in $\mathbb{E}^2$ with vector equation

$$(x, y) = (1, -2) + \lambda(1, -1).$$

We take $A = (1, -2)$ and $v = (1, -1)$. To determine a cartesian equation for r, in alternative to the procedure described at length in the previous chapter (that is removing the parameter λ), one observes that, since $\mathcal{L}(v)$ is the direction of r, and thus the vector $u = (1, 1)$ is orthogonal to v, we can write

G. Landi and A. Zampini, *Linear Algebra and Analytic Geometry for Physical Sciences*, Undergraduate Lecture Notes in Physics,
https://doi.org/10.1007/978-3-319-78361-1_15

$$\begin{aligned} P = (x, y) \in r \quad &\Longleftrightarrow \quad P - A \in \mathcal{L}(v) \\ &\Longleftrightarrow \quad (P - A) \cdot u = 0 \\ &\Longleftrightarrow \quad (x - 1, y + 2) \cdot (1, 1) = 0. \end{aligned}$$

This condition can be written as

$$x + y + 1 = 0,$$

yielding a cartesian equation Σ_r for r.

This exercise shows that, if r is a line in $\mathbb{E}^2$ whose vector equation is $r : P = A + \lambda v$, with u a vector orthogonal to v so that for the direction of r one has $S_r = \mathcal{L}(u)^\perp$, we have

$$P \in r \quad \Longleftrightarrow \quad (P - A) \cdot u = 0.$$

This expression is called the *normal equation* for the line r.

We can generalise this example to any hyperplane.

Proposition 15.1.3 *Let $H \subset \mathbb{E}^n$ be a hyperplane, with $A \in H$. If $u \in \mathbb{R}^n$ is a non zero vector orthogonal to the direction S_H of the hyperplane, that is $\mathcal{L}(u) = (S_H)^\perp$, then it holds that*

$$P \in H \quad \Longleftrightarrow \quad (P - A) \cdot u = 0.$$

Definition 15.1.4 The equation

$$\mathcal{N}_H : \quad (P - A) \cdot u = 0$$

is called the *normal equation of the hyperplane* H in $\mathbb{E}^n$. If $n = 2$, it yields the normal equation of a line; if $n = 3$, it yields the normal equation of a plane.

Remark 15.1.5 Notice that, as we already seen for a cartesian equation in the previous chapter (see the Remark 14.4.11), the normal equation $\mathcal{N}_H$ for a given hyperplane in $\mathbb{E}^n$ is not uniquely determined, since A can range in H and the vector u is given up to an arbitrary non zero scalar.

Remark 15.1.6 With a cartesian equation

$$\Sigma_H : \quad a_1 x_1 + \cdots + a_n x_n = b$$

for the hyperplane for H in $\mathbb{E}^n$, one has $S_H^\perp = \mathcal{L}((a_1, \ldots, a_n))$. This follows from the definition

$$\begin{aligned} S_H &= \{(x_1, \ldots, x_n) \in \mathbb{R}^n \; : \; a_1 x_1 + \cdots + a_n x_n = 0\} \\ &= \{(x_1, \ldots, x_n) \in \mathbb{R}^n \; : \; (a_1, \ldots, a_n) \cdot (x_1, \ldots, x_n) = 0\}. \end{aligned}$$

With A an arbitrary point in H, a normal equation for H is indeed given by

$$\mathcal{N}_H : \quad (P - A) \cdot (a_1, \ldots, a_n) = 0.$$

Exercise 15.1.7 We determine both a cartesian and a normal equation for the plane π in $\mathbb{A}^3$ whose direction is orthogonal to $u = (1, 2, 3)$ and that contains the point $A = (1, 0, -1)$. We have

$$\mathcal{N}_\pi : \quad (x - 1, y, z + 1) \cdot (1, 2, 3) = 0,$$

equivalent to the cartesian equation

$$\Sigma_\pi : \quad x + 2y + 3z + 2 = 0.$$

Exercise 15.1.8 Given the (straight) line r in $\mathbb{A}^2$ with cartesian equation

$$\Sigma_r : \quad 2x - 3y + 3 = 0$$

we look for its normal equation. We start by noticing (see the Remark 15.1.6) that the direction of r is orthogonal to the vector $u = (2, -3)$, and that the point $A = (0, 1)$ lays in r, so we can write

$$\mathcal{N}_r : \quad (P - (0, 1)) \cdot (2, -3) = 0 \quad \Leftrightarrow \quad (x, y - 1) \cdot (2, -3) = 0$$

as a normal equation for r.

From what discussed above, it is clear that there exist deep relations between cartesian and normal equations for an hyperplane in a Euclidean affine space. Moreover, as we have discussed in the previous chapter, a generic linear affine variety in $\mathbb{A}^n$ can be described as a suitable intersection of hyperplanes. Therefore it should come as no surprise that a linear affine variety can be described in a Euclidean affine space in terms of a suitable number of normal equations. The general case is illustrated by the following exercise.

Exercise 15.1.9 Let r be the line through the point $A = (1, 2, -3)$ in $\mathbb{E}^3$ which is orthogonal to the space $\mathcal{L}((1, 1, 0), (0, 1, -1))$. Its normal equation is given by

$$\mathcal{N}_r : \quad \begin{cases} (P - A) \cdot (1, 1, 0) = 0 \\ (P - A) \cdot (0, 1, -1) = 0 \end{cases},$$

that is

$$\mathcal{N}_r : \quad \begin{cases} (x - 1, y - 2, z + 3) \cdot (1, 1, 0) = 0 \\ (x - 1, y - 2, z + 3) \cdot (0, 1, -1) = 0 \end{cases}$$

yielding then the cartesian equation

$$\Sigma_r : \begin{cases} x + y - 3 = 0 \\ y - z - 5 = 0 \end{cases} .$$

15.2 Orthogonality Between Linear Affine Varieties

In the Euclidean affine space $\mathbb{E}^n$ there is the notion of orthogonality. Thus, we have:

Definition 15.2.1 One says that

(a) the lines $r, r' \subset \mathbb{E}^n$ are *orthogonal* if $v \cdot v' = 0$ for any $v \in S_r$ and any $v' \in S_{r'}$,
(b) the planes $\pi, \pi' \subset \mathbb{E}^3$ are *orthogonal* if $u \cdot u' = 0$ for any $u \in S_\pi^\perp$ and any $u' \in S_{\pi'}^\perp$,
(c) the line r with direction v is *orthogonal* to the plane π in $\mathbb{E}^3$ if $v \in S_\pi^\perp$.

Exercise 15.2.2 We consider the following lines in $\mathbb{E}^2$,

$$\begin{aligned} \Sigma_{r_1} &: \ 2x - 2y + 1 = 0, \\ \Sigma_{r_2} &: \ x + y + 3 = 0, \\ r_3 &: \ (x, y) = (1, -3) + \lambda(1, 1), \\ \mathcal{N}_{r_4} &: \ (x + 1, y - 4) \cdot (1, 2) = 0 \end{aligned}$$

with directions spanned by the vectors

$$\begin{aligned} v_1 &= (2, 2), \\ v_2 &= (1, -1), \\ v_3 &= (1, 1), \\ v_4 &= (1, -2). \end{aligned}$$

It is immediate to show that the only orthogonal pairs of lines among them are $r_1 \perp r_2$ and $r_2 \perp r_3$.

Exercise 15.2.3 Consider the lines $r, r' \subset \mathbb{E}^3$ given by

$$r : \ (x, y, z) = (1, 2, 1) + \lambda(3, 0, -1) , \qquad r' : \ \begin{cases} x = 3 + \mu \\ y = 2 - 2\mu \\ z = 3\mu \end{cases} .$$

We have $S_r = \mathcal{L}((3, 0, -1))$ and $S_{r'} = \mathcal{L}((1, -2, 3))$. Since

$$(3, 0, -1) \cdot (1, -2, 3) = 0$$

we conclude that r is orthogonal to r'.

Exercise 15.2.4 Let π be the plane in $\mathbb{E}^3$ whose cartesian equation is

$$\Sigma_\pi : \quad x - y + 2z - 3 = 0.$$

In order to find an equation for the line r through $A = (1, 2, 1)$ which is orthogonal to π we notice from the Remark 15.1.6, that it is $S_\pi^\perp = \mathcal{L}((1, -1, 2))$: we can then write

$$r : \quad (x, y, z) = (1, 2, 1) + \lambda(1, -1, 2).$$

Exercise 15.2.5 Consider in $\mathbb{E}^3$ the line given by

$$\Sigma_r : \quad \begin{cases} x - 2y + z - 1 = 0 \\ x + y = 0 \end{cases}.$$

We seek to determine:

(1) a cartesian equation for the plane π through the point $A = (-1, -1, -1)$ and orthogonal to r,
(2) the intersection between r and π.

We proceed as follows.

(1) From the cartesian equation Σ_r we have that

$$S_r^\perp = \mathcal{L}((1, -2, 1), (1, 1, 0))$$

and this subspace yields the direction S_π. Since $A \in \pi$, a vector equation for π is given by

$$\pi : \quad (x, y, z) = -(1, 1, 1) + \lambda(1, -2, 1) + \mu(1, 1, 0).$$

By noticing that $S_\pi = \mathcal{L}((1, -1, -3))$, a normal equation for π is given by

$$\mathcal{N}_\pi : \quad (P - A) \cdot (1, -1, -3) = 0$$

yielding the cartesian equation

$$\Sigma_\pi : \quad x - y - 3z - 3 = 0.$$

(2) The intersection $\pi \cap r$ is clearly given by the unique solution of the linear system

$$\Sigma_{\pi \cap r} : \quad \begin{cases} x - 2y + z - 1 = 0 \\ x + y = 0 \\ x - y - 3z - 3 = 0, \end{cases},$$

which is $P = \frac{1}{11}(6, -6, -7)$.

Exercise 15.2.6 We consider again the lines r and r' in $\mathbb{E}^3$ from the Exercise 15.2.3. We know that $r \perp r'$. We determine the plane π which is orthogonal to r' and such that $r \subset \pi$. Since $\mathcal{L}((1, -2, 3))$ is the direction of r', we can write from the Remark 15.1.6 that

$$\Sigma_\pi : \quad x - 2y + 3z + d = 0$$

with d a real parameter. The line r is in π if and only if the coordinates of every of its points $P = (1 + 3\lambda, 2, 1 - \lambda) \in r$ solve the equation Σ_π, that is, if and only if the equation

$$(1 + 3\lambda) - 2(2) + 3(1 - \lambda) + d = 0$$

has a solution for each value of λ. This is true if and only if $d = 0$, so a cartesian equation for π is

$$\Sigma_\pi : \quad x - 2y + 3z = 0.$$

Exercise 15.2.7 For the planes

$$\Sigma_\pi : \quad 2x + y - z - 3 = 0, \qquad \Sigma_{\pi'} : \quad x + y + 3z - 1 = 0$$

in $\mathbb{E}^3$ we have $S_\pi^\perp = \mathcal{L}((2, 1, -1))$ and $S_{\pi'}^\perp = \mathcal{L}((1, 1, 3))$. We conclude that π is orthogonal to π', since $(2, 1, -1) \cdot (1, 1, 3) = 0$. Notice that

$$(2, 1, -1) \in S_{\pi'} = \{(a, b, c,) \; : \; a + b + 3c = 0\},$$

that is $S_\pi^\perp \subset S_{\pi'}$. We can analogously show that $S_{\pi'}^\perp \subset S_\pi$. This leads to the following remark.

Remark 15.2.8 The planes $\pi, \pi' \subset \mathbb{E}^3$ are orthogonal if and only if $S_\pi^\perp \subset S_{\pi'}$ (or equivalently if and only if $S_{\pi'}^\perp \subset S_\pi$).

In order to recap the results we described in the previous pages, we consider the following example.

Exercise 15.2.9 Consider the point $A = (1, 0, 1)$ in $\mathbb{E}^3$ and the lines r, s with equations

$$r : \quad (x, y, z) = (1, 2, 1) + \lambda(3, 0, -1), \qquad \Sigma_s : \quad \begin{cases} x - y + z + 2 = 0 \\ x - z + 1 = 0 \end{cases}.$$

We seek to determine:

(a) the set $\mathcal{F}$ of lines through A which are orthogonal to r,
(b) the line $l \in \mathcal{F}$ which is parallel to the plane π given by $\Sigma_\pi : \; x - y + z + 2 = 0$,
(c) the line $l' \in \mathcal{F}$ which is orthogonal to s,
(d) the lines $q \subset \pi'$ with $\Sigma_{\pi'} : \; y - 2 = 0$ which are orthogonal to r.

For these we proceed as follows.

(a) A line u through A has a vector equation

$$(x, y, z) = (1, 0, 1) + \lambda(a, b, c)$$

with arbitrary direction $S_u = \mathcal{L}((a, b, c))$. Since u is to be orthogonal to r, we have the condition $(a, b, c) \cdot (3, 0, 1) = 3a - c = 0$. The set $\mathcal{F}$ is then given by the union $\mathcal{F} = \{r_\alpha\}_{\alpha \in \mathbb{R}} \cup \{\overline{r}\}$ with

$$r_\alpha : \quad (x, y, z) = (1, 0, 1) + \mu(1, \alpha, 3) \qquad \text{for} \quad a \neq 0,$$

and

$$\overline{r} : \quad (x, y, z) = (1, 0, 1) + \mu(0, 1, 0) \qquad \text{for} \quad a = c = 0.$$

(b) Since the direction S_π of the plane π is given by the subspace orthogonal to $\mathcal{L}((1, -1, 1))$, it is clear from $(0, 1, 0) \cdot (1, -1, 1) \neq 0$ that the line $\overline{r}$ is not parallel to π. This means that the line l must be found within the set $\{r_\alpha\}_{\alpha \in \mathbb{R}}$. If we impose that $(1, \alpha, 3) \cdot (1, -1, 1) = 0$, we have $\alpha = 4$, so the line l is given by

$$l : \quad (x, y, z) = (1, 0, 1) + \mu(1, 4, 3).$$

(c) A cartesian equation for s is given by solving the linear system Σ_s in terms of one free unknown. It is immediate to show that

$$s : \quad (x, y, z) = (-1 + \eta, 1 + 2\eta, \eta) = (-1, 1, 0) + \eta(1, 2, 1).$$

The condition $r_\alpha \perp s$ is equivalent to $(1, \alpha, 3) \cdot (1, 2, 1) = 0$, reading $\alpha = -2$, so we have

$$l' : \quad (x, y, z) = (1, 0, 1) + \mu(1, -2, 3).$$

This is the unique solution to the problem: we directly inspect that $\overline{r}$ is not orthogonal to s, since $(0, 1, 0) \cdot (1, 2, 1) = 2 \neq 0$.

(d) A plane π_h is orthogonal to r if and only if

$$\Sigma_{\pi_h} : \quad 3x - z + h = 0.$$

The lines q_h are then given by the intersection

$$\Sigma_{q_h} = \Sigma_{\pi_h \cap \pi'} : \quad \begin{cases} 3x - z + h = 0 \\ y - 2 = 0 \end{cases} \qquad \text{with} \quad h \in \mathbb{R}.$$

15.3 The Distance Between Linear Affine Varieties

It is evident that the distance between two points A and B on a plane is defined to be the length of the line segment whose endpoints are A and B. This definition can be consistently formulated in a Euclidean affine space.

Definition 15.3.1 Let A and B be a pair of points in $\mathbb{E}^n$. The *distance* $\mathrm{d}(A, B)$ between them is defined as

$$\mathrm{d}(A, B) = \|B - A\| = \sqrt{(B - A) \cdot (B - A)}.$$

Exercise 15.3.2 If $A = (1, 2, 0, -1)$ and $B = (0, -1, 2, 2)$ are points in $\mathbb{E}^4$, then

$$\mathrm{d}(A, B) = \|(-1, -3, 2, 3)\| = \sqrt{23}.$$

The well known properties of a Euclidean distance function are a consequence of the corresponding properties of the scalar product.

Proposition 15.3.3 *For any A, B, C points in $\mathbb{E}^n$ the following properties hold.*

(1) $\mathrm{d}(A, B) \geq 0$,
(2) $\mathrm{d}(A, B) = 0$ *if and only if* $A = B$,
(3) $\mathrm{d}(A, B) = \mathrm{d}(B, A)$.
(4) $\mathrm{d}(A, B) + \mathrm{d}(B, C) \geq \mathrm{d}(A, C)$.

In order to introduce a notion of distance between a point and a linear affine variety, we start by looking at an example. Let us consider in $\mathbb{E}^2$ the point $A = (0, 0)$ and the line r whose vector equation is $(x, y) = (1, 1) + \lambda(1, -1)$. By denoting $P_\lambda = (1 + \lambda, 1 - \lambda)$ a generic point in r, we compute

$$\mathrm{d}(A, P_\lambda) = \sqrt{2 + 2\lambda^2}.$$

It is immediate to verify that, as a function of λ, the quantity $\mathrm{d}(A, P)$ ranges between $\sqrt{2}$ and $+\infty$: it is therefore natural to consider the minimum of this range as the distance between A and r. We have then $\mathrm{d}(A, r) = \sqrt{2}$.

Definition 15.3.4 If L is a linear affine variety and A is a point in $\mathbb{E}^n$, the *distance* $\mathrm{d}(A, L)$ between A and L is defined to be

$$\mathrm{d}(A, L) = \min\{\mathrm{d}(A, B) \ : \ B \in L\}.$$

Remark 15.3.5 It is evident from the definition above that $\mathrm{d}(A, L) = 0$ if and only if $A \in L$. We shall indeed prove that, given a point A and a linear affine variety L in $\mathbb{E}^n$, there always exists a point $A_0 \in L$ such that $\mathrm{d}(A, L) = \mathrm{d}(A_0, L)$, thus showing that the previous definition is well posed.

Proposition 15.3.6 *Let L be a linear affine variety and $A \notin L$ a point in $\mathbb{E}^n$. It holds that*

$$\mathrm{d}(A, L) = \mathrm{d}(A, A_0) \quad \text{with} \quad A_0 = L \cap (A + S_L^{\perp}).$$

Here the set $A + S_L^{\perp}$ denotes the linear affine variety through A whose direction is $S_L^{\perp}$. The point A_0 is called the orthogonal projection *of A on L.*

Proof Since the linear system $\Sigma_{L \cap (A+S_L^{\perp})}$ given by the cartesian equations Σ_L and $\Sigma_{A+S_L^{\perp}}$ is of rank n with n unknowns, the intersection $L \cap (A + S_L^{\perp})$ consists of a single point that we denote by A_0.

Let B be an arbitrary point in L. We can decompose

$$A - B = (A - A_0) + (A_0 - B),$$

with $A_0 - B \in S_L$ (since both A_0 and B are in L) and $A - A_0 \in S_L^{\perp}$ (since both A and A_0 are points in the linear affine variety $A + S_L^{\perp}$). We have then

$$(A - A_0) \cdot (A_0 - B) = 0$$

and we write

$$\begin{aligned}(\mathrm{d}(A, B))^2 = \|A - B\|^2 &= \|(A - A_0) + (A_0 - B)\|^2 \\ &= \|A - A_0\|^2 + \|A_0 - B\|^2.\end{aligned}$$

As a consequence,

$$(\mathrm{d}(A, B))^2 \geq \|A - A_0\|^2 = (\mathrm{d}(A, A_0))^2$$

for any $B \in L$, and this proves the claim. □

Exercise 15.3.7 Let us compute the distance between the line $r : \; 2x + y + 4 = 0$ and the point $A = (1, -1)$ in $\mathbb{E}^2$. We start by finding the line $s_A = A + S_r^{\perp}$ through A which is orthogonal to r. The direction $S_r^{\perp}$ is spanned by the vector $(2, 1)$, so we have

$$s_A : \quad (x, y) = (1, -1) + \lambda(2, 1).$$

The intersection $A_0 = r \cap s_A$ is then given by the value of the parameter λ that solves the equation

$$2(1 + 2\lambda) + (-1 + \lambda) + 4 = 0,$$

that is $\lambda = -1$ giving $A_0 = (-1, -2)$. Therefore we have

$$\mathrm{d}(A, r) = \mathrm{d}(A, A_0) = \|(2, 1)\| = \sqrt{5}.$$

Exercise 15.3.8 Let us consider in $\mathbb{E}^3$ the point $A = (1, -1, 0)$ and the line r with vector equation $r : \ (x, y, z) = (1, 2, 1) + \lambda(1, -1, 2)$. In order to compute the distance between A and r we first determine the plane $\pi_A := A + S_r^{\perp}$. Since the direction of r must be orthogonal to π_A, from the Remark 15.1.6 the cartesian equation for π_A is given by

$$\Sigma_{\pi_A} : \quad x - y + 2z + d = 0,$$

with $d \in \mathbb{R}$. The value of d if fixed by asking that $A \in \pi_A$, that is $1 + 1 + d = 0$ giving $d = -2$. We then have

$$\Sigma_{\pi_A} : \quad x - y + 2z - 2 = 0.$$

The point A_0 is now the intersection $r \cap \pi_A$, which is given for the value of $\lambda = \frac{1}{6}$ which solves,

$$(1 + \lambda) - (2 - \lambda) + 2(1 + 2\lambda) - 2 = 0.$$

It is therefore $A_0 = (\frac{7}{6}, \frac{11}{6}, \frac{4}{3})$, with

$$\mathrm{d}(A, r) = \mathrm{d}(A, A_0) = \|(\frac{1}{6}, \frac{17}{6}, \frac{4}{3})\| = \sqrt{\frac{59}{6}}.$$

The next theorem yields a formula which allows one to compute more directly the distance $\mathrm{d}(Q, H)$ between a point Q and an hyperplane H in $\mathbb{E}^n$.

Theorem 15.3.9 *Let H be a hyperplane and Q a point in $\mathbb{E}^n$ with $\Sigma_H : a_1x_1 + \cdots + a_nx_n + b = 0$ and $Q = (x_1', \ldots, x_n')$. The distance between Q and H is given by*

$$\mathrm{d}(Q, H) = \frac{|a_1x_1' + \cdots + a_nx_n' + b|}{\sqrt{a_1^2 + \cdots + a_n^2}}.$$

Proof If we consider $X = (x_1, \ldots, x_n)$ and $A = (a_1, \ldots, a_n)$ as vectors in $\mathbb{R}^n$, using the scalar product in $\mathbb{E}^n$, the cartesian equation for H can be written as

$$\Sigma_H : \quad A \cdot X + b = 0.$$

We know that $A \in S_H^{\perp}$, so the line through A which is orthogonal to H is made of the points P such that

$$r : \quad P = Q + \lambda A.$$

The intersection point $Q_0 = r \cap H$ is given by replacing X in Σ_H with such a P, that is

$$A \cdot (Q + \lambda A) + b = 0 \quad \Rightarrow \quad A \cdot Q + \lambda A \cdot A + b = 0$$
$$\Rightarrow \quad \lambda = -\frac{A \cdot Q + b}{A \cdot A}.$$

The equation for r gives then

$$Q_0 = Q - \frac{A \cdot Q + b}{\|A\|^2} A.$$

We can now easily compute

$$\begin{aligned} \|Q - Q_0\|^2 &= \left\| \frac{A \cdot Q + b}{\|A\|^2} A \right\|^2 \\ &= \frac{|A \cdot Q + b|^2}{\|A\|^4} \|A\|^2 = \frac{|A \cdot Q + b|^2}{\|A\|^2}, \end{aligned}$$

therefore getting

$$\mathrm{d}(Q, H) = \frac{|A \cdot Q + b|}{\|A\|} = \frac{|a_1 x_1' + \cdots + a_n x_n' + b|}{\sqrt{a_1^2 + \cdots + a_n^2}},$$

as claimed. □

Exercise 15.3.10 Consider the line r with cartesian equation $\Sigma_r : \ 2x + y + 4 = 0$ and the point $A = (1, -1)$ in $\mathbb{E}^2$ as in the Exercise 15.3.7 above. From the Theorem 15.3.9 we have

$$\mathrm{d}(A, r) = \frac{|2 - 1 + 4|}{\sqrt{4 + 1}} = \sqrt{5}.$$

Exercise 15.3.11 By making again use of the Theorem 15.3.9 it is easy to compute the distance between the point $A = (1, 2, -1)$ and the plane π in $\mathbb{E}^3$ with $\Sigma_\pi : \ x + 2y - 2z + 3 = 0$. We have

$$\mathrm{d}(A, \pi) = \frac{|1 + 4 + 2 + 3|}{\sqrt{1 + 4 + 4}} = \frac{10}{3}.$$

We generalise the analysis above with a natural definition for the distance between any two linear affine varieties.

Definition 15.3.12 Let L and L' two linear affine varieties in $\mathbb{E}^n$. The *distance* between them is defined as the non negative real number

$$\mathrm{d}(L, L') = \min\{\mathrm{d}(A, A') \ : \ A \in L, \ A' \in L'\}.$$

It is evident that $\mathrm{d}(L, L') = 0$ if and only if $L \cap L' \neq \emptyset$. It is indeed possible to show that the previous definition is consistent even when $L \cap L' = \emptyset$. Moreover one can show that there exist a point $\bar{A} \in L$ and a point $\bar{A}' \in L'$, such that the minimum distance is attained for them, that is $\mathrm{d}(\bar{A}, \bar{A}') \leq \mathrm{d}(A, A')$ for any $A \in L$ and $A' \in L'$. For such a pair of points it is $\mathrm{d}(L, L') = \mathrm{d}(\bar{A}, \bar{A}')$.

In the following pages we shall study the following cases of linear varieties which do not intersect:

- lines r, r' in $\mathbb{E}^2$ which are parallel,
- planes π, π' in $\mathbb{E}^3$ which are parallel,
- a plane π and a line r in $\mathbb{E}^3$ which are parallel,
- lines r, r' in $\mathbb{E}^3$ which are parallel.

Remark 15.3.13 Consider lines r and r' in $\mathbb{E}^2$ which are parallel and distinct. Their cartesian equations are

$$\Sigma_r : \quad ax + by + c = 0, \qquad \Sigma_{r'} : \quad ax + by + c' = 0,$$

for $c' \neq c$. Let $A = (x_1', x_2') \in r$, that is $ax_1' + bx_2' + c = 0$. From the Theorem 15.3.9 it is

$$\mathrm{d}(A, r') = \frac{|ax_1' + bx_2' + c'|}{\sqrt{a^2 + b^2}} = \frac{|c' - c|}{\sqrt{a^2 + b^2}}.$$

From the Definition 15.3.12 we have $\mathrm{d}(A, A') \geq \mathrm{d}(A, r')$. Since the value $\mathrm{d}(A, r')$ we have computed does not depend on the coordinates of $A \in r$, we have that $\mathrm{d}(A, r')$ is the minimum value for $\mathrm{d}(A, A')$ when A ranges in r and A' in r', so we conclude that

$$\mathrm{d}(r, r') = \frac{|c' - c|}{\sqrt{a^2 + b^2}}.$$

Notice that, with respect to the same lines, we also have

$$\mathrm{d}(A', r) = \frac{|c' - c|}{\sqrt{a^2 + b^2}} = \mathrm{d}(A, r').$$

Exercise 15.3.14 Consider the parallel lines $r, r' \subset \mathbb{E}^2$ with cartesian equations

$$\Sigma_r : \quad 2x + y - 3 = 0, \qquad \Sigma_{r'} : \quad 2x + y + 2 = 0.$$

The distance between them is

$$\mathrm{d}(r, r') = \frac{|2 - (-3)|}{\sqrt{5}} = \sqrt{5}.$$

The distance between two parallel hyperplanes in $\mathbb{E}^n$ is given by generalising the proof of the Theorem 15.3.9.

Proposition 15.3.15 *If H and H' are parallel hyperplanes in $\mathbb{E}^n$ with cartesian equations*

$$\Sigma_H: \quad a_1x_1 + \cdots + a_nx_n + b = 0, \qquad \Sigma_{H'}: \quad a_1x_1 + \cdots + a_nx_n + b' = 0,$$

then $\mathrm{d}(H, H') = \mathrm{d}(Q, H')$, *where* Q *is an arbitrary point in* H, *and therefore it is*

$$\mathrm{d}(H, H') = \frac{|b - b'|}{\sqrt{a_1^2 + \cdots + a_n^2}}.$$

Proof We proceed as in the Theorem 15.3.9, so we set $X = (x_1, \ldots, x_n)$ and $A = (a_1, \ldots, a_n)$ and write

$$\Sigma_H: \quad A \cdot X + b = 0, \qquad \Sigma_{H'}: \quad A \cdot X + b' = 0.$$

As we argued in the Remark 15.3.13, by setting $Q = \bar{X}$ with $A\bar{X} + b = 0$, as an arbitrary point in H, we have

$$\mathrm{d}(Q, H') = \frac{|A \cdot \bar{X} + b'|}{\|A\|} = \frac{|b' - b|}{\|A\|}$$

and since such a distance does not depend on Q, we conclude that $\mathrm{d}(H, H') = \mathrm{d}(Q, H')$. □

Exercise 15.3.16 The planes

$$\Sigma_\pi: \quad x + 2y - z + 2 = 0, \qquad \Sigma_{\pi'}: \quad x + 2y - z - 4 = 0$$

are parallel and distinct. The distance between them is

$$\mathrm{d}(\pi, \pi') = \frac{|2 + 4|}{\sqrt{1 + 4 + 1}} = \sqrt{6}.$$

It is clear that not all linear affine varieties which are parallel have the same dimension. The next proposition shows a result within this situation.

Proposition 15.3.17 *Let* r *be a line and* H *an hyperplane in* $\mathbb{E}^n$, *with* r *parallel to* H. *It is*

$$\mathrm{d}(r, H) = \mathrm{d}(\bar{P}, H),$$

where $\bar{P}$ *is any point in* r.

Proof With the notations previously adopted, we have $A = (a_1, \ldots, a_n)$ and $X = (x_1, \ldots, x_n)$, we represent H by the cartesian equation

$$\Sigma_H: \quad A \cdot X + b = 0$$

and r by the vector equation

$$r: \quad P = \bar{P} + \lambda v$$

where $\bar{P} \in r$ while $v \cdot A = 0$ since r is parallel to H. From the Theorem 15.3.9 we have

$$\mathrm{d}(P, H) = \frac{|A \cdot P + b|}{\|A\|} = \frac{|A \cdot (\bar{P} + \lambda v) + b|}{\|A\|} = \frac{|A \cdot \bar{P} + b|}{\|A\|}.$$

This expression does not depend on λ: this is the reason why $\mathrm{d}(P, H) = \mathrm{d}(\bar{P}, H) = \mathrm{d}(r, H)$. □

Exercise 15.3.18 Consider in $\mathbb{E}^3$ the line r and the plane π given by:

$$\Sigma_r : \begin{cases} 2x - y + z - 2 = 0 \\ y + 2z = 0 \end{cases}, \qquad \Sigma_\pi : \quad 2x - y + z + 3 = 0.$$

Since r is parallel to π, we take the point $P = (1, 0, 0)$ in r and compute the distance between P and π. One gets

$$\mathrm{d}(r, \pi) = \mathrm{d}(P, \pi) = \frac{5}{\sqrt{6}}.$$

Exercise 15.3.19 Consider the lines r and r' in $\mathbb{E}^3$ given by the vector equations

$$r: \quad (x, y, z) = (3, 1, 2) + \lambda(1, 2, 0), \qquad r': \quad (x, y, z) = (-1, -2, 3) + \lambda(1, 2, 0).$$

Since r is parallel to r', the distance between them can be computed by proceeding as in the previous exercises, that is $\mathrm{d}(r, r') = \mathrm{d}(A, r') = \mathrm{d}(B, r)$, where A is an arbitrary point in r and B an arbitrary point in r'.

We illustrate an alternative method. We notice that, if π is a plane orthogonal to both r and r', then the distance $\mathrm{d}(r, r') = \mathrm{d}(P, P')$ where $P = \pi \cap r$ and $P' = \pi \cap r'$. We consider the plane π through the origin which is orthogonal to both r and r', and whose cartesian equation is

$$\Sigma_\pi : \quad x + 2y = 0.$$

Direct calculations show that $P = \pi \cap r = (2, -1, 2)$ and $P' = \pi \cap r' = (0, 0, 3)$, so

$$\mathrm{d}(r, r') = \mathrm{d}(P, P') = \sqrt{6}.$$

We end the section by sketching how to define the distance between skew lines in $\mathbb{E}^3$.

Remark 15.3.20 If r and r' are skew lines in $\mathbb{E}^3$, then there exist a point $P \in r$ and a point $P \in r'$ which are the intersections of the lines r and r' with the unique line s

orthogonally intersecting both r and r'. The line s is the *minimum distance line for r and r'*, and the distance $\mathrm{d}(r, r') = \mathrm{d}(P, P')$.

Exercise 15.3.21 We consider the skew lines in $\mathbb{E}^3$,

$$r: \quad (x, y, z) = \lambda(1, -1, 1), \qquad r': \quad (x, y, z) = (0, 0, 1) + \mu(1, 0, 1).$$

The subspace $N \subset E^3$ which is orthogonal to both the directions S_r and $S_{r'}$ is $N = \mathcal{L}((1, 0, -1))$. The minimum distance line s for the given r and r' has the direction $S_s = N$, and intersects r in a point P and r' in a point P'. Since $P \in r$ and $P \in r'$, there exists a value for λ and a value for μ such that $P = Q(\lambda)$ and $P' = Q'(\mu)$ with

$$Q(\lambda) + t(1, 0, -1) = Q'(\mu),$$

where ν is the parameter for s. The points $P = s \cap r$ and $P' = s \cap r'$ are then those corresponding to the values of the parameters λ and μ solving such a relation, that is

$$s: \quad \begin{cases} \lambda + \nu = \mu \\ -\lambda = 0 \\ \lambda - t = 1 + \mu \end{cases}.$$

One finds $\lambda = 0$, $\mu = \nu = -\frac{1}{2}$, so $P = (0, 0, 0)$, $P' = \frac{1}{2}(-1, 0, 1)$ and

$$\mathrm{d}(r, r') = \mathrm{d}(P, P') = \frac{1}{\sqrt{2}}.$$

15.4 Bundles of Lines and of Planes

A useful notion for several kinds of problems in affine geometry is that of bundle of lines and bundle of planes.

Definition 15.4.1 Given a point A in $\mathbb{E}^2$, the *bundle of concurrent lines* with center (or *point of concurrency*) A is the set of all the lines through A in $\mathbb{E}^2$; we shall denote it by $\mathcal{F}_A$.

The next result is immediate.

Proposition 15.4.2 *With $A = (x_0, y_0) \in \mathbb{E}^2$, the cartesian equation of an arbitrary line in the bundle $\mathcal{F}_A$ through A is given by*

$$\Sigma_{\mathcal{F}_A}: \quad \alpha(x - x_0) + \beta(y - y_0) = 0$$

for any choice of the real parameters α and β such that $(\alpha, \beta) \in \mathbb{R}^2 \setminus \{(0, 0)\}$.

Notice that the parameters α and β label a line in $\mathcal{F}_A$, but there is not a bijection between pairs (α, β) and lines in $\mathcal{F}_A$: the pairs (α_0, β_0) and $(\rho\alpha_0, \rho\beta_0)$, for $\rho \neq 0$, give the same line in $\mathcal{F}_A$.

Exercise 15.4.3 The cartesian equation for the bundle $\mathcal{F}_A$ of lines through $A = (1, -2)$ in $\mathbb{E}^2$ is

$$\Sigma_{\mathcal{F}_A} : \quad \alpha(x-1) + \beta(y+2) = 0.$$

The result described in the next proposition (whose proof we omit) shows that the bundle $\mathcal{F}_A$ can be *generated* by any pair of distinct lines concurrent in A.

Proposition 15.4.4 *Let $A \in \mathbb{E}^2$ be the unique intersection of the lines*

$$\Sigma_r : \quad ax + by + c = 0, \qquad \Sigma_{r'} : \quad a'x + b'y + c' = 0.$$

Any relation

$$\Sigma_{(\alpha,\beta)} : \quad \alpha(ax + by + c) + \beta(a'x + b'y + c') = 0$$

with $\mathbb{R}^2 \ni (\alpha, \beta) \neq (0, 0)$ is the cartesian equation for a line in the bundle $\mathcal{F}_A$ of lines with center A, and for any element s of $\mathcal{F}_A$ there exists a pair $\mathbb{R}^2 \ni (\alpha, \beta) \neq (0, 0)$ such that the cartesian equation of s can be written as

$$\Sigma_{(\alpha,\beta)} : \quad \alpha(ax + by + c) + \beta(a'x + b'y + c') = 0. \tag{15.1}$$

Definition 15.4.5 If the bundle $\mathcal{F}_A$ is given by (15.1), the distinct lines r, r' are called the *generators* of the bundle. To stress the role of the generating lines, we also write in such a case $\mathcal{F}_A = \mathcal{F}(r, r')$.

Exercise 15.4.6 The line r whose cartesian equation is $\Sigma_r : x + y + 1$ is an element in the bundle $\mathcal{F}_A$ in the Exercise 15.4.3, corresponding to the parameters $(\alpha, \beta) = (1, 1)$ or equivalently $(\alpha, \beta) = (\rho, \rho)$ with $\rho \neq 0$.

Exercise 15.4.7 Consider the following cartesian equation,

$$\Sigma_{(\alpha,\beta)} : \quad \alpha(x - y + 3) + \beta(2x + y + 3) = 0,$$

depending on a pair of real parameters $(\alpha, \beta) \neq (0, 0)$. Since the relations $x - y + 3 = 0$ and $2x + y + 3 = 0$ yield the cartesian equations for a pair of non parallel lines in $\mathbb{E}^2$, the equation $\Sigma_{(\alpha,\beta)}$ is the cartesian equation for a bundle $\mathcal{F}$ of lines in $\mathbb{E}^2$. We compute:

(a) the centre A of the bundle $\mathcal{F}$,
(b) the line $s_1 \in \mathcal{F}$ which is orthogonal to the line r_1 whose cartesian equation is $\Sigma_{r_1} : 3x + y - 1 = 0$,

(c) the line $s_2 \in \mathcal{F}$ which is parallel to the line r_2 whose cartesian equation is $\Sigma_{r_2} : x - y = 0$,

(d) the line $s_3 \in \mathcal{F}$ through the point $B = (1, 1)$.

We proceed as follows:

(a) The centre of the bundle is given by the intersection

$$\begin{cases} x - y + 3 = 0 \\ 2x + y + 3 = 0 \end{cases},$$

which is found to be $A = (-2, 1)$.

(b) We write the cartesian equation of the bundle $\mathcal{F}$,

$$\Sigma_{\mathcal{F}} : \quad (\alpha + 2\beta)x + (-\alpha + \beta)y + 3(\alpha + \beta) = 0.$$

As a consequence, the direction of an arbitrary line in the bundle $\mathcal{F}$ is spanned by the vector $v_{(\alpha,\beta)} = (\alpha + 2\beta, \beta - \alpha)$. In order for the line $s_1 \in \mathcal{F}$ to be orthogonal to r_1 we require

$$(\alpha + 2\beta, \beta - \alpha) \cdot (-1, 3) = 0 \quad \Rightarrow \quad (\alpha, \beta) = \rho(7, -2)$$

with $\rho \neq 0$. The line s has the cartesian equation $\Sigma_{(7,-2)} : x - 3y + 5 = 0$.

(c) In order for an element $s_2 \in \mathcal{F}$ to be parallel to r_2 we require that its direction coincides with the direction of r_2, which is $\mathcal{L}((1, -1))$. We impose then

$$\alpha + 2\beta = -(\beta - \alpha) \quad \Rightarrow \quad (\alpha, \beta) = \rho(1, 0)$$

with $\rho \neq 0$. So we have that s_2 is given by the cartesian equation $\Sigma_{(1,0)} : x - y + 3 = 0$. The line s_2 turns out to be indeed one of the generators of the bundle $\mathcal{F}$.

(d) We have now to require that the coordinates of B solve the equation $\Sigma_{(\alpha,\beta)}$, that is

$$(\alpha + 2\beta) + (\beta - \alpha) + 3(\alpha + \beta) = 0 \quad \Rightarrow \quad 3\alpha + 6\beta = 0,$$

giving $(\alpha, \beta) = \rho(2, -1)$ with $\rho \neq 0$. The line s_3 is therefore given by $\Sigma_{(2,-1)} : y - 1 = 0$.

Remark 15.4.8 Notice that the computations in (d) above can be generalised. If $\mathcal{F}_A$ is a bundle of lines through A, for any point $B \neq A$ there always exists a unique line in $\mathcal{F}_A$ which passes through B. We denote it as the line $r_{AB} \in \mathcal{F}_A$.

Definition 15.4.9 Let $\Sigma_r : ax + by + c = 0$ be the cartesian equation of the line r in $\mathbb{E}^2$. The set of all lines which are parallel to r is said to define a *bundle of parallel lines* or an *improper* bundle. The most convenient way to describe an improper bundle of lines is

$$\Sigma_{\mathcal{F}} : \quad ax + by + h = 0, \qquad \text{with} \quad h \in \mathbb{R}.$$

Exercise 15.4.10 We consider the line r in $\mathbb{E}^2$ given by $\Sigma_r : 2x - y + 3 = 0$. We wish to determine the lines s which are parallel to r and whose distance from r is $\mathrm{d}(s, r) = \sqrt{5}$.

The parallel lines to r are the elements r_h of the improper bundle $\mathcal{F}$ whose cartesian equation is

$$\Sigma_h : \quad 2x - y + h = 0.$$

From the Proposition 15.3.15 we have

$$\sqrt{5} = \mathrm{d}(r_h, r) = \frac{|h-3|}{\sqrt{5}} \quad \Rightarrow \quad |h - 3| = 5 \quad \Rightarrow \quad h - 3 = \pm 5.$$

The solutions of the exercise are

$$\Sigma_{r_8} : \quad 2x - y + 8 = 0, \qquad \Sigma_{r_{-2}} : \quad 2x - y - 2 = 0.$$

In a way similar to above, one has the notion of bundle of planes in a three dimensional affine space.

Definition 15.4.11 Let r be a line in $\mathbb{E}^3$. The bundle $\mathcal{F}_r$ of planes through r is the set of all planes π in $\mathbb{E}^3$ which contains r, that is $r \subset \pi$. The line r is called the *carrier* of the bundle $\mathcal{F}_r$.

Moreover, if π is a plane in $\mathbb{E}^3$, the set of all planes in $\mathbb{E}^3$ which are parallel to π gives the *(improper) bundle of parallel planes to* π.

The following proposition is the analogue of the Proposition 15.4.2.

Proposition 15.4.12 *Let r be the line in $\mathbb{E}^3$ with cartesian equation given by*

$$\Sigma_r : \quad \begin{cases} ax + by + cz + d = 0 \\ a'x + b'y + c'z + d' = 0 \end{cases}.$$

For any choice of the parameters $(\alpha, \beta) \neq (0, 0)$ the relation

$$\Sigma_{(\alpha,\beta)} : \quad \alpha(ax + by + cz + d) + \beta(a'x + b'y + c'z + d') = 0 \tag{15.2}$$

yields the cartesian equation for a plane in the bundle $\mathcal{F}_r$ with carrier line r, and for any plane π in such a bundle there is a pair $(\alpha, \beta) \neq (0, 0)$ such that the cartesian equation of π is given by (15.2).

Definition 15.4.13 If the bundle $\mathcal{F}_r$ of planes is given by the cartesian equation (15.2), the planes $\Sigma_\pi : ax + by + cz + d = 0$ and $\Sigma_{\pi'} : a'x + b'y + c'z + d' = 0$ are called the *generators* of $\mathcal{F}_r$. In such a case the equivalent notation $\mathcal{F}(\pi, \pi')$ will also be used.

Remark 15.4.14 Clearly, the bundle $\mathcal{F}_r$ is generated by any two distinct planes π, π' through r.

Exercise 15.4.15 Given the line r whose vector equation is

$$r: \quad (x, y, z) = (1, 2, -1) + \lambda(2, 3, 1),$$

we determine the bundle $\mathcal{F}_r$ of planes through r. In order to obtain a cartesian equation for r, we eliminate the parameter λ from the vector equation above, as follows

$$\begin{cases} x = 1 + 2\lambda \\ y = 2 + 3\lambda \\ \lambda = z + 1 \end{cases} \Rightarrow \quad \begin{cases} x = 1 + 2(z + 1) \\ y = 2 + 3(z + 1) \end{cases} \quad \Rightarrow \quad \Sigma_r : \begin{cases} x - 2z - 3 = 0 \\ y - 3z - 5 = 0 \end{cases}.$$

The cartesian equation for the bundle is then given by

$$\Sigma_{\mathcal{F}_r} : \quad \alpha(x - 2z - 3) + \beta(y - 3z - 5) = 0$$

with any $(\alpha, \beta) \neq (0, 0)$.

Let us next find the plane $\pi \in \mathcal{F}_r$ which passes through $A = (1, 2, 3)$. The condition $A \in \pi$ yields

$$\alpha(1 - 6 - 3) + \beta(2 - 9 - 5) = 0 \qquad \Rightarrow \qquad 2\alpha + 3\beta = 0.$$

We can pick $(\lambda, \mu) = (3, -2)$, giving $\Sigma_\pi : \ 3(x - 2z - 3) - 2(y - 3z - 5) = 0$, that is $\Sigma_\pi : \ 3x - 2y + 1 = 0$.

We also find the plane $\sigma \in \mathcal{F}_r$ which is orthogonal to $v = (1, -1, 1)$. We know that a vector orthogonal to a plane $\pi \in \mathcal{F}_r$ with equation

$$\Sigma_{\mathcal{F}_r} : \quad \alpha x + \beta y - (2\alpha + 3\beta)z - 3\alpha - 5\beta = 0,$$

is given by $(\alpha, \beta, -2\alpha - 3\beta)$. The conditions we have to meet are then

$$\begin{cases} \alpha = -\beta \\ \alpha = -2\alpha - 3\beta \end{cases} \quad \Rightarrow \quad \alpha = -\beta.$$

If we fix $(\lambda, \mu) = (1, -1)$, we have $\Sigma_\sigma : \ (x - 2z - 3) - (y - 3z - 5) = 0$, that is

$$\Sigma_\sigma : \ x - y + z + 2 = 0.$$

15.5 Symmetries

We introduce a few notions related to symmetries which are useful to solve problems in several branches of geometry and physics.

Definition 15.5.1 Consider a point $C \in \mathbb{E}^n$.

(a) Let $P \in \mathbb{E}^n$ be an arbitrary point in $\mathbb{E}^n$. The *symmetric point to P with respect to C* is the element $P' \in \mathbb{E}^n$ that belongs to the line r_{CP} passing through C and P, and such that $d(P', C) = d(P, C)$ with $P' \neq P$.
(b) Let $X \subset \mathbb{E}^n$ be a set of points. The *symmetric points to X with respect to C* is the set $X' \subset \mathbb{E}^n$ given by every point P' which is symmetric to any P in X with respect to C.
(c) Let $X \subset \mathbb{E}^n$. We say that X *is symmetric with respect to C* if $X = X'$, that is if X contains the symmetric point (with respect to C) to any of its points. In such a case, C is called a *symmetry centre* for X.

Exercise 15.5.2 In the euclidean affine plane $\mathbb{E}^2$ consider the point $C = (2, 3)$. Given the point $P = (1, -1)$, we determine its symmetric P' with respect to C. And with the line $\Sigma_r : \ 2x - y - 3 = 0$, we determine its symmetric r' with respect to C.

We consider the line r_{CP} through P and C, which has the vector equation

$$r_{CP} : \quad (x, y) = (1, -1) + \lambda(1, 4).$$

The distance between P and C is given by $\|P - C\| = \sqrt{17}$, so the point P' can be obtained by finding the value for the parameter λ such that the distance

$$\|P_\lambda - C\| = \|(-1 + \lambda, -4 + 4\lambda)\| = \sqrt{(-1 + \lambda)^2 + (-4 + 4\lambda)^2}$$

be equal to $\|P - C\|$. We have then

$$\sqrt{(-1 + \lambda)^2 + 16(-1 + \lambda)^2} = \sqrt{17} \quad \Rightarrow \quad \sqrt{17(-1 + \lambda)^2} = \sqrt{17} \quad \Rightarrow \quad \sqrt{(-1 + \lambda)^2} = 1$$

that is $\| - 1 + \lambda\| = 1$, giving $\lambda = 2$, $\lambda = 0$. For $\lambda = 0$ we have $P_{\lambda=0} = P$, so $P' = P_{\lambda=2} = (3, 7)$.

In order to determine r' we observe that $P \in r$ and we claim that, since r is a line, the set r' symmetric to r with respect to C is a line as well. It is then sufficient to write the line through P' and another point Q' which is symmetric to $Q \in r$ with respect to C. By choosing $Q = (0, -3) \in r$, it is immediate to compute, with the same steps as above, that $Q' = (4, 9)$. We conclude that $r' = r_{CQ'}$, with vector equation

$$r' : \quad (x = 3 + \lambda, \ y = 7 + 2\lambda).$$

Definition 15.5.3 Let A, B be points in $\mathbb{E}^n$. The *midpoint* M_{AB} of the line segment $\overline{AB}$ is the (unique) point of the line r_{AB} with $\|M_{AB} - A\| = \|M_{AB} - B\|$.

Notice that A is the symmetric point to B with respect to M_{AB}, and clearly B is the symmetric point to A with respect to M_{AB} with $M_{AB} = M_{BA}$. One indeed has the vector equality $A - M_{AB} = M_{AB} - B$, giving

$$M_{AB} = \frac{A + B}{2}.$$

The set H_{AB} given by the points

$$H_{AB} = \{P \in \mathbb{E}^n : \|P - A\| = \|P - B\|\}$$

can be shown to be the hyperplane passing through M_{AB} and orthogonal to the line segment $\overline{AB}$. The set H_{AB} is called the *bisecting hyperplane* of the line segment $\overline{AB}$. In $\mathbb{E}^2$ is the bisecting line of $\overline{AB}$, while in $\mathbb{E}^3$ is the *bisecting plane* of $\overline{AB}$.

Exercise 15.5.4 Consider the line segment in $\mathbb{E}^2$ whose endpoints are $A = (1, 2)$ and $B = (3, 4)$. Its midpoint is given by

$$M_{AB} = \frac{A + B}{2} = \frac{(1, 2) + (3, 4)}{2} = (2, 3).$$

A point $P = (x, y)$ belongs to the bisecting line if $\|P - A\|^2 = (x - 1)^2 + (y - 2)^2$ equates $\|P - B\|^2 = (x - 3)^2 + (y - 4)^2 = \|P_B\|^2$, which gives

$$(x - 1)^2 + (y - 2)^2 = (x - 3)^2 + (y - 4)^2 \quad \Rightarrow \quad -2x + 1 - 4y + 4 = -6x + 9 - 8y + 16,$$

that is $\Sigma_{H_{AB}} : x + y - 5 = 0$. It is immediate to check that $M \in H_{AB}$. The direction of the bisecting line is spanned by $(1, -1)$, which is orthogonal to the direction vector $B - A = (2, 2)$ spanning the direction of the line r_{AB}.

Exercise 15.5.5 Consider the points $A = (1, 2, -1)$ and $B = (3, 0, 1)$ in $\mathbb{E}^3$. The corresponding midpoint is

$$M_{AB} = \frac{A + B}{2} = \frac{(1, 2, -1) + (3, 0, 1)}{2} = (2, 1, 0).$$

The bisecting plane H_{AB} is given by the points $P = (x, y, z)$ fulfilling the condition

$$(x - 1)^2 + (y - 2)^2 + (z + 1)^2 = \|P - A\|^2 = \|P - B\|^2 = (x - 3)^2 + y^2 + (z - 1)^2$$

which gives

$$\Sigma_\pi : \quad x - y + z - 1 = 0.$$

The bisecting plane is then orthogonal to $(1, -1, 1)$, with r_{AB} having a direction vector given by $B - A = (2, -2, 2)$.

Having defined the notion of symmetry of a set in $\mathbb{E}^n$ with respect to a point, we might wonder about a meaningful definition of symmetry of a set *with respect to an arbitrary linear affine variety* in $\mathbb{E}^n$. Such a task turns out to be quite hard in general, so we focus on the easy case of defining only the notion of symmetry with respect to a hyperplane.

Firstly, a general definition.

Definition 15.5.6 Let $H \subset \mathbb{E}^n$ be a hyperplane.

(a) Let $P \in \mathbb{E}^n$ be an arbitrary point in $\mathbb{E}^n$. The *symmetric point to* P *with respect to* H is the element $P' \in \mathbb{E}^n$ such that H is the bisecting hyperplane of the line segment $\overline{PP'}$.
(b) Let $X \subset \mathbb{E}^n$ be a set of points. The *symmetric points to* X *with respect to* H is the set $X' \subset \mathbb{E}^n$ given by every point P' which is symmetric to any P in X with respect to H.
(c) Let $X \subset \mathbb{E}^n$. We say that X *is symmetric with respect to* H if $X = X'$, that is if X contains the symmetric point (with respect to H) to any of its points. In such a case, H is called a *symmetry hyperplane* for X.

Remark 15.5.7 Notice that if P' is the symmetric point to P with respect to the hyperplane H, then the line $r_{PP'}$ is orthogonal to H and $\mathrm{d}(P', H) = \mathrm{d}(P, H)$.

We finish with some examples on the simplest cases in $\mathbb{E}^2$ and $\mathbb{E}^3$.

Exercise 15.5.8 A line is a hyperplane in $\mathbb{E}^2$. Given the point $P = (1, 2)$ we determine its symmetric P' with respect to the line whose equation is $\Sigma_r : 2x + y - 2$.

We observe that if t is the line through P which is orthogonal to P, then P' is the point in t fixed by the condition $\mathrm{d}(P, r) = \mathrm{d}(P', r)$. The direction of t is clearly spanned by the vector (2, 1), so

$$t : \quad \begin{cases} x = 1 + 2\lambda \\ y = 2 + \lambda \end{cases}$$

and the points in t can be written as $Q_\lambda = (1 + 2\lambda, 2 + \lambda)$. By setting

$$\mathrm{d}(Q_\lambda, r) = \mathrm{d}(P, r) \quad \Rightarrow \quad \frac{|2(1 + 2\lambda) + (2 + \lambda) - 2|}{\sqrt{4 + 1}} = \frac{|2 + 2 - 2|}{\sqrt{4 + 1}} \quad \Rightarrow \quad |5\lambda + 2| = 2$$

we see that $Q_{\lambda=0} = P$, while $Q_{\lambda=-4/5} = P' = \frac{1}{5}(-3, 6)$.

Exercise 15.5.9 Given $P = (0, 1, -2) \in \mathbb{E}^3$, we determine its symmetric P' with respect to the hyperplane π (which is indeed a plane, since we are in $\mathbb{E}^3$) whose equation is $\Sigma_\pi : 2x + 4y + 4z - 5 = 0$.

We firstly find the line t through P which is orthogonal to π. The orthogonal subspace to π is spanned by the vector $(2, 4, 4)$ or equivalently $(1, 2, 2)$, so the line t has parametric equation

$$t : \quad \begin{cases} x = \lambda \\ y = 1 + 2\lambda \\ z = -2 + 2\lambda \end{cases} .$$

Since for the symmetric point P' it is $\mathrm{d}(P, \pi) = \mathrm{d}(P', \pi)$, we label a point Q in t by the parameter λ as $Q_\lambda = (\lambda, 1 + 2\lambda, -2 + 2\lambda)$ and impose

$$\mathrm{d}(Q_\lambda, \pi) = \mathrm{d}(P, \pi)$$
$$\Rightarrow \quad \frac{|2\lambda + 4(1+2\lambda) + 4(-2+2\lambda) - 5|}{\sqrt{36}} = \frac{|4-8-5|}{\sqrt{36}} \quad \Rightarrow \quad |18\lambda - 9| = 9.$$

We see that $P = Q_{\lambda=0}$ and $P' = Q_{\lambda=1} = P' = (1, 3, 0)$.

Exercise 15.5.10 In $\mathbb{E}^3$ let us determine the line r' which is symmetric to the line with equation $r: \ (x, y, z) = (0, 1, -2) + \mu(1, 0, 0)$ with respect to the plane π with equation $\pi: \ 2x + 4y + 4z - 5 = 0$.

The plane π is the same plane we considered in the previous exercise. Its orthogonal space is spanned by the vector $(1, 2, 2)$. By labelling a point of the line r as $P_\mu = (\mu, 1, -2)$, we find the line t_μ which passes through P_μ and is orthogonal to π. A parametric equation for t_μ is given by

$$t_\mu : \quad \begin{cases} x = \mu + \lambda \\ y = 1 + 2\lambda \\ z = -2 + 2\lambda \end{cases} .$$

We label then points Q in t_μ by writing $Q_{\lambda,\mu} = (\mu + \lambda, 1 + 2\lambda, -2 + 2\lambda)$. We require

$$\mathrm{d}(Q_{\lambda,\mu}, \pi) = \mathrm{d}(P_\mu, \pi)$$

as a condition to determine λ, since μ will yield a parameter for the line r'. We have

$$\mathrm{d}(Q_{\lambda,\mu}, \pi) = \frac{|2(\mu+\lambda) + 4(1+2\lambda) + 4(-2+2\lambda) - 5|}{\sqrt{36}}$$
$$\mathrm{d}(P_\mu, \pi) = \frac{|2\mu + 4 - 8 - 5|}{\sqrt{36}}.$$

From $\mathrm{d}(Q_{\lambda,\mu}, \pi) = \mathrm{d}(P_\mu, \pi)$ we have

$$|2\mu + 18\lambda - 9| = |2\mu - 9| \quad \Rightarrow \quad 2\mu + 18\lambda - 9 = \pm(2\mu - 9).$$

For $\lambda = 0$ we recover $Q_{\lambda=0,\mu} = P_\mu$. The other solution is $\lambda = -\frac{2}{9}\mu + 1$, giving

$$Q_{\lambda=-(2/9)\mu+1,\mu} = P'_\mu = \left(\frac{7}{9}\mu + 1, -\frac{4}{9}\mu + 3, -\frac{4}{9}\mu\right).$$

By a rescaling of the parameter μ, a vector equation for the line r' can be written as

$$r': \ (x, y, z) = (1, 3, 0) + \mu(7, -4, -4).$$

Exercise 15.5.11 Consider the set $X \subset \mathbb{E}^2$ given by

$$X = \{(x, y) \in \mathbb{E}^2 \ : \ y = 5x^2\}$$

and the line r whose cartesian equation is $\Sigma_r : x = 0$. We wish to show that r is a symmetry *axis* for X, that is X is symmetric with respect to r. We have then to prove that each point P', symmetric to any point $P \in X$ with respect to r, is an element in X.

Let us consider a generic $P = (x_0, y_0) \in X$ and determine its symmetric with respect to r. The line t through P which is orthogonal to r has the following parametric equation

$$t : \quad \begin{cases} x = x_0 + \lambda \\ y = y_0 \end{cases} .$$

A point in t is then labelled $P_\lambda = (x_0 + \lambda, y_0)$. For its distance from r we compute $\mathrm{d}(P_\lambda, r) = |x_0 + \lambda|$, while $\mathrm{d}(P, r) = |x_0|$. By imposing that these two distances coincide, we have

$$\begin{aligned} \mathrm{d}(P_\lambda, r) = \mathrm{d}(P, r) \quad &\Leftrightarrow \quad |x_0 + \lambda| = |x_0| \\ &\Leftrightarrow \quad (x_0 + \lambda)^2 = x_0^2 \\ &\Leftrightarrow \quad \lambda(2x_0 + \lambda) = 0. \end{aligned}$$

The solution $\lambda = 0$ corresponds to P, the solution $\lambda = -2x_0$ yields $P' = (-x_0, y_0)$. Such calculations do not depend on the fact that P is an element in X. If we consider only points P in X, we have to require that $y_0 = 5x_0^2$. It follows that $y_0 = 5(-x_0)^2$, that is $P' \in X$.

Chapter 16
Conic Sections

This chapter is devoted to conics. We shall describe at length their algebraic and geometric properties and their use in physics, notably for the Kepler laws for the motion of celestial bodies.

16.1 Conic Sections as *Geometric Loci*

The conic sections (or simply conics) are parabolæ, ellipses (with circles as limiting case), hyperbolæ. They are also known as geometric *loci*, that is collections of points $P(x, y) \in \mathbb{E}^2$ satisfying one or more conditions, or determined by such conditions. The following three relations, whose origins we briefly recall, should be well known

$$x^2 = 2py, \qquad \frac{x^2}{a^2} + \frac{y^2}{b^2} = 1, \qquad \frac{x^2}{a^2} - \frac{y^2}{b^2} = 1. \tag{16.1}$$

Definition 16.1.1 (*Parabolæ*) Given a straight line δ and a point F on the plane $\mathbb{E}^2$, the set (locus) of points P equidistant from δ and F is called *parabola*. The straight line δ is the *directrix* of the parabola, while the point F is the *focus* of the parabola. This is shown in Fig. 16.1.

Fix a cartesian orthogonal reference system $(O; x, y)$ for $\mathbb{E}^2$, with a generic point P having coordinates $P = (x, y)$. Consider the straight line δ given by the points with equation $y = -p/2$ and the focus $F = (0, p/2)$ (with $p > 0$). The parabola with directrix δ and focus F is the set of points fulfilling the condition

$$\mathrm{d}(P, \delta) = \mathrm{d}(P, F). \tag{16.2}$$

G. Landi and A. Zampini, *Linear Algebra and Analytic Geometry for Physical Sciences*, Undergraduate Lecture Notes in Physics,
https://doi.org/10.1007/978-3-319-78361-1_16

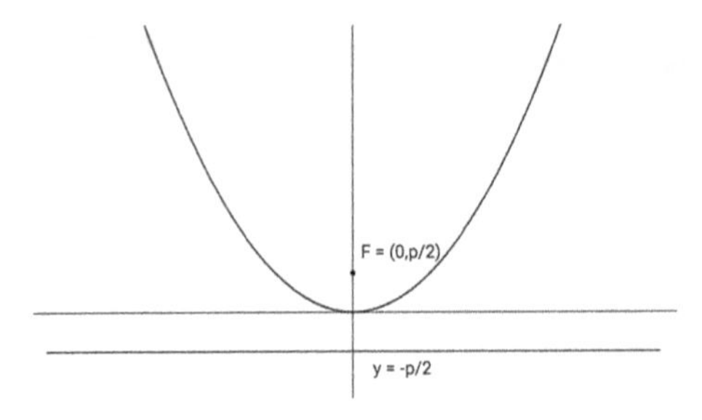

Fig. 16.1 The parabola $y = x^2/2p$

Since the point $P' = (x, -p/2)$ is the orthogonal projection of P over δ, with $\mathrm{d}(P, \delta) = \mathrm{d}(P, P')$, the condition (16.2) reads

$$\|P - P'\|^2 = \|P - F\|^2 \quad \Rightarrow \quad \|(0, y + p/2)\|^2 = \|(x, y - p/2)\|^2,$$

that is

$$(y + p/2)^2 = x^2 + (y - p/2)^2 \quad \Rightarrow \quad x^2 = 2py.$$

If C is a parabola with focus F and directrix δ then,

- the straight line through F which is orthogonal to δ is the *axis* of C,
- the point where the parabola C intersects its axis is the *vertex* of the parabola.

Definition 16.1.2 (*Ellipses*) Given two points F_1 ed F_2 on the plane $\mathbb{E}^2$, the set (locus) of points P for which the sum of the distances between P and the points F_1 and F_2 is constant is called *ellipse*. The points F_1 and F_2 are called the *foci* of the ellipse. This is shown in Fig. 16.2.

Fix a cartesian orthogonal reference system $(O; x, y)$ for $\mathbb{E}^2$, with a generic point P having coordinates $P = (x, y)$. Consider the points $F_1 = (-q, 0)$, $F_2 = (q, 0)$ (with $q \geq 0$) and k a real parameter such that $k > 2q$. The ellipse with foci F_1, F_2 and parameter k is the set of points $P = (x, y)$ fulfilling the condition

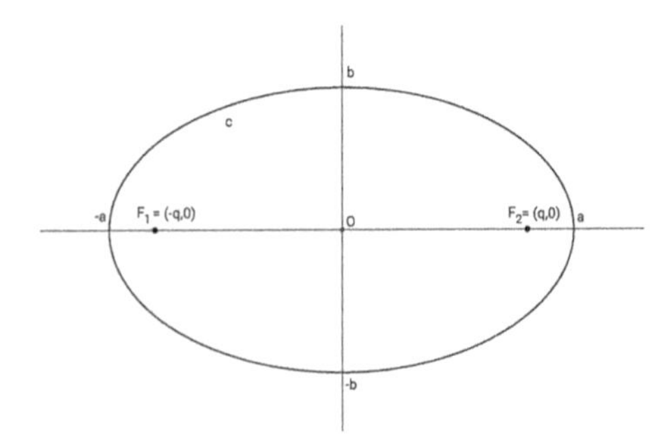

Fig. 16.2 The ellipse $x^2/a^2 + y^2/b^2 = 1$

$$d(P, F_1) + d(P, F_2) = k. \tag{16.3}$$

We denote by $A = (a, 0)$ and $B = (0, b)$ the intersection of the ellipse with the positive x-axis half-line and the positive y-axis half-line, thus $a > 0$ and $b > 0$. From $d(A, F_1) + d(A, F_2) = k$ we have that $k = 2a$; from $d(B, F_1) + d(B, F_2) = k$ we have that $2\sqrt{q^2 + b^2} = k$, so we write

$$k = 2a, \quad q^2 = a^2 - b^2,$$

with $a \geq b$. By squaring the condition (16.3) we have

$$\|(x + q, y)\|^2 + \|(x - q, y)\|^2 + 2\,\|(x + q, y)\|\,\|(x - q, y)\| = 4a^2,$$

that is

$$2(x^2 + y^2 + q^2) + 2\sqrt{(x^2 + y^2 + q^2 + 2qx)(x^2 + y^2 + q^2 - 2qx)} = 4a^2$$

that we write as

$$\sqrt{(x^2 + y^2 + q^2)^2 - 4q^2x^2} = 2a^2 - (x^2 + y^2 + q^2).$$

By squaring such a relation we have

$$-q^2x^2 = a^4 - a^2(x^2 + y^2 + q^2).$$

Since $q^2 = a^2 - b^2$, the equation of the ellipse depends on the real positive parameters a, b as follows

$$b^2x^2 + a^2y^2 = a^2b^2,$$

which is equivalent to

$$\frac{x^2}{a^2} + \frac{y^2}{b^2} = 1.$$

Notice that, if $q = 0$, that is if $a = b$, the foci F_1 ed F_2 coincide with the origin O of the reference system, and the ellipse reduces to a circle whose equation is

$$x^2 + y^2 = r^2$$

with radius $r = a = b > 0$.

If C is an ellipse with (distinct) foci F_1 and F_2, then

- the straight line passing through the foci is the *major axis* of the ellipse,
- the straight line orthogonally bisecting the segment $\overline{F_1F_2}$ is the *minor axis* of the ellipse,
- the midpoint of the segment $\overline{F_1F_2}$ is the *centre* of the ellipse,

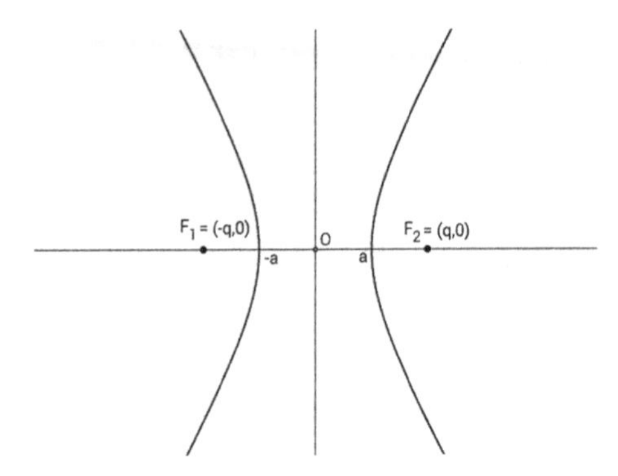

Fig. 16.3 The hyperbola $x^2/a^2 - y^2/b^2 = 1$

- the four points where the ellipse intersects its axes are the *vertices* of the ellipse,
- the distance between the centre of the ellipse and the vertices on the major axis (respectively on the minor axis) is called the *major semi-axis* (respectively *minor semi-axis*).

Definition 16.1.3 (*Hyperbolæ*) Given two points F_1 and F_2 on the plane $\mathbb{E}^2$, the set (locus) of points P for which the absolute difference of the distances $\mathrm{d}(P, F_1)$ and $\mathrm{d}(P, F_2)$ is constant, is the *hyperbola* with foci F_1, F_2. This is shown in Fig. 16.3.

Fix a cartesian orthogonal reference system $(O; x, y)$ for $\mathbb{E}^2$, with a generic point P having coordinates $P = (x, y)$. Consider the points $F_1 = (-q, 0)$, $F_2 = (q, 0)$ (with $q \geq 0$) and k a real parameter such that $k > 2q$. The hyperbola with foci F_1, F_2 and parameter k is the set of points $P = (x, y)$ fulfilling the condition

$$|\mathrm{d}(P, F_1) - \mathrm{d}(P, F_2)| = k. \tag{16.4}$$

Notice that, since $k > 0$, such a hyperbola does not intersect the y-axis, since the points on the y-axis are equidistant from the foci. By denoting by $A = (a, 0)$ (with $a > 0$) the intersection of the hyperbola with the x-axis, we have

$$k = |\mathrm{d}(A, F_1) - \mathrm{d}(A, F_2)| = \big|a + q - |a - q|\big|,$$

which yields $a < q$, since from $a > q$ it would follow that $|a - q| = a - q$, giving $k = 2q$. The previous condition then show that

$$k = |2a| = 2a.$$

By squaring the relation (16.4) we have

$$\|(x + q, y)\|^2 + \|(x - q, y)\|^2 - 2\ \|(x + q, y)\|\ \|(x - q, y)\| = 4a^2,$$

that is

$$2(x^2 + y^2 + q^2) - 2\sqrt{(x^2 + y^2 + q^2 + 2qx)(x^2 + y^2 + q^2 - 2qx)} = 4a^2$$

which we write as

$$\sqrt{(x^2 + y^2 + q^2)^2 - 4q^2x^2} = (x^2 + y^2 + q^2) - 2a^2.$$

By squaring once more, we have

$$-q^2x^2 = a^4 - a^2(x^2 + y^2 + q^2),$$

that reads

$$(a^2 - q^2)x^2 + a^2y^2 = a^2(a^2 - q^2).$$

From $a < q$ we have $q^2 - a^2 > 0$, so we set $q^2 - a^2 = b^2$ and write the previous relation as

$$-b^2x^2 + a^2y^2 = -a^2b^2,$$

which is equivalent to

$$\frac{x^2}{a^2} - \frac{y^2}{b^2} = 1.$$

If C is a hyperbola with foci F_1 and F_2, then

- the straight line through the foci is the *transverse axis* of the hyperbola,
- the straight line orthogonally bisecting the segment $\overline{F_1F_2}$ is the *axis* of the hyperbola,
- the midpoint of the segment $\overline{F_1F_2}$ is the *centre* of the hyperbola;
- the points where the hyperbola intersects its transverse axis are the *vertices* of the hyperbola,
- the distance between the centre of the hyperbola and its foci is the *transverse semi-axis* of the hyperbola.

Remark 16.1.4 The above analysis shows that, if C is a parabola with equation

$$x^2 = 2py,$$

then its directrix is the line $y = -p/2$ and its focus is the point $(0, p/2)$, while the equation

$$y^2 = 2px$$

is a parabola C with directrix $x = -p/2$ and focus $(p/2, 0)$.

If C is an ellipse with equation

$$\frac{x^2}{a^2} + \frac{y^2}{b^2} = 1$$

(and $a \geq b$), then its foci are the points $F_{\pm} = (\pm\sqrt{a^2 - b^2}, 0)$.

If C is a hyperbola with equation

$$\frac{x^2}{a^2} - \frac{y^2}{b^2} = 1$$

then its foci are the points $F_{\pm} = (\pm\sqrt{a^2 + b^2}, 0)$.

We see that the definition of a parabola requires one single focus and a straight line (not containing the focus), while the definition of an ellipse and of a hyperbola requires two distinct foci and a suitable distance k. This apparent diversity can be reconciled. If F is a point is $\mathbb{E}^2$ and δ a straight line with $F \notin \delta$, then one can consider the *locus* given by points P in $\mathbb{E}^2$ fulfilling the condition

$$\mathrm{d}(P, F) = e\,\mathrm{d}(P, \delta) \tag{16.5}$$

with $e > 0$. It is clear that, if $e = 1$, this relation defines a parabola with focus F and directrix δ. We shall show later on (in Sect. 16.4 and then Sect. 16.7) that the relation above gives an ellipse for $0 < e < 1$ and a hyperbola if $e > 1$. The parameter $e > 0$ is called the *eccentricity* of the conic.

Since symmetry properties of conics do not depend on the reference system, when dealing with symmetries or geometric properties of conics one can refer to the Eqs. (16.1).

Remark 16.1.5 With the symmetry notions given in the Sect. 15.5, the y-axis is a symmetry axis for the parabola C whose equation is $y = 2px^2$. If $P = (x_0, y_0) \in C$, the symmetric point P' to P with respect to the y-axis is $P' = (-x_0, y_0)$, which belongs to C since $2py_0 = (-x_0^2) = x_0^2$. Furthermore, the axis of a parabola is a symmetry axis and its vertex is equidistant from the focus and the directrix if the parabola.

In a similar way one shows that the axes of an ellipse or of a hyperbola, are symmetry axes and the centre is a symmetry centre in both cases. For an ellipse with equation $\alpha x^2 + \beta y^2 = 1$ or a hyperbola with equation $\alpha x^2 - \beta y^2 = 1$ the centre coincided with the origin of the reference system.

16.2 The Equation of a Conic in Matrix Form

In the previous section we have shown how, in a given reference system, a parabola, an ellipse and a hyperbola are described by one of equations in (16.1). But evidently such equations are not the most general ones for the loci we are considering, since they have particular positions with respect to the axes of the reference system.

A common feature of the Eqs. (16.1) is that they are formulated as quadratic polynomials in x and y. In the present section we study general quadratic polynomial equations in two variables.

Since to a large extent one does not make use of the euclidean structure given by the scalar product in $\mathbb{E}^n$, one can consider the affine plane $\mathbb{A}^2(\mathbb{R})$. By taking complex coordinates, with the canonical inclusion $\mathbb{R}^2 \hookrightarrow \mathbb{C}^2$, one enlarges the real affine plane to the complex one,

$$\mathbb{A}^2(\mathbb{R}) \hookrightarrow \mathbb{A}^2(\mathbb{C}).$$

Definition 16.2.1 A *conic section* (or simply a conic) is the set of points (locus) whose coordinates (x, y) satisfy a quadratic polynomial equation in the variables x, y, that is

$$a_{11}\, x^2 + 2\, a_{12}\, xy + a_{22}\, y^2 + 2\, a_{13}\, x + 2\, a_{23}\, y + a_{33} = 0 \tag{16.6}$$

with coefficients $a_{ij} \in \mathbb{R}$.

Remark 16.2.2 We notice that

(a) The equations of conics considered in the previous section are particular case of the general Eq. (16.6). As an example, for a parabola we have

$$a_{11} = 1, \quad a_{23} = -2p, \quad a_{12} = a_{22} = a_{13} = a_{33} = 0.$$

Notice also that in all the equations considered in the previous section for a parabola or an ellipse or a hyperbola we have $a_{12} = 0$.

(b) There are polynomial equations like (16.6) which do not describe any of the conics presented before: neither a parabola, nor an ellipse or a hyperbola. Consider for example the equation $x^2 - y^2 = 0$, which is factorised as $(x + y)(x - y) = 0$. The set of solutions for such an equation is the union of the two lines with cartesian equations $x + y = 0$ and $x - y = 0$.
Any quadratic polynomial equation (16.6) that can be factorised as

$$(ax + by + c)(a'x + b'y + c') = 0$$

describes the union of two lines. Such lines are not necessarily real. Consider for example the equation $x^2 + y^2 = 0$. Its set of solutions is given only by the point $(0, 0)$ in $\mathbb{A}^2(\mathbb{R})$, while in $\mathbb{A}^2(\mathbb{C})$ we can write $x^2 + y^2 = (x + \mathrm{i}y)(x - \mathrm{i}y)$, so the conic is the union of the two conjugate lines with cartesian equation $x + \mathrm{i}y = 0$ and $x - \mathrm{i}y = 0$.

Definition 16.2.3 A conic is called *degenerate* if it is the union of two lines. Such lines can be either real (coincident or distinct) or complex (in such a case they are also conjugate).

The polynomial equation (16.6) can be written in a more succinct form by means of two symmetric matrices associated with a conic. We set

$$\mathbb{R}^{2,2} \ni A = \begin{pmatrix} a_{11} & a_{12} \\ a_{12} & a_{22} \end{pmatrix}, \qquad \mathbb{R}^{3,3} \ni B = \begin{pmatrix} a_{11} & a_{12} & a_{13} \\ a_{12} & a_{22} & a_{23} \\ a_{13} & a_{23} & a_{33} \end{pmatrix}.$$

By introducing these matrices, we write the left end side of the Eq. (16.6) as

$$(x \ y \ 1) \begin{pmatrix} a_{11} & a_{12} & a_{13} \\ a_{12} & a_{22} & a_{23} \\ a_{13} & a_{23} & a_{33} \end{pmatrix} \begin{pmatrix} x \\ y \\ 1 \end{pmatrix} = a_{11}\, x^2 + 2\, a_{12}\, xy + a_{22}\, y^2 + 2\, a_{13}\, x + 2\, a_{23}\, y + a_{33}. \tag{16.7}$$

The quadratic homogeneous part of the polynomial defining (16.6) and (16.7), is written as

$$F_C(x, y) = a_{11}\, x^2 + 2\, a_{12}\, xy + a_{22}\, y^2 = (x \ y)\, A \begin{pmatrix} x \\ y \end{pmatrix}.$$

Such an F_C is a quadratic form, called the *quadratic form* associated to the conic C.

Definition 16.2.4 Let C be the conic given by the equation

$$a_{11}\, x^2 + 2\, a_{12}\, xy + a_{22}\, y^2 + 2\, a_{13}\, x + 2\, a_{23}\, y + a_{33} = 0.$$

The matrices

$$B = \begin{pmatrix} a_{11} & a_{12} & a_{13} \\ a_{12} & a_{22} & a_{23} \\ a_{13} & a_{23} & a_{33} \end{pmatrix}, \qquad A = \begin{pmatrix} a_{11} & a_{12} \\ a_{12} & a_{22} \end{pmatrix}$$

are called respectively the *matrix of the coefficients* and the *matrix of the quadratic form* of C.

Exercise 16.2.5 The matrices associated to the parabola with equation $y = 3x^2$ are,

$$B = \begin{pmatrix} 3 & 0 & 0 \\ 0 & 0 & -1/2 \\ 0 & -1/2 & 0 \end{pmatrix}, \qquad A = \begin{pmatrix} 3 & 0 \\ 0 & 0 \end{pmatrix}.$$

Remark 16.2.6 Notice that the six coefficients a_{ij} in (16.6) determine a conic, but a conic is not described by a single array of six coefficients since the equation

$$ka_{11}\, x^2 + 2\, ka_{12}\, xy + ka_{22}\, y^2 + 2\, ka_{13}\, x + 2\, ka_{23}\, y + ka_{33} = 0$$

defines the same locus for any $k \in \mathbb{R} \setminus \{0\}$.

16.3 Reduction to Canonical Form of a Conic: Translations

A natural question arises. Given a non degenerate conic with equation written as in (16.6) with respect to a reference frame, does there exist a new reference system with respect to which the equation for the conic has a form close to one of those given in (16.1)?

Definition 16.3.1 We call *canonical form* of a non degenerate conic C one of the following equations for C in a given reference system $(O; x, y)$.

(i) A *parabola* has equation

$$x^2 = 2py \qquad \text{or} \qquad y^2 = 2px. \tag{16.8}$$

(ii) A *real ellipse* has equation

$$\frac{x^2}{a^2} + \frac{y^2}{b^2} = 1 \tag{16.9}$$

while an *imaginary ellipse* has equation

$$\frac{x^2}{a^2} + \frac{y^2}{b^2} = -1. \tag{16.10}$$

(iii) A *hyperbola* has equation

$$\frac{x^2}{a^2} - \frac{y^2}{b^2} = 1 \qquad \text{or} \qquad \frac{x^2}{a^2} - \frac{y^2}{b^2} = -1. \tag{16.11}$$

A complete answer to the question above is given in two steps.

One first considers only conics whose equation, in a given reference system, $(O; x, y)$ has coefficient $a_{12} = 0$, that is conics whose equation lacks the mixed term xy. The reference system $(O'; X, Y)$ for a canonical form is obtained with a translation from $(O; x, y)$.

The general case of a conic whose equation in a given reference system $(O; x, y)$ may have the mixed term xy will require the composition of a rotation and a translation from $(O; x, y)$ to obtain the reference system $(O'; X, Y)$ for a canonical form.

Exercise 16.3.2 Let $\Gamma : y = 2x^2$ describe a parabola in the canonical form, and let us define the following translation on the plane

$$T_{(x_0, y_0)} : \quad \begin{cases} x = X + x_0 \\ y = Y + y_0 \end{cases}.$$

The equation for the conic Γ with respect to the reference system $(O'; X, Y)$ is then

$$Y = 2X^2 + 4x_0X + 2x_0^2 - y_0.$$

Exercise 16.3.3 Let $\Gamma' : x^2 + 2y^2 = 1$ be an ellipse in the canonical form. Under the translation of the previous example, the equation for Γ' with respect to the reference system $(O'; X, Y)$ is

$$X^2 + 2Y^2 + 2x_0 X + 4y_0 Y + x_0^2 + 2y_0^2 - 1 = 0.$$

Notice that, after the translation by $T_{(x_0, y_0)}$, the equations for the conics Γ and Γ' are no longer in canonical form, but both still lack the mixed term xy. We prove now, with a constructive method, that the converse holds as well.

Exercise 16.3.4 (Completing the squares) Let C be a non degenerate conic whose equation reads, with respect to the reference system $(O; x, y)$,

$$a_{11}\, x^2 + a_{22}\, y^2 + 2\, a_{13}\, x + 2\, a_{23}\, y + a_{33} = 0. \tag{16.12}$$

Since the polynomial must be quadratic, there are two possibilities. Either both a_{11} and a_{22} different from zero, or one of them is zero. We then consider:

(I) It is $a_{11} = 0$, $a_{22} \neq 0$ (the case $a_{11} \neq 0$ and $a_{22} = 0$ is analogue).
The Eq. (16.12) is then

$$a_{22}\, y^2 + 2\, a_{23}\, y + a_{33} + 2\, a_{13}\, x = 0. \tag{16.13}$$

From the algebraic identities:

$$\begin{aligned} a_{22}\, y^2 + 2\, a_{23}\, y &= a_{22}\left(y^2 + 2\,\frac{a_{23}}{a_{22}}\, y\right) \\ &= a_{22}\left[\left(y + \frac{a_{23}}{a_{22}}\right)^2 - \left(\frac{a_{23}}{a_{22}}\right)^2\right] \\ &= a_{22}\left(y + \frac{a_{23}}{a_{22}}\right)^2 - \frac{a_{23}^2}{a_{22}} \end{aligned}$$

we write the Eq. (16.13) as

$$a_{22}\left(y + \frac{a_{23}}{a_{22}}\right)^2 - \frac{a_{23}^2}{a_{22}} + a_{33} + 2\, a_{13}\, x = 0. \tag{16.14}$$

Since C is not degenerate, we have $a_{13} \neq 0$ so we write (16.14) as

$$a_{22}\left(y + \frac{a_{23}}{a_{22}}\right)^2 + 2\, a_{13}\left(x + \frac{a_{33}a_{22} - a_{23}^2}{2\, a_{22}a_{13}}\right) = 0$$

which reads

$$\left(y+\frac{a_{23}}{a_{22}}\right)^2 = -\frac{2\,a_{13}}{a_{22}}\left(x+\frac{a_{33}a_{22}-a_{23}^2}{2\,a_{22}a_{13}}\right).$$

Under the translation

$$\begin{cases} X = x + (a_{33}a_{22}-a_{23}^2)/2\,a_{22}a_{13} \\ Y = y + a_{23}/a_{22} \end{cases}$$

we get

$$Y^2 = 2\,p\,X \tag{16.15}$$

with $p = -a_{13}/a_{22}$. This is the canonical form (16.8).

If we drop the hypothesis that the conics C is non degenerate, we have $a_{13} = 0$ in the Eq. (16.13). Notice that, for the case $a_{11} = 0$ we are considering, $\det B = -a_{13}^2 a_{22}$. Thus the condition of non degeneracy can be expressed as a condition on the determinant of the matrix of the coefficients, since

$$a_{13} = 0 \qquad \Leftrightarrow \qquad \det B = -a_{13}^2 a_{22} = 0.$$

The Eq. (16.14) is then

$$\left(y+\frac{a_{23}}{a_{22}}\right)^2 = \frac{a_{23}^2 - a_{33}a_{22}}{a_{22}^2}$$

and with the translation

$$\begin{cases} X = x \\ Y = y + a_{23}/a_{22} \end{cases}$$

it reads

$$Y^2 = q \tag{16.16}$$

with $q = (a_{23}^2 - a_{33}a_{22})/a_{22}^2$.

(II) It is $a_{11} \neq 0,\ a_{22} \neq 0$.

With algebraic manipulation as above, we can write

$$a_{11}\,x^2 + 2\,a_{13}\,x = a_{11}\left(x+\frac{a_{13}}{a_{11}}\right)^2 - \frac{a_{13}^2}{a_{11}},$$

$$a_{22}\,y^2 + 2\,a_{23}\,y = a_{22}\left(y+\frac{a_{23}}{a_{22}}\right)^2 - \frac{a_{23}^2}{a_{22}}.$$

So the Eq. (16.12) is written as

$$a_{11}\left(x+\frac{a_{13}}{a_{11}}\right)^2+a_{22}\left(y+\frac{a_{23}}{a_{22}}\right)^2+a_{33}-\frac{a_{13}^2}{a_{11}}-\frac{a_{23}^2}{a_{22}}=0. \qquad (16.17)$$

If we consider the translation given by

$$\begin{cases} X = x + a_{13}/a_{11} \\ Y = y + a_{23}/a_{22} \end{cases}.$$

the conic C has the equation

$$a_{11}X^2+a_{22}Y^2=h, \qquad \text{with} \quad h=-a_{33}+\frac{a_{13}^2}{a_{11}}+\frac{a_{23}^2}{a_{22}}, \qquad (16.18)$$

and $h \neq 0$ since C is non degenerate. The coefficients a_{11} and a_{22} can be either concordant or not. Up to a global factor (-1), we can take $a_{11} > 0$. So we have the following cases.

(IIa) It is $a_{11} > 0$ and $a_{22} > 0$. One distinguish according to the sign of the coefficient h:

- If $h > 0$, the Eq. (16.18) is equivalent to

$$\frac{a_{11}}{h}X^2+\frac{a_{22}}{h}Y^2=1.$$

Since $a_{11}/h > 0$ and $a_{22}/h > 0$, we have (positive) real numbers a, b by defining $h/a_{11} = a^2$ and $h/a_{22} = b^2$. The Eq. (16.18) is written as

$$\frac{X^2}{a^2}+\frac{Y^2}{b^2}=1, \qquad (16.19)$$

which is the canonical form of a real ellipse (16.9).

- If $h < 0$, we have $-a_{11}/h > 0$ and $-a_{22}/h > 0$, we can again introduce (positive) real numbers a, b by $-h/a_{11} = a^2$ and $-h/a_{22} = b^2$. The Eq. (16.18) can be written as

$$\frac{X^2}{a^2}+\frac{Y^2}{b^2}=-1, \qquad (16.20)$$

which is the canonical form of an imaginary ellipse (16.10).

- If $h = 0$ (which means that C is degenerate), we set $1/a_{11} = a^2$ and $1/a_{22} = b^2$ with real number a, b, so to get from (16.18) the expression

$$\frac{X^2}{a^2}+\frac{Y^2}{b^2}=0. \qquad (16.21)$$

(IIb) It is $a_{11} > 0$ and $a_{22} < 0$. Again depending on the sign of the coefficient h we have:

- If $h > 0$, the Eq. (16.18) is

$$\frac{a_{11}}{h}X^2 + \frac{a_{22}}{h}Y^2 = 1.$$

Since $a_{11}/h > 0$ and $a_{22}/h < 0$, we can define $h/a_{11} = a^2$ and $-h/a_{22} = b^2$ with a, b positive real numbers. The Eq. (16.18) becomes

$$\frac{X^2}{a^2} - \frac{X^2}{b^2} = 1, \tag{16.22}$$

which the first canonical form in (16.11).

- If $h < 0$, we have $-a_{11}/h > 0$ and $-a_{22}/h < 0$, so we can define $-h/a_{11} = a^2$ and $h/a_{22} = 1/b^2$ with a, b positive real numbers. The Eq. (16.18) becomes

$$\frac{X^2}{a^2} - \frac{Y^2}{b^2} = -1, \tag{16.23}$$

which is the second canonical form in (16.11).

- If $h = 0$ (that is C is degenerate), we set $1/a_{11} = a^2$ and $-1/a_{22} = b^2$ with a, b real number, so to get from (16.18) the expression

$$\frac{X^2}{a^2} - \frac{Y^2}{b^2} = 0. \tag{16.24}$$

Once again, with B the matrix of the coefficients for C, the identity

$$\det B = a_{11}a_{22}\,h$$

shows that the condition of non degeneracy of the conic C is equivalently given by $\det B \neq 0$.

The analysis done for the cases of degenerate conics makes it natural to introduce the following definition, which has to be compared with the Definition 16.3.1.

We call canonical form of a degenerate conic C one of the following equations for C in a given reference system $(O; x, y)$.

(i) A degenerate parabola has equation

$$x^2 = q \qquad \text{or} \qquad y^2 = q. \tag{16.25}$$

(ii) A degenerate ellipse has equation

$$\frac{x^2}{a^2} + \frac{y^2}{b^2} = 0. \tag{16.26}$$

(iii) A degenerate hyperbola has equation

$$\frac{x^2}{a^2} - \frac{y^2}{b^2} = 0. \tag{16.27}$$

Remark 16.3.5 With the definition above, we have that

(i) The conic C with equation $x^2 = q$ is the union of the lines with cartesian equations $x = \pm\sqrt{q}$. If $q > 0$ the lines are real and distinct, if $q < 0$ the lines are complex and conjugate. If $q = 0$ the conic C is the y-axis counted twice. Analogue cases are obtained for the equation $y^2 = q$.

(ii) The equation $b^2\,x^2 + a^2\,y^2 = 0$ has the unique solution $(0, 0)$ if we consider real coordinates. On the complex affine plane $\mathbb{A}^2(\mathbb{C})$ the solutions to such equations give a degenerate conic C which is the union of two complex conjugate lines, since we can factorise

$$b^2\,x^2 + a^2\,y^2 \;=\; (b\,x + \mathrm{i}\,a\,y)(b\,x - \mathrm{i}\,a\,y).$$

(iii) The solutions to the equation $b^2\,x^2 - a^2\,y^2 = 0$ give the union of two real and distinct lines, since we can factorise as follows

$$b^2\,x^2 - a^2\,y^2 \;=\; (b\,x +\, a\,y)(b\,x -\, a\,y).$$

What we have studied up to now is the proof of the following theorem.

Theorem 16.3.6 *Let C be a conic whose equation, with respect to a reference system $(O; x, y)$ lacks the monomial xy. There exists a reference system $(O'; X, Y)$, obtained from $(O; x, y)$ by a translation, with respect to which the equation for the conic C has a canonical form.*

Exercise 16.3.7 We consider the conic C with equation

$$x^2 + 4y^2 + 2x - 12y + 3 = 0.$$

We wish to determine a reference system $(O'; X, Y)$ with respect to which the equation for C is canonical. We complete the squares as follows:

$$x^2 + 2x = (x + 1)^2 - 1,$$
$$4y^2 - 12y = 4\left(y - \tfrac{3}{2}\right)^2 - 9$$

and write

$$x^2 + 4y^2 + 2x - 12y + 3 = (x + 1)^2 + 4\left(y - \tfrac{3}{2}\right)^2 - 7.$$

With the translation

$$\begin{cases} X = x + 1 \\ Y = y - \frac{3}{2} \end{cases}$$

the equation for C reads

$$X^2 + 4Y^2 = 7 \quad \Rightarrow \quad \frac{X^2}{7} + \frac{Y^2}{7/4} = 1.$$

This is an ellipse with centre $(X = 0, Y = 0) = (x = -1, y = 3/2)$, with axes given by the lines $X = 0$ and $Y = 0$ which are $x = -1$ and $y = 3/2$, and semi-axes given by $\sqrt{7}$, $\sqrt{7}/2$.

16.4 Eccentricity: Part 1

We have a look now at the relation (16.5) for a particular class of examples. Consider the point $F = (a_x, a_y)$ in $\mathbb{E}^2$ and the line δ whose points satisfy the equation $x = u$, with $u \neq a_x$. The relation $\mathrm{d}(P, F) = e\,\mathrm{d}(P, \delta)$ (with $e > 0$) is satisfied by the points $P = (x, y)$ whose coordinates are the solutions of the equation

$$(y - a_y)^2 + (1 - e^2)x^2 + 2(ue^2 - a_x)x + a_x^2 - u^2e^2 = 0. \tag{16.28}$$

We have different cases, depending on the parameter e.

(a) We have already mentioned that for $e = 1$ we are describing the parabola with focus F and directrix δ. Its equation from (16.28) is given by

$$(y - a_y)^2 + 2(u - a_x)x + a_x^2 - u^2 = 0. \tag{16.29}$$

(b) Assume $e \neq 1$. Using the results of the Exercise 16.3.4, we complete the square and write

$$\begin{aligned} & (y - a_y)^2 + (1 - e^2)x^2 + 2(ue^2 - a_x)x + a_x^2 - u^2e^2 = 0 \\ \text{or} \quad & (y - a_y)^2 + (1 - e^2)\left(x + \frac{ue^2 - a_x}{1 - e^2}\right)^2 - \frac{e^2(u - a_x)^2}{1 - e^2} = 0. \end{aligned} \tag{16.30}$$

Then the translation given by

$$\begin{cases} Y = y - a_y \\ X = x + (ue^2 - a_x)/(1 - e^2) \end{cases}$$

allows us to write, with respect to the reference system $(O'; X, Y)$, the equation as

$$Y^2 + (1 - e^2)X^2 = \frac{e^2(u - a_x)^2}{1 - e^2}.$$

Depending on the value of e, we have the following possibilities.

(b1) If $0 < e < 1$, all the coefficients of the equation are positive, so we have the ellipse

$$\left(\frac{1 - e^2}{e(u - a_x)}\right)^2 X^2 + \frac{1 - e^2}{e^2(u - a_x)^2} Y^2 = 1.$$

An easy computation shows that its foci are given by

$$F_{\pm} = (\pm \frac{e^2(u - a_x)}{1 - e^2}, a_y)$$

with respect to the reference system $(O'; X, Y)$ and then clearly by

$$F_+ = (a_x, a_y), \qquad F_- = (\frac{a_x + e^2 a_x - 2ue^2}{1 - e^2}, a_y)$$

with respect to $(O; x, y)$. Notice that $F_+ = F$, the starting point.

(b2) If $e > 1$ the equation

$$\left(\frac{1 - e^2}{e(u - a_x)}\right)^2 X^2 - \frac{e^2 - 1}{e^2(u - a_x)^2} Y^2 = 1.$$

represents a hyperbola with foci again given by the points $F_{\pm}$ written before.

Remark 16.4.1 Notice that, if $e = 0$, the relation (16.28) becomes

$$(y - a_y)^2 + (x - a_x)^2 = 0,$$

that is a degenerate imaginary conic, with

$$(y - a_y + \mathrm{i}(x - a_x))(y - a_y - \mathrm{i}(x - a_x)) = 0.$$

If we fix $e^2(u - a_x)^2 = r^2 \neq 0$ and consider the limit $e \to 0$, the Eq. (16.28) can be written as

$$(x - a_x)^2 + (y - a_y)^2 = r^2.$$

This is another way of viewing a circle as a limiting case of a sequence of ellipses.

The case for which the point $F \in \delta$ also gives a degenerate conic. In this case $u = a_x$ and the Eq. (16.28) is

$$(y - a_x)^2 + (1 - e^2)(x - 2u)^2 = 0$$

which is the union of two lines either real (if $1 < e$) or imaginary (if $1 > e$).

16.5 Conic Sections and Kepler Motions

Via the notion of eccentricity it is easier to describe a fundamental relation between the conic sections and the so called *Keplerian* motions.

If $\mathbf{x}_1(t)$ and $\mathbf{x}_2(t)$ describe the motion in $\mathbb{E}^3$ of two point masses m_1 and m_2, and the only force acting on them is the mutual gravitational attraction, the equations of motions are given by

$$\begin{aligned} m_1\ddot{\mathbf{x}}_1 &= -Gm_1m_2\,\frac{\mathbf{x}_1-\mathbf{x}_2}{\|\mathbf{x}_1-\mathbf{x}_2\|^3} \\ m_2\ddot{\mathbf{x}}_2 &= -Gm_1m_2\,\frac{\mathbf{x}_2-\mathbf{x}_1}{\|\mathbf{x}_1-\mathbf{x}_2\|^3}\,. \end{aligned}$$

Here G is a constant, the gravitational constant. We know from physics that the centre of mass of this system moves with no acceleration, while for the relative motion $\mathbf{r}(t) = \mathbf{x}_1(t) - \mathbf{x}_2(t)$ the Newton equations are

$$\mu\,\ddot{\mathbf{r}}(t) = -Gm_1m_2\,\frac{\mathbf{r}}{r^3} \qquad (16.31)$$

with the norm $r = \|\mathbf{x}\|$ and $\mu = m_1m_2/(m_1+m_2)$ the so called *reduced mass* of the system. A qualitative analysis of this motion can be given as follows.

With a cartesian orthogonal reference system $(O; x, y, z)$ in $\mathbb{E}^3$, we can write $\mathbf{r}(t) = (x(t), y(t), z(t))$ and $\dot{\mathbf{r}}(t) = (\dot{x}(t), \dot{y}(t), \dot{z}(t))$ for the vector representing the corresponding velocity. From the Newton equations (16.31) the angular momentum (recall its definition and main properties from Sects. 1.3 and 11.2) with respect to the origin O,

$$\frac{\mathrm{d}\mathbf{L}_O}{\mathrm{d}t} = \mu\,\{\dot{\mathbf{r}}\wedge\dot{\mathbf{r}} + \mathbf{r}\wedge\ddot{\mathbf{r}}\} = 0,$$

is a constant of the motion, since $\ddot{\mathbf{r}}$ is parallel to $\mathbf{r}$ from (16.31). This means that both vectors $\mathbf{r}(t)$ and $\dot{\mathbf{r}}(t)$ remain orthogonal to the direction of $\mathbf{L}_O$, which is constant: if the initial velocity $\dot{\mathbf{r}}(t=0)$ is not parallel to the initial position $\mathbf{r}(t=0)$, the motion stays at any time t on the plane orthogonal to $\mathbf{L}_O(t=0)$.

We can consider the plane of the motion as $\mathbb{E}^2$, and fix a cartesian orthogonal reference system $(O; x, y)$, so that the angular momentum conservation can be written as

$$\mu\,(\dot{x}y - \dot{y}x) = l$$

with the constant l fixed by the initial conditions. We also know that the gravitational force is conservative, thus the total energy

$$\frac{1}{2}\,\mu\,\|\dot{\mathbf{r}}\|^2 - Gm_1m_2\,\frac{1}{r} = E.$$

is also a constant of the motion. It is well known that the Eq. (16.31) can be completely solved. We omit the proof of this claim, and mention that the possible trajectories of such motions are conic sections, with focus $F = (0, 0) \in \mathbb{E}^2$ and directrix δ given by the equation $x = \tilde{l}/e$ with

$$\tilde{l} = \frac{l^2}{Gm_1m_2\mu}.$$

and eccentricity parameter given by

$$e = \sqrt{1 + \frac{2\mu El^2}{(Gm_1m_2\mu)^2}}.$$

One indeed shows that

$$\frac{2\mu El^2}{(Gm_1m_2\mu)^2} > -1$$

for any choice of initial values for position and velocity.

This result is one of the reasons why conic sections deserve a special attention in affine geometry. From the analysis of the previous section, we conclude that for $E < 0$, since $0 < e < 1$, the trajectory of the motion is elliptic. If the point mass m_2 represents the Sun, while m_1 a planet in our solar system, this result gives the well observed fact that planet orbits are plane elliptic and the Sun is one of the foci of the orbit (Kepler law).

The Sun is also the focus of hyperbolic orbits ($E > 0$) or parabolic ones ($E = 0$), orbits that are travelled by comets and asteroids.

16.6 Reduction to Canonical Form of a Conic: Rotations

Let us consider two reference systems $(O; x, y)$ and $(O; X, Y)$ having the same origin and related by a rotation by an angle of α,

$$\begin{cases} x = \cos\alpha\, X + \sin\alpha\, Y \\ y = -\sin\alpha\, X + \cos\alpha\, Y \end{cases}.$$

With respect to $(O; x, y)$, consider the parabola Γ: $y = x^2$. In the rotated system $(O; X, Y)$ the equation for Γ is easily found to be

$$-\sin\alpha\, X + \cos\alpha\, Y = (\cos\alpha\, X + \sin\alpha\, Y)^2$$

$$\Rightarrow$$

$$\cos\alpha^2\, X^2 + \sin 2\alpha\, XY + \sin\alpha^2\, Y^2 + \sin\alpha\, X - \cos\alpha\, Y = 0.$$

We see that as a consequence of the rotation, there is a mixed term XY in the quadratic polynomial equation for the parabola Γ. It is natural to wonder whether such a behaviour can be reversed.

Example 16.6.1 With respect to $(O; x, y)$, consider the conic $C\colon xy = k$ for a real parameter k. Clearly, for $k = 0$ this is degenerate (the union of the coordinate axes x and y). On the other hand, the rotation to the system $(O; X, Y)$ by an angle $\alpha = \frac{\pi}{4}$,

$$\begin{cases} x = \frac{1}{\sqrt{2}}(X + Y) \\ y = \frac{1}{\sqrt{2}}(X - Y) \end{cases},$$

transforms the equation of the conic to

$$X^2 - Y^2 = 2k.$$

This is a hyperbola with foci $F_\pm = (\pm 2\sqrt{k}, 0$ when $k > 0$ or $F_\pm = (0, \pm 2\sqrt{|k|})$ when $k < 0$.

In general, if the equation of a conic has a mixed term, does there exist a reference system with respect to which the equation for the given conic does not have the mixed term?

It is clear that the answer to such a question is in the affirmative if and only if there exists a reference system with respect to which the quadratic form of the conic is diagonal. On the other hand, since the quadratic form associated to a conic is symmetric, we know from the Chap. 10 that it is always possible to diagonalise it with a suitable orthogonal matrix.

Let us first study how the equation in (16.7) for a conic changes under a general change of the reference system of the affine euclidean plane we are considering.

Definition 16.6.2 With a *rotation of the plane* we mean a change in the reference system from $(O; x, y)$ to $(O; x', y')$ that is given by

$$\begin{pmatrix} x \\ y \end{pmatrix} = P \begin{pmatrix} x' \\ y' \end{pmatrix}, \tag{16.32}$$

with $P \in \mathrm{SO}(2)$ a special orthogonal matrix, referred to as the *rotation matrix*. If we write

$$P = \begin{pmatrix} p_{11} & p_{12} \\ p_{21} & p_{22} \end{pmatrix},$$

the transformation above reads

$$\begin{cases} x = p_{11}x' + p_{12}y' \\ y = p_{21}x' + p_{22}y' \end{cases} \tag{16.33}$$

These relations give the *equations of the rotation*.

A translation from the reference system $(O; x', y')$ to another $(O'; X, Y)$ is described by the relations

$$\begin{cases} x' = X + x_0 \\ y' = Y + y_0 \end{cases} \tag{16.34}$$

where $(-x_0, -y_0)$ are the coordinates of the point O with respect to $(O'; X, Y)$ and, equivalently, (x_0, y_0) are the coordinates of the point O' with respect to $(O; x', y')$.

A *proper rigid transformation* on the affine euclidean plane $\mathbb{E}^2$ is a change of the reference system given by a rotation followed by a translation. We shall refer to a proper rigid transformation also under the name of *roto-translation*.

Let us consider the composition of the rotation given by (16.33) followed by the translation given by (16.34), so to map the reference system $(O; x, y)$ into $(O'; X, Y)$. The equation describing such a transformation are easily found to be

$$\begin{cases} x = p_{11}X + p_{12}Y + a \\ y = p_{21}X + p_{22}Y + b \end{cases} \tag{16.35}$$

where

$$\begin{cases} a = p_{11}\,x_0 + p_{12}\,y_0 \\ b = p_{21}\,x_0 + p_{22}\,y_0 \end{cases}$$

are the coordinates of O' with respect to $(O; x, y)$. The transformation (16.35) can be written as

$$\begin{pmatrix} x \\ y \\ 1 \end{pmatrix} = \begin{pmatrix} p_{11} & p_{12} & a \\ p_{21} & p_{22} & b \\ 0 & 0 & 1 \end{pmatrix} \begin{pmatrix} X \\ Y \\ 1 \end{pmatrix}, \tag{16.36}$$

and we call

$$Q = \begin{pmatrix} p_{11} & p_{12} & a \\ p_{21} & p_{22} & b \\ 0 & 0 & 1 \end{pmatrix} \tag{16.37}$$

the matrix of (associated to) the proper rigid transformation (roto-translation).

Remark 16.6.3 A rotation matrix P is special orthogonal, that is ${}^tP = P^{-1}$ and $\det(P) = 1$. A roto-translation matrix Q as in (16.37), although satisfies the identity $\det(Q) = 1$, is not orthogonal.

Clearly, with a transposition, the action (16.32) of a rotation matrix also gives $\begin{pmatrix} x & y \end{pmatrix} = \begin{pmatrix} x' & y' \end{pmatrix} {}^tP$, while the action (16.36) of a roto-translation can be written as $\begin{pmatrix} x & y & 1 \end{pmatrix} = \begin{pmatrix} X & Y & 1 \end{pmatrix} {}^tQ$.

Let us then describe how the matrices associated to the equation of a conic are transformed under a roto-translation of the reference system. Then, let us consider a conic C described, with respect to the reference system $(O; x, y)$, by

$$\begin{pmatrix} x & y & 1 \end{pmatrix} B \begin{pmatrix} x \\ y \\ 1 \end{pmatrix} = 0, \qquad F_C(x, y) = \begin{pmatrix} x & y \end{pmatrix} A \begin{pmatrix} x \\ y \end{pmatrix}.$$

Under the roto-translation transformation (16.36) the equation of the conic C with respect to the reference system $(O'; X, Y)$ is easily found to becomes

$$(X \; Y \; 1)\, {}^tQ\, B\, Q \begin{pmatrix} X \\ Y \\ 1 \end{pmatrix} = 0.$$

Also, under the same transformations, the quadratic form for C reads

$$F_C(x', y') = \begin{pmatrix} x' & y' \end{pmatrix} {}^tP\, A\, P \begin{pmatrix} x' \\ y' \end{pmatrix}$$

with respect to the reference system $(O; x', y')$ obtained from $(O; x, y)$ under the action of only the rotation P. Such a claim is made clearer by the following proposition.

Proposition 16.6.4 *The quadratic form associated to a conic C does not change for a translation of the reference system with respect to which it is defined.*

Proof Let us consider, with respect to the reference system $(O; x', y')$, the conic with quadratic form

$$F_C(x', y') = \begin{pmatrix} x' & y' \end{pmatrix} A' \begin{pmatrix} x' \\ y' \end{pmatrix} = a_{11}\,(x')^2 + 2\,a_{12}\,x'y' + a_{22}\,(y')^2.$$

Under the translation (16.34) we have $x' = X - x_0$ e $y' = Y - y_0$, that is

$$a_{11}\,X^2 + 2\,a_{12}\,XY + a_{22}\,Y^2 + \{\text{monomials of order} \leq 1\}.$$

The quadratic form associated to C, with respect to the reference system $(O'; X, Y)$, is then

$$F_C(X, Y) = a_{11}\,X^2 + 2\,a_{12}\,XY + a_{22}\,Y^2 = \begin{pmatrix} X & Y \end{pmatrix} A' \begin{pmatrix} X \\ Y \end{pmatrix},$$

with the same matrix A'. □

Given the quadratic form F_C associated to the conic C in $(O; x', y')$, we have then the following:

$$F_C(x', y') = \begin{pmatrix} x' & y' \end{pmatrix} {}^tP\, A\, P \begin{pmatrix} x' \\ y' \end{pmatrix} \quad \Rightarrow \quad F_C(X, Y) = \begin{pmatrix} X & Y \end{pmatrix} {}^tP\, A\, P \begin{pmatrix} X \\ Y \end{pmatrix}.$$

All of the above proves the following theorem.

Theorem 16.6.5 *Let C be a conic with associated matrix of the coefficients B and matrix of the quadratic form A with respect to the reference system $(O; x, y)$. If Q is the matrix of the roto-translation mapping the reference system $(O; x, y)$ to $(O'; X, Y)$, with P the corresponding rotation matrix, the matrix of the coefficients associated to the conic C with respect to $(O'; X, Y)$ is*

$$B' = {}^tQ\; B\; Q,$$

while the matrix of the canonical form is

$$A' = {}^tP\; A\; P = P^{-1}\; A\; P.$$

In light of the Definition 13.1.4, the matrices A and A' are quadratically equivalent. □

Exercise 16.6.6 Consider the conic C whose equation, in the reference system $(O; x, y)$ is

$$x^2 - 2xy + y^2 + 4x + 4y - 1 = 0.$$

Its associated matrices are

$$B = \begin{pmatrix} 1 & -1 & 2 \\ -1 & 1 & 2 \\ 2 & 2 & -1 \end{pmatrix}, \qquad A = \begin{pmatrix} 1 & -1 \\ -1 & 1 \end{pmatrix}.$$

We first diagonalise the matrix A. Its characteristic polynomial is

$$p_A(T) = |A - TI| = \begin{vmatrix} 1-T & -1 \\ -1 & 1-T \end{vmatrix} = (1-T)^2 - 1 = T(T-2).$$

The eigenvalues are $\lambda = 0$ and $\lambda = 2$ with associated eigenspaces,

$$\begin{aligned} V_0 &= \ker(f_A) = \{(x, y) \in \mathbb{R}^2 \,:\, x - y = 0\} = \mathcal{L}((1, 1)), \\ V_2 &= \ker(f_{A-2I}) = \{(x, y) \in \mathbb{R}^2 \,:\, x + y = 0\} = \mathcal{L}((1, -1)). \end{aligned}$$

It follows that the special orthogonal matrix P giving the change of the basis is

$$P = \frac{1}{\sqrt{2}} \begin{pmatrix} 1 & 1 \\ -1 & 1 \end{pmatrix}$$

and eigenvectors ordered so that $\det(P) = 1$. This rotated the reference system to $(O; x', y')$ with

$$\begin{cases} x = \frac{1}{\sqrt{2}} (x' + y') \\ y = \frac{1}{\sqrt{2}} (-x' + y') \end{cases} .$$

Without translation, the roto-translation matrix is

$$Q' = \frac{1}{\sqrt{2}} \begin{pmatrix} 1 & 1 & 0 \\ -1 & 1 & 0 \\ 0 & 0 & \sqrt{2} \end{pmatrix}$$

and from the Theorem 16.6.5, the matrix associated to C with respect to the reference system $(O; x', y')$ is $\widetilde{B} = {}^{t}Q' \, B \, Q'$. We have then

$$\widetilde{B} = \frac{1}{\sqrt{2}} \begin{pmatrix} 1 & -1 & 0 \\ 1 & 1 & 0 \\ 0 & 0 & 1 \end{pmatrix} \begin{pmatrix} 1 & -1 & 2 \\ -1 & 1 & 2 \\ 2 & 2 & -1 \end{pmatrix} \frac{1}{2} \begin{pmatrix} 1 & 1 & 0 \\ -1 & 1 & 0 \\ 0 & 0 & 1 \end{pmatrix} = \begin{pmatrix} 2 & 0 & 0 \\ 0 & 0 & 2\sqrt{2} \\ 0 & 2\sqrt{2} & -1 \end{pmatrix},$$

so that the equation for C reads

$$2(x')^2 + 4\sqrt{2}y' - 1 = 0.$$

By completing the square at the right hand side, we write this equation as

$$(x')^2 = -2\sqrt{2} \left(y' - \frac{\sqrt{2}}{8} \right).$$

With the translation

$$\begin{cases} X = x' \\ Y = y' - \frac{\sqrt{2}}{8} \end{cases}$$

we see that C is a parabola with the canonical form

$$X^2 = -2\sqrt{2}\,Y$$

and the associated matrices

$$B' = \begin{pmatrix} 1 & 0 & 0 \\ 0 & 0 & \sqrt{2} \\ 0 & \sqrt{2} & 0 \end{pmatrix}, \qquad A' = \begin{pmatrix} 1 & 0 \\ 0 & 0 \end{pmatrix}.$$

Rather than splitting the reduction to canonical form into a first step given by a rotation and a second step given by a translation, we can reduce the equation for C with respect to $(O; x, y)$ to its canonical form by a proper rigid transformation with a matrix Q encoding both a rotation and a translation. Such a composition is given by

$$\begin{cases} x = \frac{1}{\sqrt{2}}(x' + y') \\ y = \frac{1}{\sqrt{2}}(-x' + y') \end{cases} \quad \Rightarrow \quad \begin{cases} x = \frac{1}{\sqrt{2}}(X + Y + \frac{\sqrt{2}}{8}) \\ y = \frac{1}{\sqrt{2}}(-X + Y + \frac{\sqrt{2}}{8}) \end{cases}$$

which we write as

$$\begin{pmatrix} x \\ y \\ 1 \end{pmatrix} = Q \begin{pmatrix} X \\ Y \\ 1 \end{pmatrix}$$

with

$$Q = \frac{1}{\sqrt{2}} \begin{pmatrix} 1 & 1 & \sqrt{2}/8 \\ -1 & 1 & \sqrt{2}/8 \\ 0 & 0 & 1 \end{pmatrix}.$$

We end this example by checking that the matrix associated to the conic C with respect to the reference system $(O'; X, Y)$ can be computed as it is described in the Theorem 16.6.5, that is

$${}^{t}Q\,B\,Q = \begin{pmatrix} 2 & 0 & 0 \\ 0 & 0 & 2\sqrt{2} \\ 0 & 2\sqrt{2} & 0 \end{pmatrix} = 2B'.$$

We list the main steps of the method we described in order to reduce a conic to its canonical form as the proof of the following results.

Theorem 16.6.7 *Given a conic C whose equation is written in the reference system $(O; x, y)$, there always exists a reference system $(O'; X, Y)$, obtained with a roto-translation from $(O; x, y)$, with respect to which the equation for C is canonic.*

Proof Let C be a conic, with associated matrices A (of the quadratic form) and B (of the coefficients), with respect to the reference system $(O; x, y)$. Then,

(a) Diagonalise A, computing an orthonormal basis with eigenvectors $v_1 = (p_{11}, p_{21})$, $v_2 = (p_{12}, p_{22})$, given by the rotation

$$\begin{cases} x = p_{11}\,x' + p_{12}\,y' \\ y = p_{21}\,x' + p_{22}\,y' \end{cases} \tag{16.38}$$

and define

$$Q' = \begin{pmatrix} p_{11} & p_{12} & 0 \\ p_{21} & p_{22} & 0 \\ 0 & 0 & 1 \end{pmatrix}.$$

With respect to the reference system $(O; x', y')$, the conic C has matrix $B' = {}^tQ'\, B\, Q'$, and the corresponding quadratic equation, which we write as

$$\begin{pmatrix} x' & y' & 1 \end{pmatrix} B' \begin{pmatrix} x' \\ y' \\ 1 \end{pmatrix} = 0, \tag{16.39}$$

lacks the monomial term $x'y'$.

(b) Complete the square so to transform, by the corresponding translation, the reference system $(O; x', y')$ to $(O'; X, Y)$, that is

$$\begin{cases} X = x' + a \\ Y = y' + b \end{cases}. \tag{16.40}$$

From this, we can express the Eq. (16.39) for C with respect to the reference system $(O'; X, Y)$. The resulting equation is canonical for C.

(c) The equations for the roto-translation from $(O; x, y)$ to $(O'; X, Y)$ are given by substituting the translation transformation (16.40) into (16.38).

Corollary 16.6.8 *Given a degree-two polynomial equation in the variable x and y, the set (locus) of zeros of such equation is one of the following loci: ellipse, hyperbola, parabola, union of lines (either coincident or distinct).*

The proof of the Proposition 16.3.4 together with the result of the Theorem 16.6.5, which give the transformation relations for the matrices associated to a given conic C under a proper rigid transformation, allows one to prove the next proposition.

Proposition 16.6.9 *A conic C whose associated matrices are A and B with respect to a given orthonormal reference system $(O; x, y)$ is degenerate if and only if $\det B = 0$. Depending on the values of the determinant of A the following cases are possible*

$$\begin{aligned} \det A < 0 \quad &\Leftrightarrow \quad C \text{ hyperbola} \\ \det A = 0 \quad &\Leftrightarrow \quad C \text{ parabola} \\ \det A > 0 \quad &\Leftrightarrow \quad C \text{ ellipse}\,. \end{aligned}$$

The relative signs of $\det(A)$ *and* $\det B$ *determine whether the conic is real or imaginary.*

Exercise 16.6.10 As an example, we recall the results obtained in the Sect. 16.4. For the conic $\mathrm{d}(P, F) = e\,\mathrm{d}(P, \delta)$ with focus $F = (a_x, a_y)$ and directrix $\delta : x = u$, the matrix of the coefficients associated to the Eq. (16.28) is

$$B = \begin{pmatrix} 1 - e^2 & 0 & ue^2 - a_x \\ 0 & 1 & a_y \\ ue^2 - a_x & a_y & a_x^2 + a_y^2 - u^2e^2 \end{pmatrix}$$

with then $\det B = -e^2(a_x - u)^2$. We recover that the sign of $(1 - e^2)$ determines whether the conic C is an ellipse, or a parabola, or a hyperbola. We notice also that the conic is degenerate if and only if at least one of the conditions $e = 0$ or $a_x = u$ is met.

16.7 Eccentricity: Part 2

We complete now the analysis of the conics defined by the relation

$$\mathrm{d}(P, F) = e\,\mathrm{d}(P, \delta)$$

in terms of the eccentricity parameter. In Sect. 16.4 we have studied this equation with an arbitrary F and δ parallel to the y-axis, when it becomes the Eq. (16.28). In general, for a given eccentricity the previous relation depends only on the distance between F and δ. Using a suitable roto-translation as in the previous section, we have the following result.

Proposition 16.7.1 *Given a point F and a line δ in $\mathbb{E}^2$ such that $F \notin \delta$, there exists a cartesian orthogonal coordinate system $(O'; X, Y)$ with $F = O'$ and with respect to which the equation $\mathrm{d}(P, F) = e\,\mathrm{d}(P, \delta)$ (with $e > 0$) is written as*

$$Y^2 + X^2 - e^2(X - u)^2 = 0.$$

Proof Given a point F and a line $\delta \not\ni F$, it is always possible to roto-translate the starting coordinate system $(O; x, y)$ to a new one $(O'; X, Y)$ in such a way that $O' = F$ and the line δ is given by the equation $X = u \neq 0$. The result then follows from (16.28) being $a_X = a_Y = 0$. □

We know from the Sect. 16.4 that if $e = 1$, the equation represents a parabola with directrix $X = u \neq 0$ and focus $F = (0, 0)$. If $1 \neq e$, the equation represents either an ellipse $(0 < e < 1)$ or a hyperbola $(e > 1)$ with foci $F_+ = (0, 0)$ and $F_- = (-\frac{2ue^2}{1-e^2}, 0)$. Also, $e = 0$ yields the degenerate conic $X^2 + Y^2 = 0$, while $u = 0$ (that is $F \in \delta$) gives the degenerate conic $Y^2 + (1 - e^2)X^2 = 0$.

We can conclude that the Eq. (16.5) represents a conic whose type depends on the values of the eccentricity parameter. Its usefulness resides in yielding a constructive method to write the equation in canonical form, even for the degenerate cases.

We address the inverse question: given a non degenerate conic C with equation

$$a_{11}\,x^2 + 2\,a_{12}\,xy + a_{22}\,y^2 + 2\,a_{13}\,x + 2\,a_{23}\,y + a_{33} = 0$$

is it possible to determine its eccentricity and its directrix?

We give a constructive proof of the following theorem.

Theorem 16.7.2 *Given the non degenerate conic C whose equation is*

$$a_{11}\,x^2 + 2\,a_{12}\,xy + a_{22}\,y^2 + 2\,a_{13}\,x + 2\,a_{23}\,y + a_{33} = 0,$$

there exists a point F and a line δ with $F \notin \delta$ such that a point $P \in C$ if and only if

$$\mathrm{d}(P,F) \;=\; e\,\mathrm{d}(P,\delta)$$

for a suitable value $e > 0$ of the eccentricity parameter.

Proof As in the example Exercise 16.6.6 we firstly diagonalise the matrix of the quadratic form of C finding a cartesian orthogonal system $(O; x', y')$ with respect to which the equation for C is written as

$$\alpha_{11}(x')^2 + \alpha_{22}(y')^2 + 2\alpha_{13}x' + 2\alpha_{23}y' + \alpha_{33} = 0,$$

with α_{11}, α_{22} the eigenvalues of the quadratic form. This is the equation of the conic in the form studied in the Proposition 16.3.4, whose proof we now use. We have the following cases

(a) One of the eigenvalues of the quadratic form is zero, say $\alpha_{11} = 0$ (the case $\alpha_{22} = 0$ is analogous).
Up to a global (-1) factor that we can rescale, the equation for C is

$$\alpha_{22}(y')^2 + 2\alpha_{13}x' + 2\alpha_{23}y' + \alpha_{33} = 0,$$

with $\alpha_{22} > 0$ and $\alpha_{13} \neq 0$ (non degeneracy of C). Since there is no term $(x')^2$, this equation is of the form (16.28) only if $e = 1$. Thus it is of the form (16.29) written as

$$(y - a_y)^2 + 2(u - a_x)\big(x - \tfrac{1}{2}\,(u + a_x)\big) = 0.$$

The two expression are the same if and only if we have $e = 1$, and

$$a_y = -\frac{\alpha_{23}}{\alpha_{22}} \qquad \text{and} \qquad \begin{cases} u - a_x = \alpha_{13}/\alpha_{22} \\ u + a_x = (\alpha_{23}^2 - \alpha_{33}\alpha_{22})/\alpha_{13}\alpha_{22} \end{cases}.$$

These say that C is the parabola with focus and directrix given, with respect to $(O; x', y')$, by

$$F \;=\; \Big(\frac{\alpha_{23}^2 - \alpha_{33}\alpha_{22} - \alpha_{13}^2}{2\alpha_{13}\alpha_{22}},\, -\frac{\alpha_{23}}{\alpha_{22}}\Big), \qquad x' = \frac{\alpha_{13}^2 + \alpha_{23}^2 - \alpha_{33}\alpha_{22}}{2\alpha_{13}\alpha_{22}}.$$

With the translation

$$\begin{cases} X = x' + (\alpha_{33}\alpha_{22} - \alpha_{23}^2)/2\alpha_{22}\alpha_{13} \\ Y = y' + \alpha_{23}/\alpha_{22} \end{cases}$$

it can indeed be written as

$$Y^2 + 2\frac{\alpha_{13}}{\alpha_{22}} X = 0.$$

If $\alpha_{22} = 0$ and $\alpha_{11} \neq 0$ the result would be similar with the x'-axis and y' axis interchanged.

(b) Assume $\alpha_{11} \neq 0$ and $\alpha_{22} \neq 0$. We write the equation for C as in (16.17),

$$\frac{\alpha_{11}}{\alpha_{22}}\left(x + \frac{\alpha_{13}}{\alpha_{11}}\right)^2 + \left(y + \frac{\alpha_{23}}{\alpha_{22}}\right)^2 - \frac{1}{\alpha_{22}}\left(-\alpha_{33} + \frac{\alpha_{13}^2}{\alpha_{11}} + \frac{\alpha_{23}^2}{\alpha_{22}}\right) = 0, \tag{16.41}$$

and compare it with (16.30)

$$(1 - e^2)\left(x + \frac{ue^2 - a_x}{1 - e^2}\right)^2 + (y - a_y)^2 - \frac{e^2(u - a_x)^2}{1 - e^2} = 0. \tag{16.42}$$

Notice that with this choice (that the directrix be parallel to the y-axis, $x = u$) we are not treating the axes x and y in an equivalent way. We would have a similar analysis when exchanging the role of the axes x and y. The conditions to satisfy are

$$\begin{cases} 1 - e^2 = \alpha_{11}/\alpha_{22} \\ a_y = -\alpha_{23}/\alpha_{22} \end{cases} \quad \text{and} \quad \begin{cases} \frac{e^2(u-a_x)^2}{1-e^2} = \frac{h}{\alpha_{22}} \quad \text{with} \quad h = -\alpha_{33} + \frac{\alpha_{13}^2}{\alpha_{11}} + \frac{\alpha_{23}^2}{\alpha_{22}} \\ \frac{ue^2-a_x}{1-e^2} = \frac{\alpha_{13}}{\alpha_{11}} \end{cases}. \tag{16.43}$$

We see that $h = 0$ would give a degenerate conic with either $e = 0$ or $u = a_x$, that is the focus is on the directrix. As before, up to a global (-1) factor we may assume $\alpha_{22} > 0$. And as in Sect. 16.3 we have two possibilities according to the sign of α_{11}.

(b1) The eigenvalues have the same sign: $\alpha_{22} > 0$ and $\alpha_{11} > 0$. From the first condition in (16.43) we need $\alpha_{22} > \alpha_{11}$ and we get that $e < 1$. Then the last condition requires that the parameter $h > 0$ be positive. This means that C is a real ellipse. The case $\alpha_{22} < \alpha_{11}$ also results into a real ellipse but requires that the role of the axes x and y be exchanged. (The condition $\alpha_{11} = \alpha_{22}$ would give a circle and result in $e = 0$ which we are excluding.)

(b2) The eigenvalues α_{11} and α_{22} are discordant. Now the conditions (16.43) requires $e > 1$ and the parameter h to be negative. This means that C is a hyperbola of the second type in (16.11). To get the other type in (16.11), once again one needs to exchange the axes x and y.

As mentioned, the previous analysis is valid when the directrix is parallel to the y-axis. For the case when the directrix is parallel to the x-axis (the equation $y = u$), one has a similar analysis with the relations analogous to (16.43) now written as

$$\begin{cases} 1 - e^2 = \alpha_{22}/\alpha_{11} \\ a_x = -\alpha_{13}/\alpha_{11} \end{cases} \quad \text{and} \quad \begin{cases} \frac{e^2(u-a_y)^2}{1-e^2} = \frac{h}{\alpha_{11}} \\ \frac{ue^2-a_y}{1-e^2} = \frac{\alpha_{23}}{\alpha_{22}} \end{cases} \text{with } h = -\alpha_{33} + \frac{\alpha_{13}^2}{\alpha_{11}} + \frac{\alpha_{23}^2}{\alpha_{22}} \,. \tag{16.44}$$

In particular for $0 < \alpha_{22} < \alpha_{22}$ these are the data of a real ellipse, while for $\alpha_{11} > 0$ and $\alpha_{22} < 0$ (and $h < 0$) this are the data for a hyperbola of the first type in (16.11). □

In all cases above, the parameters e, u, a_x, a_y are given in terms of the conic coefficients by the relations (16.43) or (16.44). Being these quite cumbersome, we omit to write the complete solutions for these relations and rather illustrate with examples the general methods we developed.

Exercise 16.7.3 Consider the hyperbolas

$$y^2 - x^2 + k = 0, \qquad k = \pm 1 \,.$$

If $k = 1$, the relations (16.43) easily give the foci

$$F_{\pm} = (\pm\sqrt{2}, 0)$$

and corresponding directrix $\delta_{\pm}$ with equation

$$x = \pm\tfrac{\sqrt{2}}{2}.$$

On the other hand, for $k = 1$, the relations (16.44) now give the foci

$$F_{\pm} = (0, \pm\sqrt{2})$$

and corresponding directrix $\delta_{\pm}$,

$$y = \pm\tfrac{\sqrt{2}}{2}.$$

Exercise 16.7.4 Consider the C of the example Exercise 16.3.7, whose equation we write as

$$x^2 + 4y^2 + 2x - 12y + 3 \;=\; (x+1)^2 + 4(y - \tfrac{3}{2})^2 - 7 = 0.$$

It is easy now to compute that this ellipse has eccentricity $e = \frac{\sqrt{3}}{4}$ and foci

$$F_{\pm} = (-1 \pm \tfrac{\sqrt{21}}{2}, \tfrac{3}{2}).$$

The directrix $\delta_\pm$ corresponding to the focus $F_\pm$ is given by the line

$$x = -1 \pm \tfrac{2\sqrt{21}}{3}.$$

Exercise 16.7.5 Consider the conic C with equation

$$x^2 - ky^2 - 2x - 2 = 0$$

with a parameter $k \in \mathbb{R}$. By completing the square, we write this equation as

$$(x-1)^2 - ky^2 - 3 = 0.$$

Depending on the value of k, we have different cases.

(i) If $k < -1$, it is evident that C is a real ellipse with $\alpha_{11} < \alpha_{22}$, and the condition (16.43) gives eccentricity $e = \sqrt{1 + \frac{1}{k}}$, with foci

$$F_\pm = (1 \pm \sqrt{\tfrac{3(1+k)}{k}}, 0) \tag{16.45}$$

and corresponding directrix $\delta_\pm$ with equation

$$x = 1 \pm \sqrt{\tfrac{3}{k(1+k)}}\,. \tag{16.46}$$

(ii) If $-1 < k < 0$ the conic C is again a real ellipse, whose major axis is parallel to the y-axis, so $\alpha_{11} > \alpha_{22}$. Now the relations (16.44) yield eccentricity $e = \sqrt{1+k}$, with foci

$$F_\pm = (1, \pm\sqrt{-3\left(1 + \tfrac{1}{k}\right)})$$

and corresponding directrix $\delta_\pm$ given by the lines with equation

$$y = \pm\sqrt{\tfrac{3}{-k(k+1)}}\,.$$

(iii) If $k = 0$ the conic C is degenerate.
(iv) If $k > 0$, the conic C is a hyperbola. It is easy to compute the eccentricity to be $e = \sqrt{1 + \frac{1}{k}}$ (the same expression as for $k < -1$), with the foci and the directrix given by (16.45) and (16.46).

The matrix of the coefficients of this conic C is given by

$$B = \begin{pmatrix} 1 & 0 & -1 \\ 0 & -k & 0 \\ -1 & 0 & -2 \end{pmatrix},$$

with det $A = -k$ and det $B = -3k$. By the Proposition 16.6.9 we recover the listed results: C is degenerate if and only if $k = 0$; it is a hyperbola if and only if $k > 0$; an ellipse if and only if $k < 0$.

16.8 Why Conic Sections

We close the chapter by explaining where the loci on the affine euclidean plane $\mathbb{E}^2$ that we have described, the *conic* sections, get their name from. This will also be related to finding solutions to a non-linear problem in $\mathbb{E}^3$.

Fix a line γ and a point $V \in \gamma$ in $\mathbb{E}^3$. A (double) cone with axis γ and vertex V is the bundle of lines through V whose direction vectors form, with respect to γ, an angle of fixed width.

Consider now a plane $\pi \subset \mathbb{E}^3$ which does not contain the vertex of the cone. We show that, depending on the relative orientation of π with the axis of the cone, the intersection $\pi \cap \mathcal{C}$ — a *conic section* — is a non degenerate ellipse, or a parabola, or a hyperbola.

Let $(O, \mathcal{E}) = (O; x, y, z)$ be an orthonormal reference frame for $\mathbb{E}^3$, with $\mathcal{E}$ an orthonormal basis for E^3. To be definite, we take the z-axis as the axis of a cone $\mathcal{C}$, its vertex to be $V = O$ and its width an angle $0 < \theta < \pi/2$. It is immediate to see that the cone $\mathcal{C}$ is given by the points $P = (x, y, z)$ of the lines whose normalised direction vectors are

$$E^3 \ni u(\alpha) = (\sin\theta \cos\alpha, \sin\theta \sin\alpha, \cos\theta)$$

with $\alpha \in [0, 2\pi)$. The parametric equation for these lines (see the Definition 14.2.7) is then

$$r(\alpha) = \begin{cases} x = \lambda \sin\theta \cos\alpha \\ y = \lambda \sin\theta \sin\alpha \\ z = \lambda \cos\theta \end{cases} .$$

with λ a real parameter. This expression provides a vector equation for the cone $\mathcal{C}$. By eliminating the parameter, one gets a cartesian equation for $\mathcal{C}$ as given by the relation

$$\Sigma_{r(\alpha)} : \quad x^2 + y^2 - (\tan^2\theta)z^2 = 0.$$

Without loss of generality, we may intersect the cone $\mathcal{C}$ with a plane π which is orthogonal to the yz coordinate plane and meeting the z axis at the point $A = (0, 0, k > 0)$. If $\beta \in (0, \pi/2)$ is the angle between the axis of the cone (the z axis) and (its projection on) the plane π, the direction S_π of the plane is orthogonal to the normalised vector $v = (0, \cos\beta, \sin\beta)$. We know from Chap. 15 that the cartesian equation for the plane π is then

$$\Sigma_\pi : \quad (\cos\beta)y + (\sin\beta)(z - k) = 0.$$

The intersection $\mathcal{C} \cap \pi$ is then given by the solution of the system

$$\begin{cases} x^2 + y^2 - (\tan^2\theta)z^2 = 0 \\ (\cos\beta)y + (\sin\beta)(z-k) = 0 \end{cases} . \tag{16.47}$$

This is the only problem in this textbook which is formulated in terms of a system of *non-linear* equations. By inserting the second equation in the first one, elementary algebra gives, for the projection on the plane xy of the intersection $\mathcal{C} \cap \pi$, the equation,

$$x^2 + (1 - \tan^2\theta \cot^2\beta)\, y^2 + 2k\tan^2\theta \cot\beta\, y - k^2\tan^2\theta. \tag{16.48}$$

From what we have described above in this chapter, this equation represents a conic.

Its matrix of the coefficients is

$$B = \begin{pmatrix} 1 & 0 & 0 \\ 0 & 1 - \tan^2\theta\cot^2\beta & k\tan^2\theta\cot\beta \\ 0 & k\tan^2\theta\cot\beta & -k^2\tan^2\theta \end{pmatrix},$$

while the matrix of the quadratic form is

$$A = \begin{pmatrix} 1 & 0 \\ 0 & 1 - \tan^2\theta\cot^2\beta \end{pmatrix}.$$

One then computes

$$\det(A) = 1 - \tan^2\theta\cot^2\beta, \qquad \det B = -k^2\tan^2\theta\,.$$

Having excluded the cases $k = 0$ and $\tan\theta = 0$, we know from the Proposition 16.6.9 that the intersection $\mathcal{C} \cap \pi$ represents a non degenerate real conic. Some algebra indeed shows that:

$$\begin{aligned} \det(A) > 0 \quad &\Leftrightarrow \quad \tan^2\beta > \tan^2\theta \quad \Leftrightarrow \quad \beta > \theta, \\ \det(A) = 0 \quad &\Leftrightarrow \quad \tan^2\beta = \tan^2\theta \quad \Leftrightarrow \quad \beta = \theta, \\ \det(A) < 0 \quad &\Leftrightarrow \quad \tan^2\beta < \tan^2\theta \quad \Leftrightarrow \quad \beta < \theta, \end{aligned} \tag{16.49}$$

thus giving an ellipse, a parabola, a hyperbola respectively. These are shown in Figs. 16.4 and 16.5.

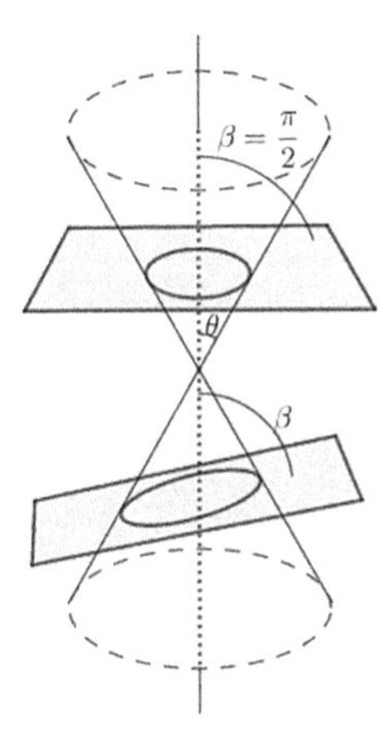

Fig. 16.4 The ellipse with the limit case of the circle

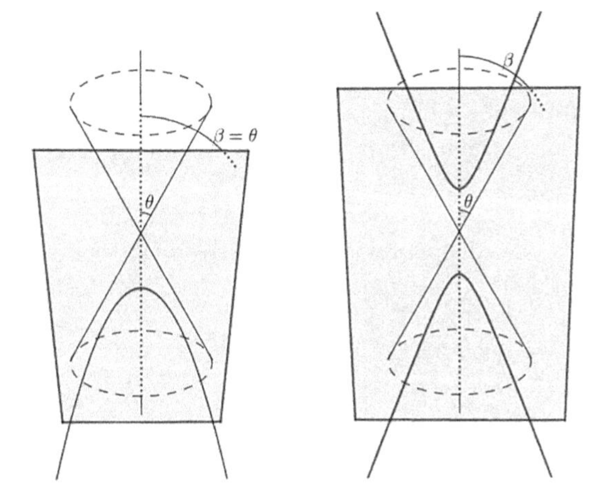

Fig. 16.5 The parabola and the hyperbola

Remark 16.8.1 As a particular case, if we take $\beta = \frac{\pi}{2}$, from (16.48) we see that $\mathcal{C} \cap \pi$ is a circle with radius $R = k \tan\theta$. On the other hand, with $k = 0$, that is π contains the vertex of the cone, one has $\det B = 0$. In such a case, the (projected) Eq. (16.48) reduces to

$$x^2 + (1 - \tan^2\theta \cot^2\beta)\, y^2 = 0.$$

Such equation represents:

2a. the union of two complex conjugate lines for $\beta > \theta$,
2b. the points $(x = 0, y)$, that is the y-axis for $\beta = \theta$,
2c. the union of two real lines for $\beta < \theta$.

We conclude giving a more transparent, in a sense, description of the intersection $\mathcal{C} \cap \pi$ by using a new reference system $(O; x', y', z')$, by a rotation around the x-axis where the plane π is orthogonal to the axis z'-axis. adapted to π. From Chap. 11, the transformation we consider is given in terms of the matrix in SO(3),

$$\begin{pmatrix} x' \\ y' \\ z' \end{pmatrix} = \begin{pmatrix} 1 & 0 & 0 \\ 0 & \sin\beta & -\cos\beta \\ 0 & \cos\beta & \sin\beta \end{pmatrix} \begin{pmatrix} x \\ y \\ z \end{pmatrix}.$$

With respect to the new reference system, the system of Eq. (16.47) becomes

$$\begin{cases} (x')^2 + \big((\sin\beta)y' + (\cos\beta)z'\big)^2 - (\tan^2\theta)\big((\sin\beta)z' - (\cos\beta)y'\big)^2 = 0 \\ z' - k\,\sin\beta = 0 \end{cases}.$$

It is then easy to see that the solutions of this system of equations are the points having coordinates $z' = k\,\sin\beta$ and (x', y') satisfying the equation

$$(x')^2 + (\sin^2\beta - \tan^2\theta\cos^2\beta)(y')^2 + 2k\cos\beta\,\sin^2\beta(1+\tan^2\theta)y' + (\cos^2\beta - \tan^2\theta\sin^2\rho)\,k^2\sin^2\beta = 0. \tag{16.50}$$

Clearly, this equation represents a conic on the plane $z' = k\,\sin\beta$ with respect to the orthonormal reference system $(O; x', y')$. Its matrix of the coefficients is

$$B = \begin{pmatrix} 1 & 0 & 0 \\ 0 & \sin^2\beta(1 - \tan^2\theta\cot^2\beta) & k\cos\beta\sin^2\beta(1+\tan^2\theta) \\ 0 & k\cos\beta\sin^2\beta(1+\tan^2\theta) & k^2\sin^2\beta\cos^2\beta(1 - \tan^2\theta\tan^2\beta) \end{pmatrix},$$

while the matrix of the quadratic form is

$$A = \begin{pmatrix} 1 & 0 \\ 0 & \sin^2\beta(1 - \tan^2\theta\cot^2\beta)) \end{pmatrix}.$$

One then computes

$$\det(A) = \sin^2\beta(1 - \tan^2\theta\cot^2\beta), \qquad \det B = -k^2\sin^2\beta\tan^2\theta.$$

With $k \neq 0$ and $\tan\theta \neq 0$, clearly also in this case the relations (16.49) are valid. And as particular cases, if we take $\beta = \pi/2$, one has that $\mathcal{C} \cap \pi$ is a circle with radius $R = k\,\tan\theta$. On the other hand, for $k = 0$, (that is π contains the vertex of the cone) so that $\det B = 0$, the Eq. (16.50) reduces to

$$(x')^2 + (\sin^2\beta - \tan^2\theta\cos^2\beta)(y')^2 = 0.$$

Such equation as before represents: the union of two complex conjugate lines for $\beta > \theta$; the points $x' = 0$ for $\beta = \theta$; the union of two real lines for $\beta < \theta$.

Remark 16.8.2 We remark that both Eqs. (16.48) and (16.50) describe the same type of conic, depending on the relative width of the angles β and θ. What differs is their eccentricity. The content of the Sect. 16.7 allows us to compute that the eccentricity of the conic in (16.48) is $e^2 = \tan^2\theta\cot^2\beta$, while for the conic in (16.50) we have $e^2 = (1 + \tan^2\theta)\cos^2\beta$.

Appendix A
Algebraic Structures

This appendix is an elementary introduction to basic notions of set theory, together with those of group, ring and field. The reader is only supposed to know about numbers, more precisely natural (containing the zero 0), integer, rational and real numbers, that will be denoted respectively by $\mathbb{N}$, $\mathbb{Z}$, $\mathbb{Q}$, $\mathbb{R}$. Some of their properties will also be recalled in the following. We shall also introduce complex numbers denoted $\mathbb{C}$ and (classes of) integers $\mathbb{Z}_p = \mathbb{Z}/p\mathbb{Z}$.

A.1 A Few Notions of Set Theory

Definition A.1.1 Given any two sets A and B, by $A \times B$ we denote their *Cartesian product*. This it is defined as the set of ordered pairs of elements from A and B, that is,

$$A \times B = \{(a, b) \mid a \in A, b \in B\}.$$

Notice that $A \times B \neq B \times A$ since we are considering ordered pairs. The previous definition is valid for sets A, B of arbitrary cardinality. The set $A \times A$ is denoted A^2.

Exercise A.1.2 Consider the set $A = \{\diamondsuit, \heartsuit, \clubsuit, \spadesuit\}$. The Cartesian product A^2 is then

$$\begin{aligned} A^2 &= A \times A \\ &= \big\{(\diamondsuit, \diamondsuit), (\diamondsuit, \heartsuit), (\diamondsuit, \clubsuit), (\diamondsuit, \spadesuit), (\heartsuit, \diamondsuit), (\heartsuit, \heartsuit), (\heartsuit, \clubsuit), (\heartsuit, \spadesuit), \\ &\qquad (\clubsuit, \diamondsuit), (\clubsuit, \heartsuit), (\clubsuit, \clubsuit)(\clubsuit, \spadesuit), (\spadesuit, \diamondsuit), (\spadesuit, \heartsuit), (\spadesuit, \clubsuit), (\spadesuit, \spadesuit)\big\}. \end{aligned}$$

Definition A.1.3 Given any set A, a *binary relation* on A is any subset of the Cartesian product $A^2 = A \times A$. If such a subset is denoted by $\mathcal{R}$, we say that the pair of elements a, b in A are *related* or *in relation* if $(a, b) \in \mathcal{R}$ and write it as $a \, \mathcal{R} \, b$.

G. Landi and A. Zampini, *Linear Algebra and Analytic Geometry for Physical Sciences*, Undergraduate Lecture Notes in Physics,
https://doi.org/10.1007/978-3-319-78361-1

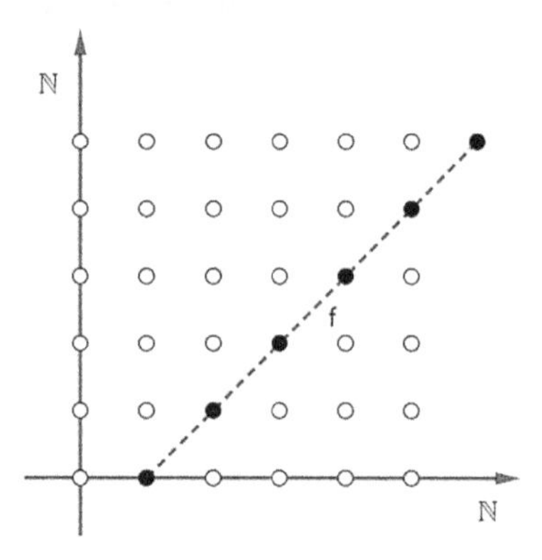

Fig. A.1 A binary relation in $\mathbb{N}^2$

Example A.1.4 Consider $A = \mathbb{N}$, the set of natural numbers, with $\mathcal{R}$ the subset of $\mathbb{N}^2 = \mathbb{N} \times \mathbb{N}$ given by the points as in the Fig. A.1. We see that $2\mathcal{R}1$, but it is not true that $1\mathcal{R}1$. One may easily check that $\mathcal{R}$ can be written by the formula

$$n\mathcal{R}m \Leftrightarrow m = n - 1, \qquad \text{for any} \quad (n, m) \in \mathbb{N}^2.$$

Definition A.1.5 A binary relation on a set A is called an *equivalence relation* if the following properties are satisfied

- $\mathcal{R}$ is *reflexive*, that is $a\mathcal{R}a$ for any $a \in A$,
- $\mathcal{R}$ is *symmetric*, that is $a\mathcal{R}b \Rightarrow b\mathcal{R}a$, for any $a, b \in A$,
- $\mathcal{R}$ is *transitive*, that is $a\mathcal{R}b$ and $b\mathcal{R}c \Rightarrow a\mathcal{R}c$ for any $a, b, c \in A$.

Exercise A.1.6 In any given set A, the equality is an equivalence relation. On the set T of all triangles, congruence of triangles and similarity of triangles are equivalence relations. The relation described in the Example A.1.4 is not an equivalence relation, since reflexivity does not hold.

Definition A.1.7 Consider a set A and let $\mathcal{R}$ be an equivalence relation defined on it. For any $a \in A$, one defines the subset

$$[a] = \{x \in A \mid x\mathcal{R}a\} \subseteq A$$

as the *equivalence class* of a in A. Any element $x \in [a]$ is called a *representative* of the class $[a]$. It is clear that an equivalence class has as many representatives as the elements it contains.

Proposition A.1.8 *With $\mathcal{R}$ an equivalence relation on the set A, the following properties hold:*

(1) If $a\,\mathcal{R}\,b$, then $[a] = [b]$.

(2) If $(a, b) \notin \mathcal{R}$*, then* $[a] \cap [b] = \emptyset$.
(3) $A = \bigcup_{a \in A} [a]$*; this is a disjoint union.*

Proof (1) One shows that the mutual inclusions $[a] \subseteq [b]$ and $[b] \subseteq [a]$ are both valid if $a \mathcal{R} b$. Let $x \in [a]$; this means $x \mathcal{R} a$. From the hypothesis $a \mathcal{R} b$, so by the transitivity of $\mathcal{R}$ one has $x \mathcal{R} b$, that is $x \in [b]$. This proves the inclusion $[a] \subseteq [b]$. The proof of the inclusion $[b] \subseteq [a]$ is analogue.
(2) Let us suppose that $A \ni x \in [a] \cup [b]$. It would mean that $x \mathcal{R} a$ and $x \mathcal{R} b$. From the symmetry of $\mathcal{R}$ we would then have $a \mathcal{R} x$, and from the transitivity this would result in $a \mathcal{R} b$, which contradicts the hypothesis.
(3) It is obvious, from (2). □

Definition A.1.9 The decomposition $A = \bigcup_{a \in A} [a]$ is called the *partition* of A associated (or corresponding) to the equivalence relation $\mathcal{R}$.

Definition A.1.10 If $\mathcal{R}$ is an equivalence relation defined on the set A, the set whose elements are the corresponding equivalence classes is denoted $A/\mathcal{R}$ and called the *quotient of* A *modulo* $\mathcal{R}$. The map

$$\pi : A \to A/\mathcal{R} \qquad \text{given by} \qquad a \mapsto [a]$$

is called *the canonical projection* of A onto the quotient $A/\mathcal{R}$.

A.2 Groups

A set has an algebraic structure if it is equipped with one or more operations. When the operations are more than one, they are required to be compatible. In this section we describe the most elementary algebraic structures.

Definition A.2.1 Given a set G, a *binary operation* $*$ on it is a map

$$* : G \times G \longrightarrow G.$$

The image of the operation between a and b is denoted by $a * b$. One also says that G is *closed*, or *stable* with respect to the operation $*$. One usually writes $(G, *)$ for the algebraic structure $*$ defined on G, that is for the set G equipped with the binary operation $*$.

Example A.2.2 It is evident that the usual sum and the usual product in $\mathbb{N}$ are binary operations.

As a further example we describe a binary operation which does not come from usual arithmetic operations in any set of numbers. Let T be an equilateral triangle whose vertices are ordered and denoted by ABC. Let R be the set of rotations on a plane under which each vertex is taken onto another vertex. The rotation that takes the vertices ABC to BCD, can be denoted by

$$\begin{pmatrix} A & B & C \\ B & C & A \end{pmatrix}.$$

It is clear that R contains three elements, which are:

$$e = \begin{pmatrix} A & B & C \\ A & B & C \end{pmatrix} \qquad x = \begin{pmatrix} A & B & C \\ B & C & A \end{pmatrix} \qquad y = \begin{pmatrix} A & B & C \\ C & A & B \end{pmatrix}.$$

The operation—denoted now $\circ$—that we consider among elements in R is the composition of rotations. The rotation $x \circ y$ is the one obtained by acting on the vertices of the triangle first with y and then with x. It is easy to see that $x \circ y = e$. The Table A.1 shows the composition law among elements in R.

$\circ$	e	x	y
e	e	x	y
x	x	y	e
y	y	e	x

(A.1)

Remark A.2.3 The algebraic structures $(\mathbb{N}, +)$ and $(\mathbb{N}, \cdot)$ have the following well known properties, for all elements $a, b, c \in \mathbb{N}$,

$$a + (b + c) = (a + b) + c, \qquad a + b = b + a,$$
$$a \cdot (b \cdot c) = (a \cdot b) \cdot c, \qquad a \cdot b = b \cdot a\,.$$

The set $\mathbb{N}$ has elements, denoted 0 and 1, whose properties are singled out,

$$0 + a = a, \qquad 1a = a$$

for any $a \in \mathbb{N}$. We give the following definition.

Definition A.2.4 Let $(G, *)$ be an algebraic structure.

(a) $(G, *)$ is called *associative* if

$$a * (b * c) = (a * b) * c$$

for any $a, b, c \in G$.

(b) $(G, *)$ is called *commutative* (or *abelian*) if

$$a * b = b * a$$

for any $a, b \in G$.

(c) An element $e \in G$ is called an *identity* (or a *neutral element*) for $(G, *)$ (and the algebraic structure is often denoted by $(G, *, e)$) if

$$e * a = a * e$$

for any $a \in G$.

(d) Let $(G, *, e)$ be an algebraic structure with an identity e. An element $b \in G$ such that

$$a * b = b * a = e$$

is called the *inverse* of a, and denoted by a^{-1}. The elements for which an inverse exists are called invertible.

Remark A.2.5 If the algebraic structure is given by a 'sum rule', like in $(\mathbb{N}, +)$, the neutral element is usually called a *zero* element, denoted by 0, with $a + 0 = 0 + a = a$. Also, the inverse of an element a is usually denoted by $-a$ and named the *opposite* of a.

Example A.2.6 It is easy to see that for the sets considered above one has $(\mathbb{N}, +, 0)$, $(\mathbb{N}, \cdot, 1)$, $(R, \circ, e)$. Every element in R is invertible (since one has $x \circ y = y \circ x = e$); the set $(\mathbb{N}, \cdot, 1)$ contains only one invertible element, which is the identity itself, while in $(\mathbb{N}, +, 0)$ no element is invertible.

From the defining relation (c) above one clearly has that if a^{-1} is the inverse of $a \in (G, *)$, then a is the inverse of a^{-1}. This suggests a way to enlarge sets containing elements which are not invertible, so to have a new algebraic structure whose elements are all invertible. For instance, one could define the set of *integer* numbers $\mathbb{Z} = \{\pm n : n \in \mathbb{N}\}$ and sees that every element in $(\mathbb{Z}, +, 0)$ is invertible.

Definition A.2.7 An algebraic structure $(G, *)$ is called a *group* when the following properties are satisfied

(a) the operation $*$ is associative,
(b) G contains an identity element e with respect to $*$,
(c) every element in G is invertible with respect to e.

A group $(G, *, e)$ is called *commutative* (or *abelian*) if the operation $*$ is commutative.

Remark A.2.8 Both $(\mathbb{Z}, +, 0)$ and $(R, \circ, e)$ are abelian groups.

Proposition A.2.9 *Let $(G, *, e)$ be a group. Then*

(i) the identity element is unique,
(ii) the inverse a^{-1} of any element $a \in G$ is unique.

Proof (i) Let us suppose that e, e' are both identities for $(G, *)$. Then it should be $e * e' = e$ since e' is an identity, and also $e * e' = e'$ since e is an identity; this would then mean $e = e'$.

(ii) Let b, c be both inverse elements to $a \in G$; this would give $a * b = b * a = e$ and $a * c = c * a = e$. Since the binary operation is associative, one has $b * (a * c) = (b * a) * c$, resulting in $b * e = e * c$ and then $b = c$. □

A.3 Rings and Fields

Next we introduce and study the properties of a set equipped with two binary operations—compatible in a suitable sense—which resemble the sum and the product of integer numbers in $\mathbb{Z}$.

Definition A.3.1 Let $A = (A, +, 0_A, \cdot, 1_A)$ be a set with two operations, called sum $(+)$ and product $(\cdot)$, with two distinguished elements called 0_A and 1_A. The set A is called a *ring* if the following conditions are satisfied:

(a) $(A, +, 0_A)$ is an abelian group,
(b) the product $\cdot$ is associative,
(c) 1_A is the identity element with respect to the product,
(d) one has $a \cdot (b + c) = (a \cdot b) + (a \cdot c)$ for any $a, b, c \in A$.

If moreover the product is abelian, A is called an *abelian* ring.

Example A.3.2 The set $(\mathbb{Z}, +, 0, \cdot, 1)$ is clearly an abelian ring.

Definition A.3.3 By $\mathbb{Z}[X]$ one denotes the set of *polynomials* in the indeterminate (or variable) X with coefficients in $\mathbb{Z}$, that is the set of formal expressions,

$$\mathbb{Z}[X] = \left\{ \sum_{i=0}^{n} a_i X^i = a_n X^n + a_{n-1} X^{n-1} + \ldots + a_1 X + a_0 \; : \; n \in \mathbb{N}, \; a_i \in \mathbb{Z} \right\}.$$

If $\mathbb{Z}[X] \ni p(X) = a_n X^n + a_{n-1} X^{n-1} + \ldots + a_1 X + a_0$ then $a_0, a_1, \ldots, a_n$ are the *coefficients* of the polynomial $p(X)$, while the term $a_i X^i$ is a *monomial* of degree i. The *degree* of the polynomial $p(X)$ is the highest degree among those of its non zero monomials. If $p(X)$ is the polynomial above, its degree is n provided $a_n \neq 0$, and one denotes $\deg p(X) = n$. The two usual operations of sum and product in $\mathbb{Z}[X]$ are defined as follows. Let $p(X), q(X)$ be two arbitrary polynomials in $\mathbb{Z}[X]$,

$$p(X) = \sum_{i=0}^{n} a_i X^i, \qquad q(X) = \sum_{i=0}^{m} b_i X^i.$$

Let us suppose $n \leq m$. One sets

$$p(X) + q(X) = \sum_{j=0}^{m} c_j X^j,$$

with $c_j = a_j + b_j$ for $0 \le j \le n$ and $c_j = b_j$ for $n < j \le m$. One would have an analogous results were $n \ge m$. For the product one sets

$$p(X) \cdot q(X) = \sum_{h=0}^{m+n} d_h X^h,$$

where $d_h = \sum_{i+j=h} a_i b_j$.

Proposition A.3.4 *Endowed with the sum and the product as defined above, the set $\mathbb{Z}[X]$ is an abelian ring, the ring of polynomials in one variable with integer coefficients.*

Proof One simply transfer to $\mathbb{Z}[X]$ the analogous structures and properties of the ring $(\mathbb{Z}, +, 0, \cdot, 1)$. Let $0_{\mathbb{Z}[X]}$ be the null polynomial, that is the polynomial whose coefficients are all equal to $0_{\mathbb{Z}}$, and let $1_{\mathbb{Z}[X]} = 1_{\mathbb{Z}}$ be the polynomial of degree 0 whose only non zero coefficient is equal to $1_{\mathbb{Z}}$. We limit ourselves to prove that $(\mathbb{Z}[X], +, 0_{\mathbb{Z}[X]})$ is an abelian group.

- Clearly, the null polynomial $0_{\mathbb{Z}[X]}$ is the identity element with respect to the sum of polynomials.
- Let us consider three arbitrary polynomials in $\mathbb{Z}[X]$,

$$p(X) = \sum_{i=0}^{n} a_i X^i, \quad q(X) = \sum_{i=0}^{m} b_i X^i, \quad r(X) = \sum_{i=0}^{p} c_i X^i.$$

We show that

$$\big(p(X) + q(X)\big) + r(X) = p(X) + \big(q(X) + r(X)\big).$$

For simplicity we consider the case $n = m = p$, since the proof for the general case is analogue. From the definition of sum of polynomials, one has

$$\begin{aligned} A(X) &= (p(X) + q(X)) + r(X) \\ &= \sum_{i=0}^{n} (a_i + b_i) X^i + \sum_{i=0}^{n} c_i X^i = \sum_{i=0}^{n} [(a_i + b_i) + c_i] X^i \end{aligned}$$

and

$$\begin{aligned} B(X) &= p(X) + (q(X) + r(X)) \\ &= \sum_{i=0}^{n} a_i X^i + \sum_{i=0}^{n} (b_i + c_i) X^i = \sum_{i=0}^{n} [a_i + (b_i + c_i)] X^i. \end{aligned}$$

The coefficients of $A(X)$ and $B(X)$ are given, for any $i = 0, \dots, n$, by

$$[(a_i + b_i) + c_i] \qquad \text{and} \qquad [a_i + (b_i + c_i)]$$

and they coincide being the sum in $\mathbb{Z}$ associative. This means that $A(X) = B(X)$.

- We show next that any polynomial $p(X) = \sum_{i=0}^{n} a_i X^i$ is invertible with respect to the sum in $\mathbb{Z}[X]$. Let us define the polynomial $p'(X) = \sum_{i=0}^{n}(-a_i)X^i$, with $(-a_i)$ denoting the inverse of $a_i \in \mathbb{Z}$ with respect to the sum. From the definition of the sum of polynomials, one clearly has

$$p(X) + p'(X) = \sum_{i=0}^{n} a_i X^i + \sum_{i=0}^{n} (-a_i)X^i = \sum_{i=0}^{n} (a_i - a_i)X^i .$$

Since $a_i - a_i = 0_{\mathbb{Z}}$ for any i, one has $p(X) + p'(X) = 0_{\mathbb{Z}[X]}$; thus $p'(X)$ is the inverse of $p(X)$.

- Finally, we show that the sum in $\mathbb{Z}[X]$ is abelian. Let $p(X)$ and $q(X)$ be two arbitrary polynomials in $\mathbb{Z}[X]$ of the same degree $\deg p(X) = n = \deg q(X)$ (again for simplicity); we wish to show that

$$p(X) + q(X) = q(X) + p(X).$$

From the definition of sum of polynomials,

$$U(X) = p(X) + q(X) = \sum_{i=0}^{n} (a_i + b_i)X^i$$

$$V(X) = q(X) + p(X) = \sum_{i=0}^{n} (b_i + a_i)X^i \ :$$

the coefficients of $U(X)$ and $V(X)$ are given, for any $i = 0, \dots, n$ by

$$a_i + b_i \qquad \text{and} \qquad b_i + a_i$$

which coincide since the sum is abelian in $\mathbb{Z}$. This means $U(X) = V(X)$.

We leave as an exercise to finish showing that $\mathbb{Z}[X]$ with the sum and the product above fulfill the conditions (b)–(d) in the Definition A.3.1 of a ring. □

Remark A.3.5 Direct computation show the following well known properties of the abelian ring $\mathbb{Z}[X]$ of polynomials. With $f(X), g(X) \in \mathbb{Z}[X]$ it holds that:

(i) $\deg(f(X) + g(X)) \leq \max\{\deg(f(X)), \deg(g(X))\}$;
(ii) $\deg(f(X) \cdot g(X)) = \deg(f(X)) + \deg(g(X))$.

It is easy to see that the set $(\mathbb{Q}, +, \cdot, 0, 1)$ of rational numbers is an abelian ring as well. The set $\mathbb{Q}$ has indeed a richer algebraic structure than $\mathbb{Z}$: any non zero element

$0 \neq a \in \mathbb{Q}$ is invertible with respect to the product. If $a = p/q$ with $p \neq 0$, then $a^{-1} = q/p \in \mathbb{Q}$.

Definition A.3.6 An abelian ring $K = (K, +, 0, \cdot, 1)$ such that each element $0 \neq a \in K$ is invertible with respect to the product $\cdot$, is called a *field*. Equivalently one sees that $(K, +, 0, \cdot, 1)$ is a field if and only if both $(K, +, 0)$ and $(K, \cdot, 1)$ are abelian groups and the product is *distributive* with respect to the sum, that is the condition (d) of the Definition A.3.1 is satisfied.

Example A.3.7 Clearly $(\mathbb{Q}, +, 0, \cdot, 1)$ is a field, while $(\mathbb{Z}, +, 0, \cdot, 1)$ is not. The fundamental example of a field for us will be the set $\mathbb{R} = (\mathbb{R}, +, 0, \cdot, 1)$ of real numbers equipped with the usual definitions of sum and product.

Analogously to the Definition A.3.3 one can define the sets $\mathbb{Q}[X]$ and $\mathbb{R}[X]$ of polynomials with rational and real coefficients. For them one naturally extends the definitions of sum and products, as well as that of degree.

Proposition A.3.8 *The set* $\mathbb{Q}[X]$ *and* $\mathbb{R}[X]$ *are both abelian rings.* □

It is worth stressing that in spite of the fact that $\mathbb{Q}$ *and* $\mathbb{R}$ *are fields, neither* $\mathbb{Q}[X]$ *nor* $\mathbb{R}[X]$ *are such since a polynomial need not admit an inverse with respect to the product.*

A.4 Maps Preserving Algebraic Structures

The Definition A.2.1 introduces the notion of algebraic structure $(G, *)$ and we have described what groups, rings and fields are. We now briefly deal with maps between algebraic structures of the same kind, which preserve the binary operations defined in them. We have the following definition

Definition A.4.1
A map $f : G \to G'$ between two groups $(G, *_G, e_G)$ and $(G', *_{G'}, e_{G'})$ is a *group homomorphism* if

$$f(x *_G y) = f(x) *_{G'} f(y) \qquad \text{for all} \quad x, y \in G.$$

A map $f : A \to B$ between two rings $(A, +_A, 0_A, \cdot_A, 1_A)$ and $(B, +_B, 0_B, \cdot_B, 1_B)$ is a *ring homomorphism* if

$$f(x +_A y) = f(x) +_B f(y), \qquad f(x \cdot_A y) = f(x) \cdot_B f(y) \qquad \text{for all} \quad x, y \in A.$$

Example A.4.2 The natural inclusions $\mathbb{Z} \subset \mathbb{Q}$, $\mathbb{Q} \subset \mathbb{R}$ are rings homomorphisms, as well as the inclusion $\mathbb{Z} \subset \mathbb{Z}[x]$ and similar ones.

Exercise A.4.3 The map $\mathbb{Z} \to \mathbb{Z}$ defined by $n \mapsto 2n$ is a group homomorphism with respect to the group structure $(\mathbb{Z}, +, 0)$, but not a ring homomorphism with respect to the ring structure $(\mathbb{Z}, +, 0, \cdot, 1)$.

To lighten notations, from now on we shall denote a sum by $+$ and a product by $\cdot$ (and more generally a binary operation by $*$), irrespectively of the set in which they are defined. It will be clear from the context which one they refers to.

Group homomorphisms present some interesting properties, as we now show.

Proposition A.4.4 *Let $(G, *, e_G)$ and $(G', *, e_{G'})$ be two groups, and $f : G \to G'$ a group homomorphism. Then,*

(i) $f(e_G) = e_{G'}$,
(ii) $f(a^{-1}) = (f(a))^{-1}$, *for any* $a \in G$.

Proof (i) Since e_G is the identity element with respect to the sum, we can write

$$f(e_G) = f(e_G * e_G) = f(e_G) * f(e_G),$$

where the second equality is valid as f is a group homomorphism. Being $f(e_G) \in G'$, it has a unique inverse (see the Proposition A.2.9), $(f(e_G))^{-1} \in G'$, that we can multiply with both sides of the previous equality, thus yielding

$$f(e_G) * (f(e_G))^{-1} = f(e_G) * f(e_G) * (f(e_G))^{-1}.$$

This relation results in

$$e_{G'} = f(e_G) * e_{G'} \quad \Rightarrow \quad e_{G'} = f(e_G).$$

(ii) Making again use of the Proposition A.2.9, in order to show that $(f(a))^{-1}$ is the inverse (with respect to the product in G') of $f(a)$ it suffices to show that

$$f(a) * (f(a))^{-1} = e_{G'}.$$

From the definition of group homomorphism, it is

$$f(a) * (f(a))^{-1} = f(a * a^{-1}) = f(e_G) = e_{G'}$$

where the last equality follows from (i).

If $f : A \to B$ is a ring homomorphism, the previous properties are valid with respect to both the sum and to the product, that is

(i') $f(0_A) = 0_B$ and $f(1_A) = 1_B$;
(ii') $f(-a) = -f(a)$ for any $a \in A$, while $f(a^{-1}) = (f(a))^{-1}$ for any invertible (with respect to the product) element $a \in A$ with inverse a^{-1}. □

If A, B are fields, a ring homomorphism $f : A \to B$ is called a *field homomorphism*. A bijective homomorphism between algebraic structures is called an *isomorphism*.

A.5 Complex Numbers

It is soon realised that one needs enlarging the field $\mathbb{R}$ of real numbers to consider zeros of polynomials with real coefficients. The real coefficient polynomial $p(x) = x^2 + 1$ has 'complex' zeros usually denoted $\pm \mathrm{i}$, and their presence leads to defining the field of complex numbers $\mathbb{C}$. One considers the smallest field containing $\mathbb{R}$, $\pm \mathrm{i}$ and all possible sums and products of them.

Definition A.5.1 The set of complex numbers is given by formal expressions

$$\mathbb{C} = \{z = a + \mathrm{i}b \mid a, b \in \mathbb{R}\}.$$

The real number a is called the *real part* of z, denoted $a = \Re(z)$; the real number b is called the *imaginary part* of z, denoted $b = \Im(z)$.

The following proposition comes as an easy exercise.

Proposition A.5.2 *The binary operations of sum and product defined in* $\mathbb{C}$ *by*

$$(a + \mathrm{i}b) + (c + \mathrm{i}d) = (a + c) + \mathrm{i}(b + d),$$
$$(a + \mathrm{i}b) \cdot (c + \mathrm{i}d) = (ac - bd) + \mathrm{i}(bc + ad)$$

make $(\mathbb{C}, +, 0_{\mathbb{C}}, \cdot, 1_{\mathbb{C}})$ *a field, with* $0_{\mathbb{C}} = 0_{\mathbb{R}} + \mathrm{i}0_{\mathbb{R}} = 0_{\mathbb{R}}$ *and* $1_{\mathbb{C}} = 1_{\mathbb{R}} + \mathrm{i}0_{\mathbb{R}} = 1_{\mathbb{R}}$. □

Exercise A.5.3 An interesting part of the proof of the proposition above is to determine the inverse z^{-1} of the complex number $z = a + \mathrm{i}b$. One easily checks that

$$(a + \mathrm{i}b)^{-1} = \frac{a}{a^2 + b^2} - \mathrm{i}\frac{b}{a^2 + b^2} = \frac{1}{a^2 + b^2}\,(a - \mathrm{i}b).$$

Again an easy exercise establishes the following proposition.

Proposition A.5.4 *Given* $z = a + \mathrm{i}b \in \mathbb{C}$ *one defines its conjugate number to be* $\bar{z} = a - \mathrm{i}b$. *Then, for any complex number* $z = a + \mathrm{i}b$ *the following properties hold:*

(i) $\bar{\bar{z}} = z$,
(ii) $\bar{z} = z$ *if and only if* $z \in \mathbb{R}$,
(iii) $z\bar{z} = a^2 + b^2$,
iv) $z + \bar{z} = 2\Re(z)$. □

Exercise A.5.5 The natural inclusions $\mathbb{R} \subset \mathbb{C}$ given by $\mathbb{R} \ni a \mapsto a + \mathrm{i}0_{\mathbb{R}}$ is a field homomorphism, while the corresponding inclusion $\mathbb{R}[x] \subset \mathbb{C}[x]$ is a ring homomorphism.

Remark A.5.6 We mentioned above that the polynomial $x^2 + 1 = p(x) \in \mathbb{R}[x]$ cannot be decomposed (i.e. cannot be factorised) as a product of degree 1 polynomials in $\mathbb{R}[x]$, that is, with real coefficients. On the other hand, the identity

$x^2 + 1 = (x - \mathrm{i})(x + \mathrm{i}) \in \mathbb{C}[x]$ shows that the same polynomial can be decomposed into degree 1 terms if the coefficients of the latter are taken in $\mathbb{C}$. This is not surprising, since the main reason to enlarge the field $\mathbb{R}$ to $\mathbb{C}$ was exactly to have a field containing the zero of the polynomial $p(x)$.

What is indeed surprising is that the field $\mathbb{C}$ contains the zeros of *any* polynomial with real coefficients. This is the result that we recall as the next theorem.

Proposition A.5.7 (Fundamental theorem of algebra) *Let* $f(x) \in \mathbb{R}[x]$ *be a polynomial with real coefficients and* $\deg f(x) \geq 1$. *Then,* $f(x)$ *has at least a zero (that is a root) in* $\mathbb{C}$. *More precisely, if* $\deg f(x) = n$, *then* $f(x)$ *has* n *(possibly non distinct) roots in* $\mathbb{C}$. *If* $z_1, \ldots, z_s$ *are these distinct roots, the polynomial* $f(x)$ *can be written as*

$$f(x) = a(x - z_1)^{m(1)}(x - z_2)^{m(2)} \cdots (x - z_s)^{m(s)},$$

with the root multiplicities $m(j)$ *for* $j = 1, \ldots s$, *such that*

$$\sum_{j=1}^{s} m(j) = n.$$

That is the polynomial $f(x)$ *it is completely factorisable on* $\mathbb{C}$. □

A more general result states that $\mathbb{C}$ is an *algebraically closed* field, that is one has the following:

Theorem A.5.8 *Let* $f(x) \in \mathbb{C}[x]$ *be a degree* n *polynomial with complex coefficients. Then there exist* n *complex (non distinct in generall) roots of* $f(x)$. *Thus the polynomial* $f(x)$ *is completely factorisable on* $\mathbb{C}$. □

A.6 Integers Modulo A Prime Number

We have seen that the integer numbers $\mathbb{Z}$ form only a ring and not a field. Out of it one can construct fields of numbers by going to the quotient with respect to an equivalence relation of 'modulo an integer'. As an example, consider the set $\mathbb{Z}_3$ of integer modulo 3. It has three elements

$$\mathbb{Z}_3 = \{[0], [1], [2]\}$$

which one also simply write $\mathbb{Z}_3 = \{0, 1, 2\}$, although one should not confuse them with the corresponding classes.

One way to think of the three elements of $\mathbb{Z}_3$ is that each one represents the equivalence class of all integers which have the same remainder when divided by 3. For instance, [2] denotes the set of all integers which have remainder 2 when divided by 3 or equivalently, [2] denotes the set of all integers which are congruent to 2 modulo 3, thus $[2] = \{2, 5, 8, 11, \dots\}$. The usual arithmetic operations determine the addition and multiplication tables for this set as show in Table A.2.

$$\begin{array}{c|c|c|c} + & 0 & 1 & 2 \\ \hline 0 & 0 & 1 & 2 \\ \hline 1 & 1 & 2 & 0 \\ \hline 2 & 2 & 0 & 1 \end{array} \quad \text{and} \quad \begin{array}{c|c|c|c} * & 0 & 1 & 2 \\ \hline 0 & 0 & 0 & 0 \\ \hline 1 & 0 & 1 & 2 \\ \hline 2 & 0 & 2 & 1 \end{array}. \tag{A.2}$$

Thus $-[1] = [2]$ and $-[2] = [1]$ and $\mathbb{Z}_3$ is an abelian group for the addition. Furthermore, $[1] * [1] = [1]$ and $[2] * [2] = [1]$ and both nonzero elements have inverse: $[1]^{-1} = [1]$ and $[2]^{-1} = [2]$. All of this makes $\mathbb{Z}_3$ a field.

The previous construction works when 3 is substituted with any *prime* number p. We recall that a positive integer p is called prime if it is only divisible by itself and by 1. Thus, for any prime number one gets the field of integers modulo p:

$$\mathbb{Z}_p = \mathbb{Z}/p\mathbb{Z} = \{[0], [1], \dots, [p-1]\}.$$

Each of its elements represents the equivalence class of all integers which have the given remainder when divided by p. Equivalently, each element denotes the equivalence class of all integers which are congruent modulo p. The corresponding addition and multiplication tables, defines as in $\mathbb{Z}$ but now taken modulo p, can be easily worked out. Notice that the construction does not work, that is $\mathbb{Z}_p$ is not a ring, if p is not a prime number: were this the case there would be divisors of zero.

Index

G. Landi and A. Zampini, *Linear Algebra and Analytic Geometry for Physical Sciences*, Undergraduate Lecture Notes in Physics,
https://doi.org/10.1007/978-3-319-78361-1

CPSIA information can be obtained
at www.ICGtesting.com
Printed in the USA
LVHW080015040119
602644LV00007B/113/P

9 783319 783604